Human Viruses in Water

Editor

Albert Bosch

*Enteric Virus Laboratory, Department of Microbiology,
University of Barcelona, Barcelona, Spain*

ELSEVIER

Amsterdam – Boston – Heidelberg – London – New York – Oxford – Paris
San Diego – San Francisco – Singapore – Sydney – Tokyo

Elsevier
Radarweg 29, PO Box 211, 1000 AE Amsterdam, The Netherlands
Linacre House, Jordan Hill, Oxford OX2 8DP, UK

First edition 2007

Notice
No responsibility is assumed by the publisher for any injury and/or damage to persons
or property as a matter of products liability, negligence or otherwise, or from any use
or operation of any methods, products, instructions or ideas contained in the material
herein. Because of rapid advances in the medical sciences, in particular, independent
verification of diagnoses and drug dosages should be made

Library of Congress Cataloguing-in-Publication Data
A catalogue record for this book is available from the Library of Congress

British Library Cataloguing in Publication Data
A catalogue record for this book is available from the British Library

ISBN: 978-0-444-52157-6
ISSN: 0168-7069

For information on all Elsevier publications
visit our website at books.elsevier.com

Working together to grow
libraries in developing countries

www.elsevier.com | www.bookaid.org | www.sabre.org

ELSEVIER   BOOK AID
            International   Sabre Foundation

Contents

Human Viruses in Water
Albert Bosch (Editor)
DOI 10.1016/S0168-7069(07)17001-4

Chapter 1

Overview of Health-Related Water Virology

Wilhelm O.K. Grabow
Department of Microbiology, University of Pretoria, South Africa

Historical perspective

Viruses play an important role in biological mechanisms by which nature maintains a balance among living organisms on earth. The role of viruses in controlling numbers of human beings is illustrated by the impact of variola viruses (smallpox) which killed an estimated 10–15 million people per year until as recently as 1967. Others include influenza viruses, which reduced numbers of humans by some 20 million during the 1918–1919 pandemic, and measles and hepatitis viruses. Eventually mankind rose above all other living organisms with intelligent and innovative resistance to biological mechanisms for controlling its proliferation at the cost of other lives on earth. The role of smallpox was finally eliminated in 1977 by eradication of the virus, and the impact of viruses like influenza, poliomyelitis and measles was restricted by vaccination. Today we have reached the point where man has severely disrupted the balance of living organisms on the planet. It is almost sad to note how nature keeps fighting a seemingly lost battle by bringing in reinforcements in the form of new viruses like the human immunodeficiency virus (HIV) and mutants of old stalwarts like influenzaviruses with new battle strategies.

Viruses affect all forms of life, from single cellular plants like bacteria and single cellular animals (protozoa), to the highest forms of plants and animals, including man. Remarkable features of virus–host relationships include the variety of mechanisms by which different viruses are transmitted from one host to the next. Some viruses are highly host specific, like HIV, while others are less host specific, like influenzaviruses. Different viruses are designed for specific modes of transmission. For instance, viruses such as HIV, rabies and haemorrhagic fever viruses, are designed for direct inoculation of contaminated body fluids from an infected host into the tissue or blood stream of a new host. Viruses like influenza and measles are designed for airborne transmission and inhalation of air containing the viruses by a new host. Then there is the large group of enteric viruses that primarily infect the intestinal tract and are typically transmitted by the faecal-oral route often involving the ingestion of food and water contaminated with the viruses. However, there are no rules and regulations cast in concrete. There are exceptions to all these principles, allowing viruses to exploit alternative options when it serves their best interest.

The earliest record of diseases caused by enteric viruses may well be the report in the Babylonian Talmud that hepatitis was common in the fifth century BC (Zuckerman, 1983). It would appear that the most likely cause of the hepatitis referred to was hepatitis A and/or E viruses, both of which enteric viruses typically transmitted by food and water (Grabow, 2002; Chapter 3).

Impact of human viruses in water

Health impact

The global impact of waterborne and water-related diseases is difficult to assess. This is due to a lack of data, many variables and shortcomings in epidemiological studies and the interpretation of results. Additional reasons include the difficulty to confirm the source of infections, sub-clinical infections and secondary transmission of infections, increased susceptibility to infections in certain communities due to undernourishment and immune incompetence, reduced susceptibility in others due to immunity and geographical and seasonal distribution of diseases (Gerba et al., 1996a).

Consequently estimates of the health impact of these diseases vary. The following gives an indication.

Infectious diarrhoea or gastroenteritis is the most frequent, non-vector, water-related health outcome, in both the developed and developing world. Diarrhoea causes approximately 2.2 million deaths per year, mostly among children under the age of five, and while water is not solely responsible, water sanitation and hygiene are extremely important factors in this death toll (Prüss-Üstün and Fewtrell, 2004). The World Health Organization (WHO) estimated that every year there are 1.7 million deaths related to unsafe water, sanitation and hygiene, mainly through infectious diarrhoea. Some 4 billion cases of diarrhoea annually account for over 82 million disability adjusted life years (DALYs), representing 5.7% of the global burden of disease and placing diarrhoeal diseases as the third highest cause of morbidity and sixth highest cause of mortality (Prüss and Havelaar, 2001). In addition, waterborne disease is a major threat to millions who live in underdeveloped and informal conditions, or are displaced or otherwise affected by conflicts and disasters. Although the developing world is hardest hit by waterborne diseases, developed countries are also affected. For instance, the largest outbreak of a waterborne disease on record with some 403,000 cases of cryptosporidiosis occurred in

1993 in Milwaukee, a highly developed modern city in the USA (MacKenzie et al., 1994).

There is reason to believe that the health impact of waterborne diseases, and particularly those caused by viruses, tends to be underestimated (Regli et al., 1991; Gerba et al., 1996a). For instance, mortality data do not reflect the large number of infected individuals who suffer from clinical manifestations that range from mild unreported discomfort to non-fatal severe illness, with far-reaching socio-economic implications (Pegram et al., 1998). Waterborne and water-related diseases are associated with exposure to water environments in many ways. These include treated waters like those used for drinking and recreation in swimming pools and related facilities, and in food processing and other industrial activities, as well as untreated waters used for drinking, recreation and agricultural purposes such as crop irrigation and animal husbandry.

Expressed in terms easier to understand, data on waterborne diseases have been calculated for purposes of comparison as equivalent to a jumbo jet with 400 children and 100 adults on board crashing with no survivors every half hour around the clock (see Grabow, 1996). This illustration is based on authentic estimates that some 50,000 people die each day in the world due to waterborne and water-related diseases.

Viruses are a major cause of waterborne and water-related diseases. Extreme examples include the outbreak of 300,000 cases of hepatitis A and 25,000 cases of viral gastroenteritis in 1988 in Shanghai caused by shellfish harvested from a sewage-polluted estuary (Halliday et al., 1991). In 1991, an outbreak of 79,000 cases of hepatitis E in Kanpur was ascribed to polluted drinking water (Ray et al., 1991).

Socio-economic impact

Although the mortality of many waterborne diseases is relatively low, the socio-economic impact even of non-fatal infections is immense. Undetected diarrhoeal illnesses are common but generally not severe; their significance is often unrecognised and many illnesses are unreported. The societal cost of the so-called "mild gastrointestinal illnesses" is several orders of magnitude higher than the costs associated with acute hospitalised cases. In the US, the annual cost to society of gastrointestinal infectious illnesses was estimated as $19,500 million dollars (1985 US dollars) for cases with no consultation by physician, $2750 million dollars for those with consultations, and only $760 million dollars for those requiring hospitalisation. From the data collected during the Canadian studies and based on reported symptoms in the US (population of 300 million individuals) the estimate of the cost of waterborne illness ranges from US$269 to 806 million for medical costs and US$40 to 107 million for absences from work. These figures illustrate the enormous economic costs of endemic gastrointestinal illnesses, even in societies where waterborne disease is not perceived to be a problem (Payment, 2006b).

The socio-economic costs of epidemics and outbreaks of diseases with more severe illness and higher mortality rates such as cholera, typhoid fever and shigellosis are much higher (Pegram et al., 1998).

Viruses associated with waterborne transmission

Viruses predominantly associated with waterborne transmission are members of the group of enteric viruses that primarily infect cells of the gastrointestinal tract, and are excreted in the faeces of infected individuals. The viruses concerned are highly host specific, which implies that their presence in water environments is sound evidence of human faecal pollution. In some cases different strains of a viral species, or even different species of a viral genus, may infect animals. The extent of the host specificity of enteric viruses is such that it is used as a valuable tool to distinguish between faecal pollution of human and animal origin, or to identify the origin of faecal pollution. The hepatitis E virus may be the only meaningful exception to this rule, having strains which seem to infect both humans and certain animals, complying with the definition of a zoonosis (Grabow, 2002; Maluquer de Motes et al., 2004; Chapter 3). The following is a summary of typical human enteric viruses:

Adenoviruses

The family Adenoviridae consists of the genus *Mastadenovirus* associated with mammals, and three other genera associated with a spectrum of animals including birds and reptiles. The 51 antigenic types of human adenoviruses (HAds) consist of a double-stranded DNA genome in a non-enveloped icosahedral capsid with diameter about 80 nm and unique fibres. HAds cause a wide range of infections with a spectrum of clinical manifestations in the gastrointestinal, respiratory and urinary tracts, as well as the eyes. Relevant examples include types 40 and 41, which are an important cause of childhood gastroenteritis, and types 3, 4 and 7 associated with pharyngo-conjunctival fever commonly known as "swimming pool conjunctivitis". HAds are excreted by infected individuals in high numbers (enteric HAds in numbers of $10^{11} \, g^{-1}$ of faeces) and occur in relatively large numbers in faecally polluted waters, often outnumbering cytopathogenic enteroviruses. They are relatively resistant to unfavourable conditions, notably ultraviolet light. Apart from types 40 and 41, most HAds are readily detectable by cell culture propagation. In view of these features they have been suggested as useful indicators for enteric viruses. The US EPA has included HAds in its drinking water Candidate Contaminant List (CCL) as a group of high-priority viruses for water research, together with the groups of entero-, rota- and caliciviruses. For further details see EPA (1989), Muniain-Mujika et al. (2003), WHO (2004), Van Heerden et al. (2005a,b).

Astroviruses

The family Astroviridae contains eight types of human astroviruses (HAstVs) consisting of a single-stranded RNA genome in a 28 nm diameter non-enveloped icosahedral capsid, which displays a characteristic Star of David surface image under the electron microscope. HAstVs are excreted in substantial numbers in the faeces of infected individuals and are readily detectable in polluted water environments. The virus is a common cause of gastroenteritis, predominantly in children. HAstVs do not readily produce a cytopathogenic effect (CPE) in cell cultures, but their nucleic acid is detectable by molecular techniques after amplification in cell cultures. For further details see Nadan et al. (2003), WHO (2004).

Caliciviruses

The family Caliciviridae contains the genera *Norovirus* (Norwalk-like viruses) and *Sapovirus* (Sapporo-like viruses), which typically infect humans (HuCVs), as well as two other genera associated with infections of a wide variety of animals including mammals, fish, reptiles and insects. HuCVs consist of a single-stranded RNA genome in a non-enveloped 35–40 nm diameter icosahedral capsid, which under optimal conditions displays 32 calicle-like (cup-like) structures on the surface. HuCVs are exceptionally difficult to detect. They do not even seem to infect available cell culture systems since no viral RNA is detectable in cell cultures exposed to the viruses. Much of the initial research on the viruses was, therefore, carried out in human volunteers. Today progress is accomplished by means of molecular techniques. Noroviruses (NoVs) may be excreted in numbers of $10^{10} g^{-1}$ of stool or more. They are highly infectious and the most common cause of gastroenteritis associated with water and food in all age groups. However, the most frequent routes of transmission are person-to-person contact and the inhalation of contaminated aerosols and dust particles, as well as airborne particles of vomitus. Although clinical symptoms typically including vomiting, abdominal cramps, headache and muscular pain are relatively mild and rarely last for more than 3 days, the socio-economic impact is enormous. HuCVs are notorious for outbreaks on cruise ships, at holiday resorts, hotels, schools and hospitals, causing interruptions with far-reaching implications. Only about 40% of infected cases present with diarrhoea. Since cases with vomiting in the absence of diarrhoea are common, the infection is also known as "winter-vomiting disease", even though there is no meaningful seasonal trend. The immune response is poor and immunological protection short-lived, many infections are sub-clinical, and reinfection of individuals by the same HuCV strain is common. Secondary spread occurs often. For further details see Graham et al. (1994), Monroe et al. (2000), Duizer et al. (2004), Maunula et al. (2004), WHO (2004), Chan et al. (2006), Chapter 2.

Enteroviruses

The family Picornaviridae includes the genus *Enterovirus* of which the following species infect humans: poliovirus (PV1-3), coxsackievirus A (CVA1-24 with no type 23), coxsackievirus B (CVB1-6), echovirus (EV1-35 with no types 10 and 28), and enterovirus (EV68-71). Other species of the genus infect animals, for instance, the bovine (ECBO) group of enteroviruses. Enteroviruses are among the smallest known viruses and consist of a single-stranded RNA genome in a non-enveloped icosahedral capsid with a 20–30 nm diameter. Some members of the genus are readily detectable by CPE in cell cultures, notably polio-, coxsackie B-, echo- and enteroviruses. Members of the genus *Enterovirus* are among the most common causes of human infection, with an estimated 30 million infections per year in the USA. The viruses are associated with a broad spectrum of diseases ranging from mild febrile illness to myocarditis; meningoencephalitis; poliomyelitis; haemorrhagic conjunctivitis; herpangina; Bornholm disease; hand, foot and mouth disease; diabetes mellitus and neonatal multi-organ failure. Chronic infections are associated with conditions such as polymyositis, dilated cardiomyopathy and chronic fatigue syndrome. There is reason to believe that the health implications of enterovirus infections are not fully understood, particularly in terms of the long-term effects of chronic infections that are not readily evident from epidemiological data. Most infections, particularly in children, are asymptomatic, but still result in excretion of large numbers of the viruses that may cause clinical disease in others. Poliomyelitis, which caused severe mortality and suffering for a long time, has almost been eradicated by vaccination, but mutants of live vaccine strains are of concern. Since enteroviruses are excreted in large numbers by many people, they are detected in large numbers in raw and treated water supplies worldwide by techniques commonly used for water analysis. They tend to be outnumbered, at least at times, only by adenoviruses. In view of their common presence, resistance to treatment and disinfection processes, and easy detection of some members, they are widely used in water quality assessment, control and monitoring.

The family Picornaviridae also includes the genus *Hepatovirus* with only one species, the hepatitis A virus (HAV). This virus, of which there is only one antigenic type, shares all the basic features of other picornaviruses, including primary infection of cells of the gastrointestinal tract. From here HAV readily spreads via the blood stream to the liver where it may cause serious damage known as acute hepatitis with jaundice a typical clinical symptom. HAV is highly infectious and one of the best-known waterborne diseases with well-defined records of outbreaks and cases. As with many other picornaviruses, up to 90% of infected individuals, particularly children, display no clinical symptoms of infection, but they do excrete the virus, which may cause clinical disease in others. Although the mortality is generally less than 1%, recovery is a slow process that may keep patients incapacitated for 6 weeks or longer, which has substantial burden of disease implications. Immunity acquired by natural infection is typically lifelong, but not vaccine-derived immunity, which may constitute risks for immunised individuals later in life. The virus is not detectable by conventional cell culture systems, but molecular detection of the viral RNA is well established. For further details see Bosch et al. (1991), Grabow et al. (1999, 2001), Grabow (2002), WHO (2004), Pavlov et al. (2005), Chapters 3 and 4.

Hepatitis E virus

The hepatitis E virus (HEV) has some unique genetic and epidemiological properties, which rendered classification into existing families and genera inappropriate. After various efforts of classification over many years failed, the virus with one antigenic type only has eventually been classified into its own exclusive genus *Hepevirus*, in its own family Hepeviridae. HEV and HAV share many clinical and epidemiological features, to the extent that they were identified as different viruses only relatively recently. HEV causes acute hepatitis and typical waterborne outbreaks very much like HAV, but there are differences. For instance, the incubation period is longer for HEV. Particularly important is that HEV has a mortality rate of up to 25% in pregnant women. Although HEV seems to occur in most parts of the world at least in animals, clinical disease and outbreaks in humans tend to have a specific geographical distribution, with high incidence in developing countries of India, Pakistan, Mexico and some parts of Africa. As mentioned earlier, outbreaks with tens of thousands of cases are on record for these areas. HAV, on the other hand, causes unprecedented clinical disease in non-immune populations all over the world. Immunity derived by natural infection is lifelong. One of the unique features of HEV is that it appears to be the only known enteric virus that is a typical zoonosis. There seems to be meaningful evidence that the same strains of HEV infect both humans and at least certain animals, notably swine, cattle, goats and rodents. This unfortunately implies that some animals may serve as reservoir for HEV strains that infect humans. For further details see Grabow (2002), WHO (2004), Chapter 3.

Rotaviruses

The genera *Rotavirus* and *Orthoreovirus* of the family Reoviridae are associated with water quality. The family has other genera that are irrelevant. Viruses belonging to this family consist of a double-stranded RNA genome in a non-enveloped icosahedral capsid with a diameter of 60–80 nm. The capsid has a characteristic double layer with spikes between the layers giving it the appearance of a wheel (Latin "rota").

Species of the genus *Rotavirus* are known as rotaviruses (RVs). They are divided into seven antigenic groups, A–G, each of which is subdivided into a number of subgroups. Certain members of groups A–C, predominantly group A, are associated with human infections (HRVs), while the rest infect a variety of animals including calves, swine, dogs, mice and monkeys. The stool of infected individuals may contain HRVs in numbers of $10^{12}\,g^{-1}$. No other enteric viruses are excreted in numbers this high. In addition, HRVs are highly infectious, and if infections are not treated in time, the mortality rate is high. Consequently it is not surprising that HRVs are the most important single cause of infantile death in the world. Typically, HRVs are the cause of 50–60% of hospitalised cases of children with acute gastroenteritis. The burden of disease of HRV infections is, therefore, extremely high. Despite faecal excretion in exceptionally high numbers, and confirmation that waterborne transmission may occur, the predominant route of transmission is by personal contact, droplets, aerosols and airborne particles. HRVs are not readily detectable by CPE in cell cultures. HRVs recovered from water have successfully been identified by infection of cell cultures followed by detection of the replicated RNA by molecular techniques. Some RVs, notably monkeys strains, are readily detected by CPE in cell cultures, and are used as models for research on HRVs with regard to behaviour in water treatment and disinfection, as well as HRV vaccine production.

Species of the genus *Orthoreovirus* are known as reoviruses. The name is derived from "Respiratory Enteropathogenic Orphan" virus. These are typical "orphan" viruses, referring to viruses that have not been associated to meaningful extent with any disease. There seem to be indications of an association with gastroenteritis under circumstances. The three antigenic types infect humans and a variety of animals, including cattle, mice, chimpanzees and monkeys, typically without indications of disease. They seem to replicate in the respiratory tract of healthy individuals and are excreted in large numbers in the faeces of many humans and animals. Reoviruses are, therefore, commonly detected in water environments, often outnumbering other viruses. They are readily detected by CPE in cell cultures, although it takes longer for the CPE to become visible than with enteroviruses. Since reoviruses are readily detectable, occur in relatively high numbers in faecally polluted water and constitute no health risk, they are often used as indicators for other viruses. For further details see Hopkins et al. (1984), Gerba et al. (1996b), WHO (2004), Van Zyl et al. (2006), Chapters 2–4.

The above six groups of enteric viruses are the best known and commonly associated with water quality because they are excreted in faeces and their detection in sewage-polluted water environments is well established. However, a variety of other viruses are also excreted in faeces. These include parvoviruses (Family Parvoviridae) and corona- and toroviruses (Family Coronaviridae), all of which are at least suspected of being associated with gastroenteritis under circumstances. Viruses excreted in urine, notably polyomaviruses (Bofill-Mas et al., 2001), are also relevant.

Viruses excreted in faeces: transmission by water

The excretion in faeces and detection in water does not necessarily imply that viruses are predominantly, typically or even to meaningful extent transmitted by water. For enteric adeno-, all entero- and astroviruses there is little if any meaningful evidence of waterborne transmission. Waterborne transmission of rotaviruses has been confirmed, but it is not a common route of transmission. Likewise, more than 30 years ago coxsackievirus infections have once been associated with bathing in polluted lake water, but water and food are not recognised as important vehicles for the transmission of these viruses. The same may apply to polyomaviruses excreted in urine, for which waterborne transmission has not yet been confirmed. Even noroviruses, the most common cause of waterborne disease, are not predominantly transmitted by water. The same applies to hepatitis A and E viruses. Virtually without exception viruses transmitted by the faecal-oral route are predominantly transmitted by routes other than food and water. The most important mechanisms involve personal contact and the transfer of viruses in droplets or the inhalation of viruses in aerosols or airborne particles. For instance, at the height of poliomyelitis epidemics some decades ago, swimming pools were closed not for fear of waterborne transmission of the viruses, but to restrict transmission among bathers by other routes. Likewise, to this day schools and related gatherings are closed at times of outbreaks to prevent the spread by personal contact of typical enteric viruses such as coxsackievirus A16 and enterovirus 71, the aetiological agents of hand, foot and mouth disease.

Obviously, the exposure to any viable viruses in water constitutes a certain risk of infection. Although in most cases the risk may be considered negligible, under circumstances it may take on catastrophic dimensions. Appropriate caution is therefore essential, and recommended guidelines to control the risk of waterborne viral infections should be strictly adhered to at all times (WHO, 2004). It should be noted that without exception the absence of commonly used faecal indicator bacteria, such as coliforms, is not reliable evidence of the absence of any enteric viruses, notably in water treated and disinfected for human consumption.

Avian influenza and SARS

It is feared that new strains of influenzaviruses may cause pandemics similar to the one that killed some 20 million people in 1918–1919. For a number of reasons, the global impact may be much larger this time (Webster, 1994). The mutation rate of influenzaviruses is exceptionally high because the single-strand RNA genome consists of eight segments that facilitates recombination among different strains (antigenic shift) in addition to point mutations (antigenic drift). Another important factor is that influenzaviruses have animal hosts, notably birds such as waterfowl and domestic chickens, and pigs, all of which occur in exceptionally large numbers in certain parts of the world. This promotes high-rate multiplication of influenzaviruses with abundant opportunities for mutations and recombinations among human and animal strains. Currently, concerns are that the highly virulent avian influenza A strain H5N1, which recently emerged, may undergo a mutation or recombination changing the host specificity to also infect humans. Human influenzaviruses replicate primarily in the respiratory tract, but avian strains primarily in the gastrointestinal tract of birds. This implies that the viruses are faecally excreted in large numbers into the water on which dense populations of waterfowl occur. Consequently it is feared that water may play an important role in the transmission of influenzaviruses among waterfowl and potentially also from waterfowl to humans. In assessment of the risks involved it should be taken into account that viruses excreted in faeces are not necessarily transmitted to meaningful extent by water as has been explained earlier. There is little information on the mode of transmission of influenzaviruses among waterfowl. Water may possibly not play a particularly important role in the transmission of influenzaviruses among waterfowl because the virus seems to spread equally rapid among birds with restricted exposure to faecally polluted water such as poultry in breading batteries and ostriches in high-density farming units. The most important route of transmission may be direct contact and inhalation of droplets or aerosols, as is the predominant route of influenzavirus transmission among humans.

Severely acute respiratory syndrome (SARS) caused by an apparently new strain of coronavirus was diagnosed for the first time in patients in the Guangdong Province of China in November 2002. The virus spread rapidly and proved highly virulent. By 10 April 2003, 2781 cases with 111 deaths had been reported from 17 countries on 3 continents. Since the virus resembles coronaviruses known to be part of the intestinal viral flora of many people, and it was detected in the stool of patients, an association with waterborne transmission was considered possible. However, as in the case of avian influenzaviruses, waterborne transmission has not yet been confirmed.

Both influenza- and coronaviruses have a typical envelope, which is a distinct difference from naked enteric viruses typically associated with waterborne transmission. Viral envelopes have a high lipid content, which renders them vulnerable to detergents, oxidising agents such as chlorine commonly used for water disinfection and other unfavourable environmental conditions. Since the viral receptor sites are located on the envelope, any damage to the envelope renders the virus noninfectious. Consequently enveloped viruses are not as resistant as typical enteric viruses to water treatment and disinfection processes.

The potential risk of infection associated with respiratory viruses such as influenza and SARS in water environments cannot be ignored. However, there is sound reason to believe that treatment and disinfection processes recommended for the acceptable control of enteric viruses (WHO, 2004) will also accommodate enveloped viruses with a substantial safety margin.

Virological analysis of water

Virological analysis of water is required for a number of purposes. These include research on the incidence and behaviour of viruses in water environments, assessment of the presence of viruses and the risk of infection, evaluation of the efficiency of treatment and disinfection processes and routine quality monitoring to test the compliance of water quality with guidelines and specifications. These analyses generally consist of the following basic components:

- Recovery of small numbers of viruses from large volumes of water.
- Detection of the recovered viruses.
- Confirmation of the infectivity, or potential health risk, of the viruses detected.

In view of the fundamental nature of viruses, their size and composition, and their mode of replication in specific host cells, each of these components constitutes major challenges. The following is an introductory summary of these challenges, with further details in other chapters of this book.

Recovery of viruses from water

A wide variety of procedures has been described for the recovery of viruses from water. The most commonly applied techniques are based on adsorption–elution methods using negatively or positively charged filters, ultrafiltration or extraction. The efficiency of recovery (EOR) depends on a number of variables, including the volume, turbidity and pH of the water samples under investigation. In some studies, an EOR of 50% was considered optimal under the conditions concerned, while other studies claimed higher levels of efficiency. The effect of the recovery procedure on the viability of viruses is also important. A reliable indication of the EOR of recovery procedures is of fundamental importance for purposes such as monitoring the compliance of water with quality guidelines and assessment of infection risks. Available evidence confirms that currently available methods are in need of improvement with regard to efficiency, cost and meaningful data on EOR for viruses (Grabow et al., 2001; Vivier et al., 2002; Maunula et al., 2004; Chapter 9).

Detection of viruses

Research on viruses in water started in the 1940s. One of the pioneers in the field was Joseph L. Melnick in the USA. He started his work on the detection of polioviruses in the East River where it flowed through New York City. He used vervet monkeys to detect the polioviruses. After ingestion the viruses infect the epithelial cells of the gastrointestinal tract of the monkey. These cells release replicated viruses into the blood stream of the monkey and via this route they reach central nervous cells in the brain and spinal cord. These cells are highly susceptible to infection by polioviruses. Replication of polioviruses in these cells causes a typical CPE, which results in readily detectable paralysis of the monkey. In addition, the damaged cells are clearly visible by microscopic analysis of brain and spinal cord autopsy specimens. Monkeys inoculated with test samples were observed daily over a few weeks for signs of paralysis. Those that displayed paralysis were sacrificed for microscopic analysis of brain and spinal cord specimens.

Melnick expressed the detection of polioviruses in river water by this method in "monkey infectious doses" of polioviruses. Despite this very tedious, time-consuming and labour intensive procedure, in which large numbers of vervet monkeys

W.O.K. Grabow

had to pay the ultimate toll, Melnick made some fundamentally import observations on the incidence and behaviour of viruses in water (Melnick, 1976).

The next major step forward in the development of techniques for the detection of viruses was the establishment of procedures for the laboratory cultivation of mammalian cells. This implied that cell cultures could be infected with viruses and the CPE caused by virus replication was readily detectable by microscopic analysis of the cells. During the 1950s, the cell culture detection of viruses became established as a routine procedure for the virological analysis of water. Cell cultures retained the role of fundamentally important tools in research on viruses to this day, and will probably carry on playing that role for a long time to come.

Despite attractive features, cell cultures have important shortcomings for the detection of waterborne viruses. This is due to the exceptional host specificity of enteric viruses. A small selection of these viruses, notably polio-, some coxsackie-, some echo-, some entero- and some adenoviruses, as well as reoviruses, readily infect cells in culture and cause a distinctive CPE. Monkey kidney cells are exceptionally susceptible to most of these viruses and are commonly used for their detection. A number of other cell cultures of animal and human origin, each with their own advantages and disadvantages, are also used.

Unfortunately, however, the great majority of the wide spectrum of enteric viruses fails to infect available cell cultures with the production of a detectable CPE. The reasons are not altogether clear, but are probably related to the loss of features required for viral replication under *in vitro* laboratory conditions. This is illustrated by viruses that successfully infect cell cultures, replicate their nucleic acid and produce capsid components, but fail to assemble complete virions and produce a CPE. These viruses are readily detectable by confirming the presence of their nucleic acid using molecular techniques. However, some enteric viruses seem to even fail to infect cell cultures, which may be due to the absence of viral adsorption sites on the cells. This includes the large group of noroviruses. Since noroviruses are so difficult to detect, much of the early information on these viruses was derived from research in which human volunteers were used to detect the viruses by infection and clinical symptoms of disease (Graham et al., 1994).

The next major breakthrough in the development of methods unfolded in the1960s when molecular techniques for the detection of viral nucleic acid were established (Metcalf et al., 1995). These techniques are based on the detection of the nucleic acid of viruses by a diversity of procedures including gene probe hybridisation, the polymerase chain reaction (PCR) and reverse transcriptase-PCR (RT-PCR). Important benefits of these techniques include the ability to detect any virus for which the nucleotide sequence of the nucleic acid is known. In addition, the techniques are highly specific and sensitive. Generally, they yield results in a shorter period of time than the isolation of viruses by cell culture propagation. Unfortunately, they also have shortcomings. Among these is the need for special procedures to distinguish between viable and non-viable viruses. Also, it is difficult to obtain quantitative data on the numbers of viruses detected by molecular techniques. The tests are relatively complicated and require well-trained staff and appropriate laboratory facilities. False positive results due to contamination of test specimens, and laboratory environments contaminated with amplicons of viral nucleic acid, are major risks. Research on the improvement and modification of molecular techniques is a high priority in many laboratories worldwide. Recent progress includes assessment of viability by RT-PCR detection of m-RNA (Ko et al., 2003), and enumeration of viruses by real-time quantitative RT-PCR (Choi and Jiang, 2005; Fuhrman et al., 2005; Ko et al., 2005; Van Heerden et al., 2005b).

Confirmation of viability and infectivity

Viability of viruses is confirmed when they infect cell cultures and produce a CPE. Viability may also be accepted for viruses, which infect cell cultures and replicate their nucleic acid but fail to complete the viral multiplication cycle to release complete virions and produce a visible CPE (Pintó et al., 1996; Reynolds et al., 1997; Grabow et al., 1999; Van Zyl et al., 2006). At least sometimes these incomplete multiplication cycles even go as far as the production of capsid components. Failure to complete multiplication cycles may be due to a number of factors. Among others, it is known that viral infection of host cells and multiplication in the *in vitro* conditions is less successful than under *in vivo* conditions in the natural host. Another important factor is that the transformed cell cultures generally used for laboratory work have lost a variety of their original features, including functions required by at least some viruses for normal multiplication (Grabow et al., 1992). Apart from the detection of m-RNA mentioned earlier, confirmation of the viability of viruses detected by molecular techniques may be carried out by infecting cell cultures followed by molecular detection of the nucleic acid of viruses, which failed to produce a CPE. This offers a sensitive and highly specific procedure for the qualitative detection of viable viruses (Reynolds et al., 1997; Grabow et al., 1999, 2001; Fuhrman et al., 2005). It has been confirmed that at least some viruses detected in water environments by means of conventional molecular techniques are non-viable (Sobsey et al., 1998; Fuhrman et al., 2005).

The ultimate confirmation of infectivity is by infection of the natural host and the production of clinical disease. In the case of human viruses this is not practical for routine laboratory work, although human volunteers have been used in research on some viruses (Graham et al., 1994; see Grabow, 2002). Today the use of laboratory animals such as monkeys has largely been phased out. Remaining work with live animals is restricted to, for instance, chicken embryos for influenzaviruses.

Viruses confirmed viable by cell culture infection and certain molecular techniques are, therefore, considered as infectious, at least for practical purposes. In this assumption it is taken into account that most enteric viruses infect the natural host more readily than available cell culture systems. Consequently the number of detected viruses confirmed viable by laboratory techniques is almost certainly an underestimate of the true number of viable and infectious viruses present (Grabow et al., 1999).

Although presently available techniques for the detection of viruses still have many shortcomings, they made it possible to obtain valuable information on the incidence and behaviour of viruses in water, and on the risk of infection they constitute. However, in combination the cumulative effect of all the variables in viral detection methods mentioned almost certainly results in a major underestimate of the true number of viruses in water under investigation. This has implications for water quality management, risk assessment, practical safety guidelines and specifications, and routine quality monitoring of water supplies.

Quality guidelines

Infectious diseases are the most important concern about the quality of water intended for human use (Craun et al., 1994). WHO (2004) emphasises that "The potential health consequences of microbial contamination are such that its control must always be of paramount importance and must never be compromised." Over many years a traditional approach to the management of safe drinking water was established and based predominantly on guidelines and specifications for (EPA, 1989; WHO, 1996; EC, 1998; SABS, 2001):

- Raw water quality in terms of bacterial indicators of faecal pollution.
- Efficiency of treatment and disinfection.
- Final quality in terms of the absence of faecal indicator bacteria, notably coliforms.

Pathogenic micro-organisms were rarely or only vaguely referred to in these guidelines and specifications. This is largely because it would be impractical to include recommendations for the wide variety of pathogens that may be transmitted by water. Major considerations included cost, time of analysis, expertise and lack of suitable techniques. This applied in particular to viruses that required complicated and expensive techniques for detection, while many viruses of prime importance remained to be identified and characterised. In addition, not enough information was available on the wide spectrum of viruses involved to define meaningful and practical guidelines for routine quality monitoring. Some recommendations mention an unqualified and undefined "absence of viruses", which is basically meaningless because it is practically impossible, and unnecessary, to sterilize water supplies. Some of the recommendations are a little more specific and refer to enteroviruses or enteric viruses. In practice these terms referred to the small group of enteric viruses that cause a cytopathogenic effect in certain cell cultures, because that was the only practical virus detection technology available at the time. Tests for these viruses are basically restricted to an indicator function because very few of them are typically associated with waterborne diseases.

Despite shortcomings, this approach to quality testing has played a valuable role in water quality management and will undoubtedly carry on doing so for a long time to come (see Chapters 11 and 12). The principles concerned also played a fundamental role in the establishment of a wide spectrum of water treatment and disinfection processes for preparing acceptably safe water supplies (see Chapter 6). Today, it is possible to even directly reclaim safe drinking water from wastewater.

However, as expertise and technology for the detection of pathogens and waterborne diseases were refined, shortcomings in the above approach to the management of drinking water safety were disclosed. For instance, according to Payment et al. (1991) at least in certain situations as much as 35% of household infectious gastroenteritis may be caused by drinking water supplies which meet conventional quality specifications. Cases and outbreaks of waterborne disease associated with conventionally treated drinking water have been reported in many epidemiological studies (Hejkal et al., 1982; Zmirou et al., 1987; Bosch et al., 1991; MacKenzie et al., 1994; Payment et al., 1997; Payment, 2006b). Shortcomings of widely accepted guidelines for the microbiological safety of drinking water are also confirmed by the detection of pathogens, notably viruses, in supplies that have been treated according to specifications and comply with limits for faecal indicator bacteria (Keswick et al., 1984; Payment et al., 1985; Rose et al., 1986; Regli et al., 1991; Moore et al., 1994; Grabow et al., 2001; Vivier et al., 2004). For instance, in routine monitoring of drinking water supplies which complied with all specifications, enteroviruses were detected in 17%, adenoviruses in 4% and hepatitis A virus in 3% of 413 samples analysed over 2 years (Grabow et al., 2001). All viruses detected were confirmed viable by replication of nucleic acid in cell cultures. There is sound reason to believe that these data, as in most other studies, represent an underestimate of the true number of viruses present. The routine monitoring included conventional tests for coliform bacteria and heterotrophic plate counts, as well as tests for somatic and F-RNA coliphages using presence–absence tests on 500 ml samples. By definition the detection of any viruses fail the widely accepted drinking water quality guidelines and specifications referred to earlier.

Further shortcomings of the above approach to drinking water quality management were disclosed when guidelines based on a quantifiable acceptable risk of infection were established (Haas et al., 1999; Chapter 8). In 1989, the US Environmental Protection Agency (EPA) defined one infection per 10 000 consumers per year as an acceptable risk for drinking water (EPA, 1989; Macler, 1993; Macler and Regli, 1993). This definition of an acceptable risk of infection has since been used worldwide at least as a guideline for drinking water quality. Another approach to the definition of an acceptable risk for drinking water has been defined by the WHO (2004). This is based on the burden of disease constituted by a water supply, and a limit of 10–6 DALYs has been suggested. Risk assessment analyses carried out on data for viable viruses in drinking water supplies which complied with all requirements for treatment and disinfection reveal that these supplies exceeded the recommended level of acceptable risk of infection (Regli et al., 1991; Haas et al., 1993; Crabtree et al., 1997; Grabow et al., 2001; Van Heerden et al., 2005b). Available data suggest that the number of drinking water supplies worldwide that comply with this level of an acceptable risk may be rather restricted (Grabow et al., 2001). The same seems to apply to the virological quality of swimming pool water

(Van Heerden et al., 2005a) and other water environments used for recreational purposes (Fuhrman et al., 2005).

The above data and considerations outline shortcomings and controversies in traditional water quality management practices of the past. Shortly, there is sound evidence that many drinking water supplies, which comply with specifications for raw water quality, treatment, disinfection and faecal indicators on the one hand, exceed recommendations for viruses and an acceptable risk of infection at least for viruses on the other hand. This situation is confusing to water supply utilities, water quality authorities and others concerned because now there is evidence that traditional specification for water treatment, disinfection and indicator monitoring fail to produce acceptably safe drinking water. Resolution may be approached by one or both of the following options:

- Tighten the specifications for treatment and disinfection to obtain water that complies with the recommended absence of viruses and an acceptable risk of infection. This will have major financial and practical implications for the water industry (Clark et al., 1993; Regli et al., 1993).
- Retain specifications for treatment and disinfection unchanged by relaxing recommendations for the absence of viruses and an acceptable risk of infection. This may prove unacceptable from a public health point of view (WHO, 2004).

Another approach (WHO, 2004) is summarised in the next section.

Shortcomings of faecal bacteria like *Escherichia coli* as indicators for the potential presence and survival of enteric viruses in water environments were already noted by Melnick in his pioneering work during the 1940s, and are not surprising. Reasons are based on major differences in the composition, structure, size and resistance to unfavourable conditions including water treatment and disinfection processes. Other reasons for the absence of a direct correlation in numbers of viruses and faecal indicators include differences in excretion by the general population and infected individuals in terms of numbers, seasonal incidence, epidemics and geographic distribution (Grabow, 1996; Chapter 5). The same applies to the value of faecal bacteria as indicators for many other pathogens, notably protozoan parasites.

Despite limitations as indicators for the potential presence and behaviour of viruses, faecal bacteria have a long history of valuable indicators of faecal pollution, and of the efficiency of water treatment and disinfection processes (WHO, 2004). A variety of more resistant indicators is widely used to supplement the shortcomings of faecal bacteria for selected purposes. More resistant indicators with valuable features include spore-forming bacteria and bacteriophages. The latter proved particularly useful as indicators for human viruses because they share many fundamentally relevant features (Grabow, 2001). The benefits of indicator organisms are due to be fully utilised in future water quality management strategies (Chapters 6, 7, 11, 12).

An important shortcoming of the above approach to water quality monitoring is endpoint analysis of grab samples. Basically this implies that by the time results are available for tests carried out on samples of the final product, the water is already in the distribution system and drunk by any number of consumers. It is often too late then to take remedial or preventive measures (Payment, 2006b).

Water safety plans

Despite progress in technology and expertise, waterborne diseases keep having far-reaching public health and socio-economic implications worldwide. In addition, new challenges emerge all the time (Ford and Colwell, 1996). These are due to factors such as an escalating world population of humans and domestic animals that increase faecal pollution of water resources while the demand for potable water that has to be derived from these sources increases. Also, the cycle of water reuse is getting shorter which results in a selection for organisms more resistant to treatment and disinfection processes, like viruses and protozoan parasites. This is reflected by the epidemiology of waterborne diseases (Craun, 1991). Another factor of concern is the ongoing appearance of new pathogens, mutants of pathogens and the re-appearance of pathogens due to changing conditions and selective pressures driven by closer contact in escalating populations, rapid and frequent movement of people and animals all over the globe, and changing lifestyle and standards of living (Nel and Markotter, 2006).

The challenges of keeping up with the escalating demand for water of acceptable quality, and the shortcomings of traditional approaches to water quality control outlined in the previous section, prompted research on new strategies for water quality management (Fewtrell and Bartram, 2001; Payment, 2006a). Combined inputs from experts all over the world led to the establishment of a Framework for Safe Drinking Water (WHO, 2004) based on the principles of HACCP (Hazard Assessment and Critical Control Points). Basically the strategy implies that the quality of water is controlled at a selected set of critical control points (CPs) in a multiple barrier drinking water treatment system. Typically, CPs are monitored by testing water quality using physico-chemical and microbiological parameters. However, other parameters may also be used, such as observational monitoring of livestock barriers and the integrity of groundwater sources. Raw water sources may be seen as the first barrier with its own set of control measures for quality protection. Final disinfection at a treatment plant is an important CP, but quality control commences throughout the distribution system. Risk assessment is used to define the quality of the final product, as well as the efficiency of each CP to accomplish the desired final quality. Once a functional system has been established it has benefits such as:

- Routine quality monitoring does not require complicated, expensive and time-consuming analysis of grab samples collected from the final product; routine monitoring is carried out by practical, reliable, cost-effective and rapid or

continuous physico-chemical and microbial indicator analysis of CPs; this includes the elimination of tests for viruses and other pathogens.
• Breakdown and failure at CPs is detected in time to take remedial or preventive measures before the water is released into the distribution system.
• The quality of the final product is based on quantifiable acceptable risks derived from assessment of infection risk and burden of disease data.

Although the principles of the new strategy for water quality management are clearly defined (WHO, 2004, 2006), the application in practice, referred to by terms such as "water safety plans", is at least in certain respects still in a developmental stage and subject to ongoing research. This includes a need for more data to reliably and effectively monitor the removal and inactivation of viruses at CPs. Available information on the efficiency of a number of treatment processes with regard to viruses may be considered sufficient to initiate the implementation of water safety plans. However, more details are required to comply with the ultimate objectives (WHO, 2004). For instance, many of the available data are based on research in which viruses were detected by CPE in cell cultures. As has been pointed out earlier, this does not take into account viruses that have been damaged to the extent that they fail to produce a CPE but remain viable and at least potentially infectious. Also, most of the tests were carried out by recovery procedures with restricted and poorly defined efficiency and detected by techniques with questionable reliability in many cases. Most of the data are restricted to readily detectable viruses such as poliovirus, which may indeed serve as reliable indicators for other viruses, but are themselves not associated with waterborne diseases to meaningful extent. Little information is available on viruses of primary importance such as caliciviruses. More accurate data on the behaviour and survival of viruses in treatment processes have to be compared to those of practical indicators in order to establish appropriate routine monitoring procedures for CPs. Definition of ultimate goals for water quality, from which the efficiency of CPs is calculated, requires assessment of risk of infection and burden of diseases at least for representative (model) viruses such as rota, coxsackie B or hepatitis A, on which meaningful data required for these estimates are available.

Another important aspect of the framework for safe drinking water that seems to require attention is the recommendation to design goals on the basis of health-based targets. It is recommended that each country should have its own realistic targets designed by national authorities that take into consideration variables such as relevance to local conditions including economic, environmental, social and cultural factors. Considerations would have to include public health priorities and burden of disease, as well as financial, technical and institutional resources, and possibly also factors such as susceptibility to infection of communities with compromised immune status due to undernourishment or AIDS. Accomplishing these objectives may not prove easy in a world where over more than 100 years practices and perceptions based on specifications for treatment, disinfection and end-point monitoring for coliforms, got cast in concrete. This seems to be confirmed by slow progress along these lines. The acceptable risk for drinking water of one infection per 10,000 consumers per year was recommended in the USA in 1989. In 2001, this same acceptable risk was accepted as a standard in The Netherlands (Netherlands, 2001) and there are indications that other countries may be moving into the same direction. However, as far as can be established, no country has yet accepted an alternative level of infection risk as an official guideline or standard for drinking water. Many countries, particularly in the developing world, may not find it easy to define their own health-based targets for drinking water based on a risk of infection or burden of disease as recommended. Reasons include lack of expertise and relevant information. It also seems unlikely that countries will readily accept a health-based target that differs from that in countries such as the USA and The Netherlands because it would basically be a political decision with implications for the image of the country, which affects international relations, trade, tourism and many other aspects.

The acceptable risk of infection for drinking water recommended by the US EPA is used as a valuable unofficial guideline worldwide, and as an indication of what such a figure might look like. However, it possibly has the disadvantage of creating false expectations in the mind of many people. The EPA recommendation may be feasible in a developed country such as the USA. However, in many parts of the world, without the financial resources and infrastructure of the USA, this may be unrealistic. In fact, the guideline may not even be particularly realistic in the USA because, as mentioned earlier, many drinking water supplies in the USA seem to fail the recommendation (Regli et al., 1991; Haas et al., 1993; Crabtree et al., 1997).

It should be noted that the guideline specifies "infection" without reference to clinical disease. This is intentional to cover sub-clinical infections, which may result in secondary infection of others in whom it may cause clinical disease (Macler, 1993). In the case of enteric viruses typically associated with waterborne transmission, the great majority of infections are sub-clinical, which are difficult to monitor because there is no reliable relation between infections with or without clinical disease. The relation also varies for different communities and situations, and different viruses. These shortcomings are largely eliminated by health-based targets based on burden of disease. However, on their part these may prove more difficult to apply in practice and to take into consideration the significance of sub-clinical infections.

Challenges regarding the implementation of the WHO framework for safe drinking water may be accomplished by taking the process forward step-by-step. This seems to be the approach followed in Australia where apparently a preventive risk management plan, including performance targets and technology targets indirectly related to health outcomes, has been introduced (Australian Drinking Water Guidelines, 2004). This seems to succeed in changing general thinking from end-point testing to a preventive risk management plan, which is a major step forward (Sinclair and Rizak, 2004). The next step might be to develop the best way of implementing appropriate health-based targets in the management strategy.

According to unconfirmed reports a number of other countries are making progress by first introducing the principles of water safety plans and management of water quality by monitoring critical control points in multiple barrier systems rather than traditional end-point analysis.

Efficiency of treatment and disinfection processes

Viruses constitute special challenges in water treatment and disinfection because of their unique structure, composition and size. Basically, viruses typically associated with waterborne transmission consist of nucleic acid neatly wrapped up in a small protein capsid, which is exceptionally resistant to unfavourable environmental conditions, including those in water treatment and disinfection processes. These particles are specifically designed for transmission by the faecal-oral route via water and food environments. Viruses can only multiply inside metabolically active host cells and not in any water environment like many bacteria. As a result of these features, the behaviour and survival of viruses in water environments differs substantially in many respects from that of much larger organisms like coliform bacteria commonly used as indicators of faecal pollution.

Details on the behaviour and survival of the wide spectrum of viruses in water environments, including water treatment and disinfection processes, are restricted because it is relatively difficult, expensive and time-consuming to detect these viruses and confirm their infectivity. WHO (2004) contains a useful summary of available data on the relative survival of bacteria, viruses and protozoa (cysts and oocysts) in selected water treatment and disinfection processes. The table clearly illustrates the difference in resistance of these groups of organisms to drinking water treatment and disinfection processes. Viruses are substantially more resistant than bacteria to disinfectants like chlorine, monochloramine, chlorine dioxide, ozone and ultravioet light, and are less readily removed by processes such as sand filtration and membrane filtration. The data for viruses in this summary refer predominantly to readily detectable viruses such as vaccine strains of poliovirus detected by CPE in cell cultures. Although polioviruses are recognised as exceptionally resistant and serve as sound indicators for the survival of other viruses, the data do not cover viruses damaged to the extent that they fail to produce a CPE in cell culture, but are still viable and at least potentially infectious, as discussed earlier. For further details see Chapter 6.

Data on the efficiency of treatment and disinfection processes confirm that it is indeed possible to accomplish goals for drinking water treatment such as a 4-log reduction in unqualified numbers of viruses (EPA, 1989) and an acceptable risk of infection or burden of disease as discussed earlier. However, in practice this requires suitable facilities, treatment processes and meticulous management.

Future challenges

Pioneer Louis Pasteur said, "It is characteristic of science and progress that they continually open new fields to our vision". This is true today as much as it was at his time more than 100 years ago. Although in recent times, particularly since the 1940s, major progress has been made in knowledge about viruses in water, many questions remain unanswered. Among these are a better understanding and appreciation of the public health impact of viruses. This includes the potential health implications of latent infections of entero- and other viruses, long-term effects of viruses such as polyoma associated with cancer and the health effects of many viruses that have not yet been disclosed. The challenges are complex and increase in complexity with the emergence of new viruses and mutants of existing viruses, a process driven by factors such as escalating populations of humans and related animals. Examples include avian influenzaviruses and other respiratory agents like SARS, which are associated with water to an extent that remains to be clearly defined. At the same time, major progress is being made with new strategies for water-quality management. However, the optimisation and application in practice of approaches such as water safety plans based on HACCP principles with health-based quality targets require more information on viruses. There is sound reason to believe that new tools for the detection of viruses such as molecular techniques, assessment of the health impact of viruses based on risk assessment and burden of diseases, and refined expertise on epidemiological surveillance and interpretation of results are due to ensure exciting progress in the future (Chapter 13).

References

Australian Drinking Water Quality Guidelines. Australian National Health and Medical Research Council, and Agriculture and xxesource Management Council of Australia and New Zealand; Canberra: Commonwealth of Australia; 2004.

Bofill-Mas S, Formiga-Cruz M, Clemente-Casares P, Calafell F, Girones R. Potential transmission of human polyomaviruses through the gastrointestinal tract after exposure to virions or viral DNA. J Virol. 2001; 75: 10290–10299.

Bosch A, Lucena F, Diez JM, Gajardo R, Blasi M, Jofre J. Waterborne viruses associated with hepatitis outbreak. J Am Water Wks Ass. 1991; 83: 80–83.

Chan MCW, Sung JJY, Lam RKY, Chan PKS, Lee NLS, Lai RWM, Leung WK. Fecal viral load and norovirus-associated gastroenteritis. Emer Infect Dis. 2006; 12: 1278–1280.

Clark RM, Hurst CJ, Regli S. Costs and benefits of pathogen control in drinking water. In: Safety of Water Disinfection: Balancing Chemical and Microbial Risks (Craun GF, editor). Washington, DC: International Life Sciences Institute; 1993; pp. 181–198.

Choi S, Jiang SC. Real-time PCR quantification of human adenoviruses in urban rivers indicates genome prevalence but low infectivity. Appl Environ Microbiol 2005; 71: 7426–7433.

Crabtree KB, Gerba CP, Rose JB, Haas CN. Waterborne adenovirus: a risk assessment. Water Sci Technol. 1997; 35: 1–6.

Craun GF. Causes of waterborne outbreaks in the United States. Water Sci Technol. 1991; 24: 17–20.

Craun GF, Regli S, Clark RM, Bull RJ, Doull J, Grabow W, Marsh GM, Okun DA, Sobsey MD, Symons JM. Balancing chemical and microbial risks of drinking water disinfection, Part II Managing the risks. Aqua 1994; 43: 207–218.

Duizer E, Schwab KJ, Neill FH, Atmar RL, Koopmans MPG, Estes MK. Laboratory efforts to cultivate noroviruses. J Gen Virol. 2004; 85: 79–85.

EC. Council Directive 98/83/EC of 3 November 1998 on the quality of water intended for human consumption. Off J Eur Comm. 1998; L330: 32–54.

EPA. Maximum contaminant level goals for microbiological contaminants. Fed Reg Part II/ Part 141.52. USA: Environmental Protection Agency; 1989; 27527 pp.

Fewtrell L, Bartram J, editors. Water Quality: Guidelines, Standards and Health. Assessment of Risk and Risk Management for Water-Related Infectious Disease. World Health Organization Water Series. London: IWA Publishing; 2001.

Ford T.E., Colwell R.R. A Global Decline in Microbiological Safety of Water: A Call for Action. Washington, DC: American Academy of Microbiology; 1996; 40 pp.

Fuhrman JA, Xiaolin L, Noble RT. Rapid detection of enteroviruses in small volumes of natural waters by real-time quantitative reverse transcriptase PCR. Appl Environ Microbiol. 2005; 71: 4523–4530.

Gerba CP, Rose JB, Haas CN. Sensitive populations—Who is at the greatest risk? Int J Food Microbiol. 1996a; 30: 113–123.

Gerba CP, Rose JB, Haas CN, Crabtree KD. Waterborne rotavirus: a risk assessment. Water Res. 1996b; 30: 2929–2940.

Grabow WOK. Waterborne diseases: update on water quality assessment and control. Water SA 1996; 22: 193–202.

Grabow WOK. Bacteriophages: update on application as models for viruses in water. Water SA 2001; 27: 251–268.

Grabow WOK. Enteric hepatitis viruses. In: Guidelines for Drinking-Water Quality. 2nd ed. Addendum: Microbiological agents in drinking water. Geneva: World Health Organization; 2002; pp. 18–39, 142 pp.

Grabow WOK, Botma KL, de Villiers JC, Clay CG, Erasmus B. Assessment of cell culture and polymerase chain reaction procedures for the detection of polioviruses in wastewater. Bull Wrld Hlth Org. 1999; 77: 973–980.

Grabow WOK, Puttergill DL, Bosch A. Propagation of adeno-virus types 40 and 41 in the PLC/PRF/5 primary liver carcinoma cell line. J Virol Methods 1992; 37: 201–208.

Grabow WOK, Taylor MB, de Villiers JC. New methods for the detection of viruses: call for review of drinking water quality guidelines. Water Sci Technol. 2001; 43: 1–8.

Graham DY, Jiang X, Tanaka T, Opekun AR, Madore HP, Estes MK. Norwalk virus infection of volunteers: new insights based on improved assays. J Infect Dis. 1994; 170: 34–43.

Haas CN, Rose JB, Gerba CP. Quantitative Microbial Risk Assessment. New York: John Wiley & Sons; 1999.

Haas CN, Rose JB, Gerba CP, Regli S. Risk assessment of virus in drinking water. Risk Analysis 1993; 13: 545–552.

Halliday ML, Kang L-Y, Zhout T-K, Hu M-D, Pan Q-C, Fu T-Y, Huang Y-S, Hu S-L. An epidemic of hepatitis A attributable to the ingestion of raw clams in Shanghai, China. J Infect Dis. 1991; 164: 852–859.

Hejkal TW, Keswick B, LaBelle RL, Gerba CP, Sanchez Y, Dreesman G, Hafkin B, Melnick JL. Viruses in community water supply associated with an outbreak of gastroenteritis and infectious hepatitis. J Am Water Wks Ass. 1982; 74: 318–321.

Hopkins RS, Gaspard GB, Williams Jr. FP, Karlin RJ, Cukor G, Blacklow NR. A community waterborne gastroenteritis outbreak: evidence for rotavirus as the agent. Am J Publ Hlth. 1984; 74: 263–265.

Keswick BH, Gerba CP, Dupond HL, Rose JB. Detection of enteric viruses in treated drinking water. Appl Environ Microbiol. 1984; 47: 1290–1294.

Ko G, Cromeans TL, Sobsey MD. Detection of infectious adenovirus in cell culture by mRNA reverse transcription-PCR. Appl Environ Microbiol. 2003; 69: 7377–7384.

Ko G, Jothikumar N, Hill VR, Sobsey MD. Rapid detection of infectious adenoviruses by mRNA real-time RT-PCR. J Virol Methods 2005; 127: 148–154.

MacKenzie WR, Hoxie NJ, Proctor ME, Gradus MS, Blair KA, Peterson DE, Kazmierczak JJ, Addiss DG, Fox KR, Rose JB, Davis JP. A massive outbreak in Milwaukee of cryptosporidium infection transmitted through the public water supply. New Engl J Med. 1994; 331: 161–167.

Macler BA. Acceptable risk and US microbial drinking water standards. In: Safety of Water Disinfection: Balancing Chemical and Microbial Risks (Craun GF, editor). Washington, DC: International Life Sciences Institute; 1993; pp. 619–623.

Macler BA, Regli S. Use of microbial risk assessment in setting US drinking water standards. Int J Food Microbiol. 1993; 18: 245–256.

Maluquer de Motes C, Clemente-Casares P, Hundesa A, Martin M, Girones R. Detection of bovine and porcine adenoviruses for tracing the source of fecal contamination. Appl Environ Microbiol. 2004; 70: 1448–1454.

Maunula L, Kalso S, Von Bonsdorff CH, Ponka A. Wading pool water contaminated with both norovirus and astroviruses as the source of a gastroenteritis outbreak. Epidemiol Infect. 2004; 132: 737–743.

Melnick JL. Viruses in water. In: Viruses in Water (Berg G, Bodily HL, Lennette EH, Melnick JL, Metcalf TG, editors). Washington, DC: American Public Health Association; 1976; pp. 3–11.

Metcalf TG, Melnick JL, Estes MK. Environmental virology: from detection of virus in sewage and water by isolation to identification by molecular biology—a trip over 50 years. Ann Rev Microbiol. 1995; 49: 461–487.

Monroe SS, Ando T, Glass R. Introduction: Human enteric caliciviruses—An emerging pathogen whose time has come. J Infect Dis. 2000; 181(Suppl 2): S249–S251.

Moore AC, Herwaldt BL, Craun GF, Calderon RL, Highsmith AK, Juranek DD. Waterborne disease in the United States, 1991 and 1992. J Am Water Works Assoc. 1994; 86: 87–99.

Muniain-Mujika I, Calvo M, Lucena F, Girones R. Comparative analysis of viral pathogens and potential indicators in shellfish. Int J Food Microbiol. 2003; 83: 75–85.

Nadan S, Walter JE, Grabow WOK, Mitchell DK, Taylor MB. Molecular characterization of astroviruses by reverse transcriptase PCR and sequence analysis: comparison of clinical and environmental isolates from South Africa. Appl Environ Microbiol. 2003; 69: 747–753.

Nel LH, Markotter W. New and emerging waterborne infectious diseases. In: Water and Public Health (Grabow, WOK, editor). Encyclopedia of Life Support Systems (EOLSS),

Developed under the auspices of the UNESCO, Oxford, UK: Eolss Publishers; 2006. [http://www.eolss.net].

Netherlands. Artikel 31/4, Waterleidingbesluit, Bijlage A: Kwaliteitseisen, Tabel 1 Microbiologische parameters. In: Staatsblad van het Koninkrijk der Nederlanden. The Netherlands: Centrale Directie Juridische Zaken; 2001.

Pavlov DN, Van Zyl WB, Van Heerden J, Grabow WOK, Ehlers MM. Prevalence of vaccine-derived polioviruses in sewage and river water in South Africa. Water Res. 2005; 39: 3309–3319.

Payment P. Goals of water treatment and disinfection: reduction in morbidity and mortality. In: Water and Public Health. (Grabow, WOK, editor). Encyclopedia of Life Support Systems (EOLSS), Developed under the auspices of the UNESCO. Oxford, UK: Eolss Publishers; 2006a. [http://www.eolss.net].

Payment P. Diseases associated with drinking water supplies that meet treatment and indicator specifications. In: Water and Public Health. (Grabow, WOK, editor). Encyclopedia of Life Support Systems (EOLSS), Developed under the auspices of the UNESCO. Oxford, UK: Eolss Publishers; 2006b. [http://www.eolss.net].

Payment P, Richardson L, Siemiatycki J, Dewar R, Edwardes M, Franco E. A randomized trial to evaluate the risk of gastrointestinal disease due to consumption of drinking water meeting current microbiological standards. Am J Publ Hlth. 1991; 81: 703–708.

Payment P, Siemiatycki J, Richardson L, Renaud G, Franco E, Prévost M. A prospective epidemiological study of gastrointestinal health effects due to the consumption of drinking water. Int J Environ Hlth Res. 1997; 7: 5–31.

Payment P, Tremblay M, Trudel M. Relative resistance to chlorine of poliovirus and coxsackievirus isolates from environmental sources and drinking water. Appl Environ Microbiol. 1985; 49: 981–983.

Pegram GC, Rollins N, Espey Q. Estimating the costs of diarrhoea and epidemic dysentery in KwaZulu-Natal and South Africa. Water SA 1998; 24: 11–20.

Pintó RM, Abad FX, Gajardo R, Bosch A. Detection of infectious astroviruses in water. Appl Environ Microbiol. 1996; 62: 1811–1813.

Prüss A, Havelaar A. The global burden of disease study and applications in water, sanitation and hygiene. In: Water Quality: Guidelines, Standards and Health World Health Organization Water Series (Fewtrell L, Bartram J, editors). London: IWA publishing; 2001; pp. 41–59.

Prüss-Üstün A, Fewtrell L. Burden of disease: current situation and trends. In: Water and Public Health. (Grabow, WOK, editor). Encyclopedia of Life Support Systems (EOLSS), Developed under the auspices of the UNESCO. Oxford, UK: Eolss Publishers; 2004. [http://www.eolss.net].

Ray R, Aggarwal R, Salunke PN, Mehrotra NN, Talwar GP, Naik SR. Hepatitis E virus genome in stools of hepatitis patients during large epidemic in north India. Lancet 1991; 338: 783–784.

Regli S, Berger P, Macler B, Haas C. Proposed decision tree for management of risks in drinking water: consideration for health and socio-economic factors. In: Safety of Water Disinfection: Balancing Chemical and Microbial Risks (Craun GF, editor). Washington, DC: International Life Sciences Institute; 1993; pp. 39–80.

Regli S, Rose JB, Haas CN, Gerba CP. Modelling the risk from giardia and viruses in drinking water. J Am Water Wks Ass. 1991; 92: 76–84.

Reynolds KS, Gerba CP, Pepper IL. Rapid PCR based monitoring of infectious entero-viruses in drinking water. Water Sci Technol. 1997; 35: 423–427.

Rose JB, Gerba CP, Singh SN, Toranzos GA, Keswick B. Isolating viruses from finished water. Res Technol. 1986; 78: 56–61.

SABS. South African Standard. Specification: Drinking water. SABS-241:2001. 5th ed. Pretoria: South African Bureau of Standards; 2001.

Sinclair M, Rizak S. Drinking water quality management: the Australian Framework. J Toxicol Environ Health, Part A 2004; 67: 1567–1579.

Sobsey MD, Battigelli DA, Shin GA, Newland S. RT-PCR amplification detects inactivated viruses in water and wastewater. Water Sci Technol. 1998; 38: 91–94.

Van Heerden J, Ehlers MM, Grabow WOK. Detection and risk assessment of adenoviruses in swimming pool water. J Appl Microbiol. 2005a; 99: 1256–1264.

Van Heerden J, Ehlers MM, Heim A, Grabow WOK. Prevalence, quantification and typing of adenoviruses detected in river and treated drinking water in South Africa. J Appl Microbiol. 2005b; 99: 234–242.

Van Zyl WB, Page NA, Grabow WOK, Steele AD, Taylor MB. Molecular epidemiology of group A rotaviruses in water sources and selected raw vegetables in southern Africa. Appl Environ Microbiol. 2006; 72: 4554–4560.

Vivier JC, Ehlers MM, Grabow WOK. Detection of enteroviruses in treated drinking water. Water Res. 2004; 38: 2699–2705.

Vivier JC, Ehlers MM, Grabow WOK, Havelaar AH. Assessment of the risk of infection associated with coxsackie B viruses in drinking water. Water Sci Technol: Water Supply 2002; 2: 1–8.

Webster RG. While awaiting the next pandemic of influenza A. Bri Med J. 1994; 309: 1179–1180.

WHO. Guidelines for drinking-water quality: Volume 2—Health criteria and other sup-porting information. Second Edition. Geneva: World Health Organization; 1996; 973 pp.

WHO. Guidelines for Drinking-water Quality. Recommendations. vol. 1. 3rd ed. Geneva: World Health Organization; 2004; 515 pp.

WHO. Guidelines for Drinking-water Quality. First Addendum to the Third Edition. Recommendations. vol. 1. Geneva: World Health Organization; 2006; 68 pp.

Zmirou D, Ferley JP, Collin JF, Charrel M, Berlin J. A follow-up study of gastro-intestinal diseases related to bacteriologically substandard drinking water. Am J Publ Hlth. 1987; 77: 582–584.

Zuckerman AJ. The history of viral hepatitis from antiquity to the present. In: Viral Hepatitis: Laboratory and Clinical Science (Deinhardt F, Deinhardt J, editors). New York: Marcel Dekker; 1983; pp. 3–32.

Human Viruses in Water
Albert Bosch (Editor)
© 2007 Elsevier B.V. All rights reserved
DOI 10.1016/S0168-7069(07)17002-6

Chapter 2

Waterborne Gastroenteritis Viruses

Kellogg Schwab

Johns Hopkins Bloomberg School of Public Health, Baltimore, MD

Introduction

Many human viruses capable of causing gastroenteritis can be transmitted via water. Gastroenteritis is a communicable disease of sudden onset characterized by fever, abdominal cramps, nausea, vomiting, diarrhea and headache (Cheng et al., 2005). In developed areas of the world viral gastroenteritis is predominantly self-limiting with very low mortality. In developing areas of the world however, viral gastroenteritis can cause significant mortality, predominantly due to dehydration compounded by lack of rehydration therapy (Cheng et al., 2005). In developed areas of the world, the over-whelming majority of waterborne viral gastroenteritis illness originates from ingestion of human calicivirus (predominantly noroviruses), with rotavirus, astrovirus, adeno-virus, emerging viruses and enteroviruses (the latter two described in detail elsewhere in this book) also contributing to the cases of waterborne gastroenteritis. In devel-oping areas of the world, rotavirus is a major contributor to childhood morbidity and mortality with the aforementioned viruses also causing considerable morbidity. Little information is known about the magnitude of norovirus in these areas of the world as very few studies have been conducted.

There are numerous physical, chemical and biological factors that influence virus persistence in the environment. Viruses that cause waterborne gastroenteritis are highly infectious, small (25–100 nm), contain non-enveloped protein capsids and can persist for weeks to months in the aquatic environment (Hurst, 1991; Gordon and Toze, 2003; Wetz et al., 2004). Viral capsids are comprised of one or more structural proteins that have charged surfaces due to the ionic functional groups of the acidic and basic amino acids comprising the virion proteins. In the ambient pH (6–8) of most aqueous environments, viral protein capsids are neg-atively charged with this charge facilitating virion association to particles that can

subsequently affect both viral transport and resistance to environmental degradation and chemical inactivation (Gerba, 1984). The extremely small size of waterborne gastroenteritis viruses facilitates their transport though groundwater with
travel distances of hundreds to thousands of meters from sources of pollution
(Gerba, 1984). The combination of small size, environmental resistance, facile
transport and high infectivity of waterborne gastroenteritis viruses results in a
potent vector for significant waterborne gastrointestinal morbidity.

Calicivirus

The family *Caliciviridae* is comprised of four genera: *Norovirus, Sapovirus, Lago-
virus* and *Vesivirus*. Viruses in the latter two genera infect only animals. The
Norovirus and *Sapovirus* genera contain human and animal virus strains, but cross-
species transmission of strains within these genera has not been recognized to occur
(Atmar and Estes, 2006). Human noroviruses are the leading cause of non-bacterial
gastroenteritis worldwide (Atmar and Estes, 2006). Similar to other *Caliciviridae*,
norovirus contain a positive-sense, single-stranded RNA genome of approximately
7600 nucleotides in length (Jiang et al., 1990). Noroviruses are genetically and
antigenically diverse. Based on sequence information of the non-structural and
structural genome, the norovirus genus is divided into five genogroups (genogroup
I (GGI) to GGV), which can be divided further into genetic clusters, each represented by a prototype virus (Zheng et al., 2006). Three of these, GGI, GGII and
GGIV, contain human strains, whereas GGIII contains bovine strains and GGV
contains mouse strains. Unfortunately, to date there is no cell-culture system or
small-animal model for human noroviruses so little is known about the replication
of human noroviruses (Duizer et al., 2004). The recent report of a murine norovirus amenable to cell culture should facilitate information on the replication of
noroviruses (Wobus et al., 2004, 2006).
  Susceptibility to norovirus infection involves genetic resistance to infection and
acquired immunity. Resistance of an individual to Norwalk virus infection correlates with the individual's histo-blood group antigen expression. Secretor-
negative individuals have non-functional *FUT2* genes and do not express a
fucosyltransferase enzyme that is responsible for making H type-1 antigen that
appears on the surface of epithelial cells (Hutson et al., 2004). Secretor-negative
persons who do not express this antigen on their epithelial cells are uniformly
resistant to Norwalk virus infection following experimental challenge (Lindesmith
et al., 2003). Acquired immunity also plays a role in susceptibility to infection.
Short-term resistance to reinfection occurs following rechallenge with the same
strain 6–14 weeks later. Immunity is strain specific, in that infection can be induced
following challenge with a serologically distinct strain (Atmar and Estes, 2006). The
mechanism of immunity is unclear. In some studies absence of serum antibody was
correlated with resistance to infection, whereas in other studies, decreased infection
frequencies were associated with higher serum antibody levels (Schwab et al., 2000).
If an antibody is the mediator of immunity, this apparent paradox may be due to
differences in the populations studied. In studies where an antibody was correlated

with protection from infection, higher antibody levels may be due to recent or repeated infection of the population, whereas in studies where absence of an antibody was associated with resistance to infection, the absence of antibody is likely due to genetic mechanisms that prevent infection, and thus, adaptive immune responses (Atmar and Estes, 2006). A significant contributing factor for the high number of norovirus cases is the very low-infectious dose. Unfortunately, due to the complexity of establishing human volunteer feeding studies, few studies have examined this thoroughly. Moe and colleagues (Lindesmith et al., 2003) have infected volunteers with less than 10^4 viral genomes (measured as reverse transcription-polymerase chain reaction (RT-PCR) units) of Norwalk virus. One RT-PCR unit is estimated to be approximately 10 virions. These researchers have estimated that the infectious dose is less than 10–100 virions based on experimental human challenge studies.

The lack of a cell culture or animal-infectivity model has limited studies on the stability of noroviruses to environmental degradation. Human volunteer studies have shown that Norwalk virus is resistant to inactivation following treatment with chlorine levels frequently found in drinking water, and Norwalk virus is more resistant to inactivation by chlorine than poliovirus 1, human rotavirus (Wa), simian rotavirus (SA11) or f2 bacteriophage (Keswick et al., 1985). Norwalk virus retained infectivity for volunteers following either exposure to pH 2.7 for 3 h at room temperature, treatment with 20% ether at 4°C for 18 h, or incubation at 60°C for 30 min (Dolin et al., 1972). The development of newer assays including enzyme-linked immunosorbant assays (ELISA) and RT-PCR has increased the accuracy of disease surveillance for noroviruses to the point that the majority of gastroenteritis cases which otherwise might have been considered of "unknown etiology" are now attributable to noroviruses with an estimated 23 million cases per year in the United States (Mead et al., 1999). Numerous outbreaks originating from contaminated drinking water as well as recreational water have been reported (Blanton et al., 2006). Outbreaks of waterborne norovirus have been associated with private wells, small water systems, community water systems, groundwater contamination and even ice (Taylor et al., 1981; Kaplan et al., 1982; Wilson et al., 1982; Cannon et al., 1991; Lawson et al., 1991; McAnulty et al., 1993; Khan et al., 1994; Payment et al., 1994; Beller et al., 1997; Kim et al., 2005; Maunula et al., 2005). Several outbreaks have been associated with recreational waters (Baron et al., 1982; Kappus et al., 1982; Koopman et al., 1982; Gray et al., 1997; Hoebe et al., 2004) and outbreaks have also occurred when food has been washed with contaminated water. One such outbreak involved over 1,500 cadets and staff at the U.S. Air Force Academy. In that instance, celery had been soaked for 1 h in water from a hose used earlier in the day to clear clogged drains in the kitchen after sewage had backed up (Warner et al., 1991).

Rotavirus

Rotaviruses are the main etiological agent of viral gastroenteritis in infants and young children. Each year, rotavirus causes approximately 111 million episodes of gastroenteritis, 25 million clinic visits, 2 million hospitalizations and over

600,000 deaths in children less than 5 years of age with children in the poorest countries accounting for over 80% of these deaths (Parashar et al., 2003, 2006). Among children younger than 2 years, nearly half of all the cases of diarrhea requiring admission into a hospital can be attributed to rotavirus infection. Rotavirus is increasingly recognized as a cause of infectious diarrhea in adults as well as children (Anderson and Weber, 2004). Rotaviruses are a genus of the *Reoviridae* family and possess a genome of 11 segments of double-stranded RNA. The genome codes for six structural viral proteins (VPs; termed VP1, VP2, VP3, VP4, VP6 and VP7) and six non-structural proteins (NSPs; termed NSP1–NSP6) (Estes, 2001). Rotaviruses are triple-layered icosahedral particles approximately 75 nm in diameter with the RNA segments residing within the core. The core is surrounded by an inner capsid, composed mostly of VP6, the primary group antigen, and includes the epitope detected by most common diagnostic assays. Seven distinct groups of rotavirus (named A–G) have been shown to infect various animal species. Of these, only groups A, B and C have been reported as human pathogens (Estes, 2001). Group A is the primary pathogen worldwide and is the group detected by commercially available assays. Additional subgroups and serotypes can be identified by further characterization of VP4, VP6 and VP7 antigens (Kapikian et al., 2001). Group B appears to be limited to causing epidemic infection is Asia and the Indian subcontinent, whereas group C rotavirus causes endemic infections that frequently go unrecognized (Kapikian et al., 2001). Protective immunity to rotavirus is of short duration and reinfection occurs throughout life due to the great number of serotypes (Estes, 2001).

Rotaviruses are shed in extremely high numbers (up to $10^{10} g^{-1}$) from the feces of infected individuals and can persist in the environment for extended periods of time (Carter, 2005) resulting in the potential for recreational and drinking water contamination. There have been several waterborne outbreaks of rotavirus (Hopkins et al., 1984; Hung et al., 1984; Hrdy, 1987; Gerba et al., 1996; Kukkula et al., 1997; Villena et al., 2003) resulting in illness in both children and adults. Gerba et al. (1996) developed a risk assessment for waterborne rotavirus that estimated the public-health impacts from exposure to human rotavirus in drinking and recreational waters. By incorporating human volunteer data and reports of waterborne rotavirus outbreaks, they concluded that rotavirus illness is most severe for the very young, the elderly, and the immunocompromised with case fatality rates in the United States being 0.01% in the general population, 1% in the elderly and up to 50% in the immunocompromised (Gerba et al., 1996).

Astroviruses

Astroviruses have been isolated from humans as well as numerous animal species. Human astroviruses, classified in genus *Mamastrovirus* (human and animal astrovirus) within the family *Astroviridae*, are small (27–43 nm in diameter), non-enveloped viruses that have a single-stranded polyadenylated positive-sense RNA genome approximately 6.8–7.2 kb in length (Matsui and Greenberg, 2001). These viruses were given the name astro (astron: "star" in Greek) because of the characteristic five- or six-point star shape they often display when viewed by electron microscope (EM). To date, eight human astrovirus serotypes have been described with serotype 1 being the most globally prevalent, while the second most common serotype differs depending on the country. Serotypes 6 and 7 have rarely been detected (Guix et al., 2005). In most species, astroviruses are found in association with gastroenteritis although extraintestinal manifestations are also observed in avian species (Matsui and Greenberg, 2001). In humans, after a 1–4 day incubation period, the clinical symptoms of astrovirus infection present as self-limiting watery diarrhea that typically lasts for 2–3 days associated with vomiting, fever, anorexia and abdominal pain (Matsui and Greenberg, 2001). Protracted diarrhea and viral shedding has been observed as well as documented asymptomatic infections. While the disease is most often mild and does not result in severe dehydration, it can be more serious in immunocompromised children and adults, and elderly institutionalized patients (Matsui and Greenberg, 2001). Studies have indicated that astroviruses are common causes of diarrhea in children worldwide, and that most children are infected during their first two years of life (Herrmann et al., 1991; Glass et al., 1996). Astroviruses have been reported to be second only to rotaviruses as a cause of hospitalization for childhood viral gastroenteritis (Herrmann et al., 1991). Human volunteer studies have indicated that astrovirus, unlike norovirus, is of relatively low pathogenicity in adults (Glass et al., 1996).

Astroviruses were first observed by EM in stool specimens from infants with gastroenteritis. However, this method was relatively insensitive and studies based on EM detection indicated that astroviruses were a rare cause of gastroenteritis. For example, hospital-based EM studies indicated that incidence rates of astrovirus infection never exceed 4% (Herrmann et al., 1991). The development of more advanced methods of detection such as ELISA, nucleic acid sequence-based amplification (NASBA) and RT-PCR have revealed that astroviruses are a common cause of viral gastroenteritis in children worldwide and that adults can also become infected if exposed to large doses of astrovirus(Guix et al., 2005).

A major advance in the ability of investigators to study astroviruses came as the result of the finding that with the addition of trypsin into the assay, astroviruses could be propagated in the CaCo2 continuous line of colon carcinoma cells (Pinto et al., 1994). Researchers have now begun to investigate how infectious astroviruses respond to environmental degradation and drinking water disinfectants using both cell culture and integrated cell culture-reverse transcription-PCR (ICC-RT-PCR) procedures (Abad et al., 1997; Taylor et al., 1997; Brinker et al., 2000).

In addition to rotaviruses and noroviruses, astroviruses are now recognized as important etiologic agents of viral gastroenteritis in all age groups. However, astrovirus is not routinely screened for in stool or environmental samples, and data on the health impact of waterborne astrovirus are lacking. A recent study evaluated the potential impact of astrovirus in drinking water by assessing the relationship between incidence of gastroenteritis and astrovirus RNA prevalence in public drinking water systems in France (Gofti-Laroche et al., 2003). The study found that 12% (8/68) of the analyzed water samples were positive for astrovirus and that the presence of astrovirus RNA was associated with a significant increased risk of gastroenteritis, suggesting a role for waterborne astrovirus in the endemic level of gastoenteritis in the general population. Astroviruses have also been associated with an outbreak of gastroenteritis at a wading pool in Scandinavia (Maunula et al., 2004).

There have been limited studies on the effectiveness of water treatment in removing or inactivating astroviruses. Human astroviruses are similar in size, morphology and nucleic acid composition with other enteric viruses such as poliovirus, hepatitis A virus and caliciviruses and thus astroviruses most likely respond to treatment processes in a similar fashion. A recent study used ICC-RT-PCR to measure the persistence of astrovirus suspended in dechlorinated tap water and free chlorine (Abad et al., 1997). In dechlorinated tap water, the reduction of astrovirus infectivity following 60 days was 2 $\log_{10}$ units at 4°C and 3.2 $\log_{10}$ units at 20°C, and following 90 days there was 3.3 and 5 $\log_{10}$ units reduction at 4°C and 20°C, respectively. Following a 2-h incubation in the presence of 0.5 or 1 mg free chlorine per liter, residual astrovirus infectivity was still found with a 2.4 and 4 $\log_{10}$ reduction in titer, respectively.

Adenovirus

Adenoviruses are non-enveloped, icosohedral, double-stranded DNA viruses ranging in size between 80 and 110 nm in diameter. Adenoviruses are widespread in nature, infecting birds (genus *Aviadenovirus*) and many mammals (genus *Mastadenovirus*). They are generally species-specific, and few cross-species infections have been documented (Horwitz, 2001). Fifty-one different human adenovirus serotypes, organized into six subgroups (A–F), have been identified based on their responses to neutralizing antibodies and their ability to agglutinate red blood cells (Horwitz, 2001). Adenoviruses cause a variety of clinical manifestations including gastroenteritis, acute respiratory disease, pneumonia, epidemic keratoconjunctivitis, meningoencephalitis, hepatitis, myocarditis, acute febrile pharyngitis and pharyngoconjunctival fever (Horwitz, 2001). Adenoviruses can establish latent and persistent infections with viral shedding for weeks, and regardless of the site of primary infection, the virus is often shed from the gut in high numbers (Yates et al., 2006). Infection with any adenovirus may be accompanied by diarrhea and the viruses can be excreted even if diarrhea is not present (Spigland et al., 1966; Fox et al., 1969). It has been reported that fecal–oral transmission may account for most adenovirus infections in young children, irrespective of the primary site of infection. In addition, adenoviruses that are not traditionally considered "enteric" can cause diarrhea, thus the potential exists for waterborne transmission of all adenovirus serotypes, not just the "enteric" adenoviruses Ad40 and Ad41 (Spigland et al., 1966; Yates et al., 2006).

Adenoviruses are readily inactivated by most chemical disinfectants in routine use to treat drinking water, including free chlorine, chlorine dioxide and ozone, with inactivation kinetics similar to other viruses (Thurston-Enriquez et al., 2003a; Ballester and Malley, 2004; Thurston-Enriquez et al., 2005a,b). Several researchers have evaluated the effectiveness of UV disinfection on adenoviruses. Most of these studies were conducted using low-pressure UV sources at 254 nm denoted as UVC, (i.e. short wave or germicidal, < 280 nm) (Yates et al., 2006). These reports consistently demonstrated that adenoviruses are much more resistant to UVC disinfection than are any of the other viruses, parasites or vegetative bacteria studied (Gerba et al., 2002; Thompson et al., 2003; Thurston-Enriquez et al., 2003b; Ko et al., 2005; Nwachuku et al., 2005; Yates et al., 2006). Exposure to UVC causes damage to adenoviral DNA in the form of cyclobutane–pyrimidine dimers and pyrimidine–pyrimidone dimers (Jagger, 1967). In the presence of these dimers, the virus genome cannot be replicated because of the inhibition of polymerase activity. However, the UV-exposed viruses can still attach to and infect susceptible host cells (Rainbow and Mak, 1973). Once the viral genome is inside the host cell, the dimers can be repaired by nuclear excision repair, restoring the ability of the virus to be replicated (Rainbow and Mak, 1973).

As with the other human enteric viruses amenable to cell culture, infectivity assays have been used to detect adenoviruses from environmental samples. Although adenoviruses are culturable in several cell lines including fibroblasts, PLC/PRF/5, A549, CaCo2, HeLa and HEK 293 cells, many adenoviruses, such as Ad40 and Ad41, are difficult to culture and do not produce clear and consistent cytopathogenic effects (CPE). Some of the adenoviruses are very slow growing, necessitating multiple passages in fresh cells over several weeks to obtain CPE. Direct antigen detection by immunofluorescence techniques, enzyme immunoassay or latex agglutination has been used for clinical diagnosis of adenovirus infection (Horwitz, 2001) but these methods are usually too insensitive for detection of the low concentrations of adenoviruses usually present in environmental samples. Many researchers have used molecular methods targeting adenovirus nucleic acid to detect adenoviruses in various environmental media including swimming pools (Papapetropoulou and Vantarakis, 1998), coastal waters (Jiang et al., 2001), surface waters (Chapron et al., 2000) and sewage and shellfish (Pina et al., 1998). Many instances of recreational waterborne outbreaks attributed to adenovirus have been reported over the years (Foy et al., 1968; D'Angelo et al., 1979; Martone et al., 1980; Turner et al., 1987). However, assessing the health risk posed by adenoviruses in water is hampered by a lack of information of the infectious dose of enterically transmitted adenoviruses and the recovery efficiencies of adenovirus isolation methods (Crabtree et al., 1997; Heerden et al., 2005).

Conclusion

In conclusion, there are numerous viruses that can contribute to waterborne gastroenteritis resulting in significant morbidity and mortality worldwide. These

34 K. Schwab

viruses predominantly cause self-limiting gastroenteritis that is usually resolved in a few days by immunocompetent individuals. Waterborne viral illness in immunosuppressed individuals or children that are malnourished or have multiple infections can result in chronic, severe gastroenteritis that is potentially life threatening if appropriate rehydration therapy is not initiated. The lack of standardized viral recovery and detection techniques has severely limited our understanding of the magnitude of the potential public health threat of waterborne viruses that cause gastroenteritis. Further research in this field of study is warranted.

References

Abad FX, Pinto RM, Villena C, Gajardo R, Bosch A. Astrovirus survival in drinking water. Appl Environ Microbiol 1997; 63(8): 3119–3122.

Anderson EJ, Weber SG. Rotavirus infection in adults. Lancet Infect Dis 2004; 4(2): 91–99.

Atmar RL, Estes MK. The epidemiologic and clinical importance of norovirus infection: gastroenterol. Clin North Am 2006; 35(2): 275–290.

Ballester NA, Malley J. Sequential disinfection of adenovirus type 2 with UV-chlorine-chloramine. J Am Water Works Assoc 2004; 96(10): 97–103.

Baron RC, Murphy FD, Greenberg HB, Davis CE, Bregman DJ, Gary GW, Hughes JM, Schonberger LB. Norwalk gastrointestinal illness: an outbreak associated with swimming in a recreational lake and secondary person-to-person transmission. Am J Epidemiol 1982; 115: 163–172.

Beller M, et al. Outbreak of viral gastroenteritis due to a contaminated well. International consequences. JAMA 1997; 278(7): 563–568.

Blanton LH, Adams SM, Beard RS, Wei G, Bulens SN, Widdowson MA, Glass RI, Monroe SS. Molecular and epidemiologic trends of caliciviruses associated with outbreaks of acute gastroenteritis in the United States, 2000–2004. J Infect Dis 2006; 193(3): 413–421.

Brinker JP, Blacklow NR, Herrmann JE. Human astrovirus isolation and propagation in multiple cell lines. Arch Virol 2000; 145(9): 1847–1856.

Cannon RO, et al. A multistate outbreak of Norwalk virus gastroenteritis associated with consumption of commercial ice. J Infect Dis 1991; 164: 860–863 [published erratum appears in J Infect Dis 1992; 166(3): 698].

Carter MJ. Enterically infecting viruses: pathogenicity, transmission and significance for food and waterborne infection. J Appl Microbiol 2005; 98(6): 1354–1380.

Chapron CD, Ballester NA, Fontaine JH, Frades CN, Margolin AB. Detection of astroviruses, enteroviruses, and adenovirus types 40 and 41 in surface waters collected and evaluated by the information collection rule and an integrated cell culture-nested PCR procedure. Appl Environ Microbiol 2000; 66(6): 2520–2525.

Cheng AC, McDonald JR, Thielman NM. Infectious diarrhea in developed and developing countries. J Clin Gastroenterol 2005; 39(9): 757–773.

Crabtree KD, Gerba CP, Rose JB, Haas CN. Waterborne adenovirus: a risk assessment. Water Sci Technol 1997; 35(11–12): 1–6.

D'Angelo LJ, Hierholzer JC, Keenlyside RA, Anderson LJ, Martone WJ. Pharyngoconjunctival fever caused by adenovirus type 4: report of a swimming pool-related outbreak with recovery of virus from pool water. J Infect Dis 1979; 140(1): 42–47.

Dolin R, Blacklow NR, DuPont H, Buscho RF, Wyatt RG, Kasel JA, Hornick R, Chanock RM. Biological properties of Norwalk agent of acute infectious nonbacterial gastroenteritis. Proc Soc Exp Biol Med 1972; 140: 578–583.

Duizer E, Schwab KJ, Neill FH, Atmar RL, Koopmans MP, Estes MK. Laboratory efforts to cultivate noroviruses. J Gen Virol 2004; 85(1): 79–87.

Estes MK. Rotaviruses and their replication. In: Fields Virology (Knipe DM, Howley PM, editors). Baltimore, MD: Lippincott Williams and Wilkins; 2001; pp. 1747–1785.

Fox JP, Brandt CD, Wassermann FE, Hall CE, Spigland I, Kogon A, Elveback LR. The virus watch program: a continuing surveillance of viral infections in metropolitan New York families. VI. Observations of adenovirus infections: virus excretion patterns, antibody response, efficiency of surveillance, patterns of infections, and relation to illness. Am J Epidemiol 1969; 89(1): 25–50.

Foy HM, Cooney MK, Hatlen JB. Adenovirus type 3 epidemic associated with intermittent chlorination of a swimming pool. Arch Environ Health 1968; 17(5): 795–802.

Gerba CP. Applied and theoretical aspects of virus adsorption to surfaces. Adv Appl Microbiol 1984; 30: 133–168.

Gerba CP, Gramos DM, Nwachuku N. Comparative inactivation of enteroviruses and adenovirus 2 by UV light. Appl Environ Microbiol 2002; 68(10): 5167–5169.

Gerba CP, Rose JB, Haas CN, Crabtree KD. Waterborne rotavirus: a risk assessment. Water Res 1996; 30(12): 2929–2940.

Glass RI, et al. The changing epidemiology of astrovirus-associated gastroenteritis: a review. Arch Virol Suppl 1996; 12: 287–300.

Gofti-Laroche L, Gratacap-Cavallier B, Demanse D, Genoulaz O, Seigneurin JM, Zmirou D. Are waterborne astrovirus implicated in acute digestive morbidity (E.MI.R.A. study)? J Clin Virol 2003; 27(1): 74–82.

Gordon C, Toze S. Influence of groundwater characteristics on the survival of enteric viruses. J Appl Microbiol 2003; 95(3): 536–544.

Gray JJ, Green J, Cunliffe C, Gallimore CI, Lee JV, Neal K, Brown DW. Mixed genogroup SRSV infections among a party of canoeists exposed to contaminated recreational water. J Med Virol 1997; 52(4): 425–429.

Guix S, Bosch A, Pinto RM. Human astrovirus diagnosis and typing: current and future prospects. Lett Appl Microbiol 2005; 41(2): 103–105.

Heerden J, Ehlers MM, Vivier JC, Grabow WO. Risk assessment of adenoviruses detected in treated drinking water and recreational water. J Appl Microbiol 2005; 99(4): 926–933.

Herrmann JE, Taylor DN, Echeverria P, Blacklow NR. Astroviruses as a cause of gastroenteritis in children. N Engl J Med 1991; 324(25): 1757–1760.

Hoebe CJ, Vennema H, de Roda Husman AM, van Duynhoven YT. Norovirus outbreak among primary schoolchildren who had played in a recreational water fountain. J Infect Dis 2004; 189(4): 699–705.

Hopkins RS, Gaspard GB, Williams Jr. FP, Karlin RJ, Cukor G, Blacklow NR. A community waterborne gastroenteritis outbreak: evidence for rotavirus as the agent. Am J Public Health 1984; 74: 263–265.

Horwitz MS. Adenoviruses. In: Fields Virology (Knipe DM, Howley PM, editors). Baltimore, MD: Lippincott Williams and Wilkins; 2001; pp. 2301–2326.

Hrdy DB. Epidemiology of rotaviral infection in adults. Rev Infect Dis 1987; 9(3): 461–469.

Hung T, et al. Waterborne outbreak of rotavirus diarrhoea in adults in China caused by a novel rotavirus. Lancet 1984; 1(8387): 1139–1142.

Hurst CJ. Presence of enteric viruses in freshwater and their removal by the conventional drinking water treatment process. Bull World Health Organ 1991; 69(1): 113–119.

Hutson AM, Atmar RL, Estes MK. Norovirus disease: changing epidemiology and host susceptibility factors. Trends Microbiol 2004; 12(6): 279–287.

Jagger J. Introduction to Research in Ultraviolet Photobiology. Englewood Cliffs, NJ: Prentice-Hall, Inc.; 1967.

Jiang X, Graham DY, Wang KN, Estes MK. Norwalk virus genome cloning and characterization. Science 1990; 250: 1580–1583.

Jiang S, Noble R, Chu W. Human adenoviruses and coliphages in urban runoff-impacted coastal waters of Southern California. Appl Environ Microbiol 2001; 67(1): 179–184.

Kapikian AZ, Hoshino Y, Chanock RM. Rotaviruses. In: Fields Virology (Knipe DM, Howley PM, editors). Baltimore, MD: Lippincott Williams and Wilkins; 2001; pp. 1787–1833.

Kaplan JE, Goodman RA, Schonberger LB, Lippy EC, Gary GW. Gastroenteritis due to Norwalk virus: an outbreak associated with a municipal water system. J Infect Dis 1982; 146: 190–197.

Kappus KD, Marks JS, Holman RC, Bryant JK, Baker C, Gary GW, Greenberg HB. An outbreak of Norwalk gastroenteritis associated with swimming in a pool and secondary person-to-person transmission. Am J Epidemiol 1982; 116: 834–839.

Keswick BH, Satterwhite TK, Johnson PC, DuPont HL, Secor SL, Bitsura JA, Gary GW, Hoff JC. Inactivation of Norwalk virus in drinking water by chlorine. Appl Environ Microbiol 1985; 50: 261–264.

Khan AS, et al. Norwalk virus-associated gastroenteritis traced to ice consumption aboard a cruise ship in Hawaii: application of molecular-based assays. J Clin Microbiol 1994; 32: 318–322.

Kim SH, Cheon DS, Kim JH, Lee DH, Jheong WH, Heo YJ, Chung HM, Jee Y, Lee JS. Outbreaks of gastroenteritis that occurred during school excursions in Korea were associated with several waterborne strains of norovirus. J Clin Microbiol 2005; 43(9): 4836–4839.

Ko G, Cromeans TL, Sobsey MD. UV inactivation of adenovirus type 41 measured by cell culture mRNA RT-PCR. Water Res 2005; 39(15): 3643–3649.

Koopman JS, Eckert EA, Greenberg HB, Strohm BC, Isaacson RE, Monto AS. Norwalk virus enteric illness acquired by swimming exposure. Am J Epidemiol 1982; 115: 173–177.

Kukkula M, Arstila P, Klossner ML, Maunula L, Bonsdorff CH, Jaatinen P. Waterborne outbreak of viral gastroenteritis. Scand J Infect Dis 1997; 29(4): 415–418.

Lawson HW, Braun MM, Glass RI, Stine SE, Monroe SS, Atrash HK, Lee LE, Englender SJ. Waterborne outbreak of Norwalk virus gastroenteritis at a southwest US resort: role of geological formations in contamination of well water. Lancet 1991; 337: 1200–1204.

Lindesmith L, Moe C, Marionneau S, Ruvoen N, Jiang X, Lindblad L, Stewart P, LePendu J, Baric R. Human susceptibility and resistance to Norwalk virus infection. Nat Med 2003; 9(5): 548–553.

Martone WJ, Hierholzer JC, Keenlyside RA, Fraser DW, D'Angelo LJ, Winkler WG. An outbreak of adenovirus type 3 disease at a private recreation center swimming pool. Am J Epidemiol 1980; 111(2): 229–237.

Matsui SM, Greenberg HB. Astroviruses. In: Fields Virology (Knipe DM, Howley PM, editors). Baltimore, MD: Lippincott Williams and Wilkins; 2001; pp. 875–894.

Maunula L, Kalso S, von Bonsdorff CH, Ponka A. Wading pool water contaminated with both noroviruses and astroviruses as the source of a gastroenteritis outbreak. Epidemiol Infect 2004; 132(4): 737–743.

Maunula L, Miettinen IT, von Bonsdorff CH. Norovirus outbreaks from drinking water. Emerg Infect Dis 2005; 11(11): 1716–1721.

McAnulty JM, Rubin GL, Carvan CT, Huntley EJ, Grohmann G, Hunter R. An outbreak of Norwalk-like gastroenteritis associated with contaminated drinking water at a caravan park. Aust J Public Health 1993; 17(1): 36–41.

Mead PS, Slutsker L, Dietz V, McCaig LF, Bresee JS, Shapiro C, Griffin PM, Tauxe RV. Food-related illness and death in the United States. Emerg Infect Dis 1999; 5(5): 607–625.

Nwachuku N, Gerba CP, Oswald A, Mashadi FD. Comparative inactivation of adenovirus serotypes by UV light disinfection. Appl Environ Microbiol 2005; 71(9): 5633–5636.

Papapetropoulou M, Vantarakis AC. Detection of adenovirus outbreak at a municipal swimming pool by nested PCR amplification. J Infect 1998; 36(1): 101–103.

Parashar UD, Gibson CJ, Bresse JS, Glass RI. Rotavirus and severe childhood diarrhea. Emerg Infect Dis 2006; 12(2): 304–306.

Parashar UD, Hummelman EG, Bresee JS, Miller MA, Glass RI. Global illness and deaths caused by rotavirus disease in children. Emerg Infect Dis 2003; 9(5): 565–572.

Payment P, Franco E, Fout GS. Incidence of Norwalk virus infections during a prospective epidemiological study of drinking water-related gastrointestinal illness. Can J Microbiol 1994; 40(10): 805–809.

Pina S, Puig M, Lucena F, Jofre J, Girones R. Viral pollution in the environment and in shellfish: human adenovirus detection by PCR as an index of human viruses. Appl Environ Microbiol 1998; 64(9): 3376–3382.

Pinto RM, Diez JM, Bosch A. Use of the colonic carcinoma cell line CaCo-2 for in vivo amplification and detection of enteric viruses. J Med Virol 1994; 44(3): 310–315.

Rainbow AJ, Mak S. DNA damage and biological function of human adenovirus after u.v.-irradiation. Int J Radiat Biol Relat Stud Phys Chem Med 1973; 24(1): 59–72.

Schwab KJ, Estes MK, Atmar RL. Norwalk and other human caliciviruses: molecular characterization, epidemiology and pathogenesis. In: Microbial Foodborne Diseases: Mechanisms of Pathogenicity and Toxin Synthesis (Cary JW, Linz JE, Stein MA, Bhatnagar C, editors). Lancaster, PA: Technomic Publishing Company, Inc.; 2000; pp. 460–493.

Spigland I, Fox JP, Elveback LR, Wassermann FE, Ketler A, Brandt CD, Kogon A. The Virus Watch program: a continuing surveillance of viral infections in metropolitan New York families. II. Laboratory methods and preliminary report on infections revealed by virus isolation. Am J Epidemiol 1966; 83(3): 413–435.

Taylor JW, Gary Jr. GW, Greenberg HB. Norwalk-related viral gastroenteritis due to contaminated drinking water. Am J Epidemiol 1981; 114: 584–592.

Taylor MB, Grabow WO, Cubitt WD. Propagation of human astrovirus in the PLC/PRF/5 hepatoma cell line. J Virol Methods 1997; 67(1): 13–18.

Thompson S, Jackson JL, Suva-Castillo M, Yanko WA, Jack Z, Kuo J, Chen CL, Williams FP, Schnurr DP. Detection of infectious human adenoviruses in tertiary-treated and ultraviolet disinfected wastewater. Water Environ Res 2003; 75(2): 163–170.

Thurston-Enriquez JA, Haas CN, Jacangelo J, Gerba CP. Chlorine inactivation of adenovirus type 40 and feline calicivirus. Appl Environ Microbiol 2003a; 69(7): 3979–3985.

Thurston-Enriquez JA, Haas CN, Jacangelo J, Gerba CP. Inactivation of enteric adenovirus and feline calicivirus by chlorine dioxide. Appl Environ Microbiol 2005a; 71(6): 3100–3105.

Thurston-Enriquez JA, Haas CN, Jacangelo J, Gerba CP. Inactivation of enteric adenovirus and feline calicivirus by ozone. Water Res 2005b; 39(15): 3650–3656.

Thurston-Enriquez JA, Haas CN, Jacangelo J, Riley K, Gerba CP. Inactivation of feline calicivirus and adenovirus type 40 by UV radiation. Appl Environ Microbiol 2003b; 69(1): 577–582.

Turner M, Istre GR, Beauchamp H, Baum M, Arnold S. Community outbreak of adenovirus type 7a infections associated with a swimming pool. South Med J 1987; 80(6): 712–715.

Villena C, et al. A large infantile gastroenteritis outbreak in Albania caused by multiple emerging rotavirus genotype. Epidemiol Infect 2003; 131(3): 1105–1110.

Warner RD, Carr RW, McCleskey FK, Johnson PC, Elmer LM, Davison VE. A large nontypical outbreak of Norwalk virus. Gastroenteritis associated with exposing celery to nonpotable water and with *Citrobacter freundii*. Arch Intern Med 1991; 151: 2419–2424.

Wetz JJ, Lipp EK, Griffin DW, Lukasik J, Wait D, Sobsey MD, Scott TM, Rose JB. Presence, infectivity, and stability of enteric viruses in seawater: relationship to marine water quality in the Florida Keys. Mar Pollut Bull 2004; 48(7-8): 698–704.

Wilson R, Anderson LJ, Holman RC, Gary GW, Greenberg HB. Waterborne gastroenteritis due to the Norwalk agent: clinical and epidemiologic investigation. Am J Public Health 1982; 72: 72–74.

Wobus CE, et al. Replication of Norovirus in cell culture reveals a tropism for dendritic cells and macrophages. PLoS Biol 2004; 2(12): e432.

Wobus CE, Thackray LB, Virgin HW. Murine norovirus: a model system to study norovirus biology and pathogenesis. J Virol 2006; 80(11): 5104–5112.

Yates MV, Malley J, Rochelle P, Hoffman R. Effect of adenovirus resistance on UV disinfection requirements: a report on the state of adenovirus science. J Am Water Works Assoc 2006; 98(6): 93–106.

Zheng DP, Ando T, Fankhauser RL, Beard RS, Glass RI, Monroe SS. Norovirus classification and proposed strain nomenclature. Virology 2006; 346(2): 312–323.

Human Viruses in Water
Albert Bosch (Editor)
© 2007 Elsevier B.V. All rights reserved
DOI 10.1016/S0168-7069(07)17003-8

Chapter 3

Enteric Hepatitis Viruses

Rosa M. Pintó[a], Juan-Carlos Saiz[b]

[a]*Enteric Virus Laboratory, Department of Microbiology, School of Biology, University of Barcelona, Diagonal 645, 08028 Barcelona, Spain*
[b]*Laboratory of Zoonotic and Environmental Virology, Department of Biotechnology, Instituto Nacional de Investigación Agraria y Alimentaria (INIA). Ctra. Coruña km. 7.5, 28040 Madrid, Spain*

Background

The term "jaundice" was used as early as in the ancient Greece when Hippocrates described an illness probably corresponding to a viral hepatitis. However, it was not until the beginning of the twentieth century when a form of hepatitis was associated to an infectious disease occurring in epidemics and the term "infectious hepatitis" was established. Later on in the early 1940s two separate entities were defined: "infectious" and "serum" hepatitis, and from 1965 to nowadays the different etiological agents of viral hepatitis have been identified. Although all viral hepatitis are infectious the aforementioned terms refer to the mode of transmission, corresponding the "infectious" entity to those hepatitis transmitted through the fecal-oral route and the "serum" hepatitis to those transmitted parenterally. Thus, the infectious or enteric hepatitis include two types: hepatitis A and hepatitis E, which will be reviewed here.

Hepatitis A

Natural course and epidemiology of hepatitis A

Hepatitis A infection may develop asymptomatically. This type of subclinical infection is most common among young children (under 5), while in older children and in the adulthood the infection usually proceeds with symptoms (Previsani et al., 2004). In this latter case, the clinical course of hepatitis A is indistinguishable from that of other types of acute viral hepatitis. The clinical case definition for hepatitis A is an acute illness with moderate onset of symptoms (fever, malaise, anorexia, nausea, abdominal discomfort, dark urine) and jaundice, and elevated serum bilirubin and aminotransferases levels later on. The incubation period of hepatitis A ranges from 15 to 50 days and clinical illness usually does not last longer than 2 months, although 10–15% of patients have prolonged or relapsing signs and symptoms for up to 6 months (Sjogren et al., 1987; Glikson et al., 1992). In fact, with the advent of new highly sensitive techniques even in normal clinical courses a high and long lasting viremia has been detected (Costafreda et al., 2006), with the peak (up to 10^7 genome copies/ml of sera) occurring at 2 weeks after the onset of symptoms and lasting up to an average of 6 weeks after the start of symptoms (Bower et al., 2000; Costafreda et al., 2006). However, there is no evidence of chronicity of the infection. Hepatitis A infection may occasionally produce fulminant hepatitis, mainly among patients with underlying chronic liver diseases (Akriviadis and Redeker, 1989; Previsani et al., 2004).

The distribution patterns of hepatitis A in different geographical areas of the world are closely related to their socioeconomic development (Gust, 1992; Hollinger and Emerson, 2001; Previsani et al., 2004). The endemicity is low in developed regions and high in underdeveloped countries. The epidemiological pattern has important implications on the average age of exposure and hence, as above stated, on the severity of the clinical disease. Since hepatitis A infection induces a life-long immunity (Hollinger and Emerson, 2001), severe infections among adults are rare in highly endemic regions where most children are infected early in life. In contrast, in low endemic areas the disease occurs mostly in adulthood, mainly as a consequence of traveling to endemic regions, or as food or waterborne outbreaks, and hence the likelihood of developing severe symptomatic illness is high. An epidemiological shift, from intermediate to low prevalence, has been noticed in recent decades in many countries, particularly in southern Europe, including Spain, Italy and Greece (Germinario and Lopalco, 1999; Salleras, 1999; Van Damme et al., 1999). Consequently, the Mediterranean basin as a whole should no longer be considered as an endemic area (Previsani et al., 2004; Pintó et al., 2006). Additionally, some other countries from eastern Europe (Cianciara, 2000; Tallo et al., 2003) have also described significant declines in the incidence of hepatitis A.

General features of hepatitis A virus (HAV)

The etiological agent of hepatitis A is the hepatitis A virus (HAV) which belongs to genus *Hepatovirus* within family *Picornaviridae*, and as such it consists of a non-envoloped icosaedral capsid of around 30 nm in diameter containing a positive ssRNA genomic molecule of 7.5 Kb (Fauquet et al., 2005). The genome contains a single open reading frame (ORF) encoding a polyprotein of around 2225 amino acids (aa) preceded by a 5' non-coding region (5'NCR) that makes around 10% of the total genome, and followed by a much shorter 3'NCR that contains a poly(A) tract (Baroudy et al., 1985; Cohen et al., 1987). This genome is uncapped but covalently linked to a small viral protein (VPg) (Weitz et al., 1986). The singly translated polyprotein is subsequently cleaved into 11 proteins through a cascade of proteolytic events brought about mainly by the viral 3C protease (Schultheiss et al., 1994; Schultheiss et al., 1995). However, although the general genomic organization and the expression pattern of HAV are very similar to those of most picornaviruses (Hollinger and Emerson, 2001; Agol, 2002), many differences exist which deserve a special attention.

What makes hepatitis A virus such a special picornavirus

The genetic distance between the genus *Hepatovirus* and the other genera of the family reflects not merely a difference in the nucleotide and amino acid composition but a difference in the molecular and biological characteristics of HAV. From the genomic and proteomic points of view, several interrelated key issues must be brought up. First of all, the structure of the 5'NCR and its internal ribosome entry site (IRES). It is likely that picornavirus IRES has evolved by gradual addition of domains and elements that improved its function in ribosome recruitment or otherwise conferred regulation to the process of viral protein synthesis in a specific cell environment (Ehrenfeld and Teterina, 2002). The HAV IRES is unique among picornaviruses and constitutes the type III model (Braun et al., 1994; Ehrenfeld and Teterina, 2002), which shows a very low efficiency in directing translation (Whetter et al., 1994). Second, HAV encodes only a protease, 3C, while other picornavirus code for additional proteases such as the L protease, in genus *Aphtovirus*, or the 2A protease in *Enterovirus* and *Rhinovirus* genera (Leong et al., 2002). L and 2A proteases, when present, play a crucial role in the primary cleavages of the viral polyprotein while in those genera lacking these proteases, such as *Hepatovirus* and *Paraechovirus*, both primary and secondary cleavages are conducted by the 3C protease. But what is most important is that these additional proteases are involved in the cellular protein shutoff induction (Leong et al., 2002). Since picornaviruses utilize a mechanism of translation that is cap-independent and IRES-dependent, the inhibition of non-essential cap-dependent cellular translation could be advantageous to the virus. In doing so, the cellular translation machinery is utilized almost exclusively for the production of viral proteins (Kuechler et al., 2002). An early event preceding the shutoff of host cell protein synthesis is the cleavage of the cellular translation initiation factor eIF4G, and evidence exists supporting that the enzymes responsible of such a cleavage are 2A and L proteases in enteroviruses and rhinoviruses, and aphtoviruses, respectively (Kuechler et al., 2002). An immediate consequence of the lack of any of these proteolytic activities in HAV is its incapacity to induce cellular shutoff which otherwise is directly related with its requirement for an intact uncleaved eIF4G factor for the formation of the initiation of translation complex (Borman et al., 1997; Jackson, 2002). An intriguing evolutionary question remains to be solved regarding the selection of such an inefficient IRES in HAV. An explanation has been suggested which accounts for the constraints caused by the need to accommodate the dual functions of translation and replication in adjacent regions of the 5'NCR, generating inadvertent consequences for either function (Ehrenfeld and Teterina, 2002). What has been described up to now denotes that HAV must inefficiently compete for the cellular translational machinery and thus it presents a unique translation strategy. This points out to the third difference between HAV and other picornavirus members: the codon usage. HAV presents a higher codon usage bias compared to other members of its family, which conveys in the adaptation to use abundant and rare codons (Sánchez et al., 2003b). In fact, 14 aa families contain rare codons, defined in terms of their frequencies, making a total of 22 used rare codons. But what is more surprising is that the HAV codon usage has evolved to be complementary to that of human cells, never adopting as abundant codons those abundant for the host cell, and even in some instances using these latter as rare codons. This disparity, unique to HAV, has been interpreted as a subtle strategy to avoid, as much as possible, competition for the cellular tRNAs in the absence of a precise mechanism of inducing shutoff of cellular protein synthesis (Sánchez et al., 2003b). As stated before, a consequence of this special codon bias is an increase in the number of rare codons used by HAV. Overall this increment is the result of the addition to the cellular rare codons, also used as rare by the virus, of those most abundant cellular codons that being unavailable for the virus are used at low frequencies. Altogether, the HAV codon usage may contribute to its slow replication and to its low yields. It has been largely documented, (Robinson et al., 1984; Sørensen et al., 1989; Chou and Lakatos, 2004) the role of rare codons in the control of translation speed, in the sense that clusters of rare codons would induce a transient stop of the translational complex in order to seek for a suitable tRNA present at a very low concentration among the pool of tRNAs. A function of these ribosome stallings has been suggested to be the assurance of the proper folding of the nascent protein (Adzhubei et al., 1996; Gavrilin et al., 2000; Evans et al., 2005). Such a function may be postulated for HAV, where highly conserved clusters of rare codons strategically located at the carboxi-ends of the structured elements have been reported (Sánchez et al., 2003b). A certain contribution of the codon usage to the low variability of the HAV capsid has been proposed taking into account that 15% of its surface residues are encoded by such functional rare codons (Sánchez et al., 2003b). This low capsid variability indeed correlates with a very low antigenic variability: a single serotype exists, this being another striking difference with other picornaviruses. The low capsid variability should rely on negative selection acting against potential newly arising proteins, since the viral population replicates as a quasispecies (Sánchez et al., 2003a). The quasispecies analysis revealed a dynamics of mutation selection at and around the rare codons, confirming a seminal role of the codon usage on HAV evolution (Sánchez et al., 2003a).

Other important differences exist between HAV and other picornaviruses at the morphogenetic/structural level. The role of both ends (amino-VP4 and carboxi-2A) of the capsid polyprotein in the virion assembly is still controversial (Probst et al., 1999), and while there is no agreement on the requirement of VP4 for the maturation of pentamers into capsids (Probst et al.,1999; Martin and Lemon, 2006), a complete consensus exists on the necessity of 2A for pentamer formation (Probst et al., 1999; Martin and Lemon, 2006). The ulterior removal of 2A in the mature virion must be performed by a host cell protease (Graff et al., 1999; Martin et al., 1999), although the mature 2A protein has never been identified directly in infected cells.

The X-ray crystallographic structure has not yet been solved, due to the low viral yields obtained by *in vitro* replication. However, recent 3D images of HAV produced by cryoelectron microscopy (Holland Cheng, unpublished results) have revealed important data being the most intriguing the lack of a well-defined canyon around the fivefold axis of symmetry. The pit region of many picornaviruses contains the receptor binding residues (Rieder and Wimmer, 2002) playing, thus, an important biological role. A human HAV receptor (huhavr-1) has been identified (Feigelstock et al., 1998), which contains both an amino terminal Ig-like domain followed by a mucin-like domain (Silberstein et al., 2003). Huhavr-1 has been detected in several human tissues, including the liver. Alternatively the asialoglyco-protein receptor, to which IgA binds to, has been described to enable HAV internalization provided that the virus is complexed with such immunoglobulin (Dotzauer et al., 2000). However, whichever is the receptor, the capsid region involved in such an interaction remains to be elucidated. In contrast, the capsid region interacting with the glycophorin A of the human erythrocytes is indeed located around the putative pit area (Sánchez et al., 2004a). The capsid structure, however, is such that it tolerates this interaction only to occur at acid conditions, being impaired at neutral biological conditions. Erythrocyte glycoproteins may function as decoy receptors attracting pathogens to the erythrocyte and keeping them away from target tissues (Gagneux and Varki, 1999), and hence the actual capsid conformation that allows escaping from erythrocyte attachment may constitute an advantage for a viremic infectious agent whose target organ is the liver. In fact, pathogenesis is in part determined by the spread of the virus to the target tissues (Rieder and Wimmer, 2002). In this context, key factors for the viral biological cycle and infection outcome are a high stability to the acid pH of the stomach during the entry phase, a safe viremic phase, and resistance to the action of detergents, particularly biliary salts, during the exit phase. This extreme resistant phenotype of HAV explains its high persistence in the environment (Abad et al., 1994a,b) and its transmission by contaminated foods and drinking water (Reid and Robinson, 1987; Rosemblum et al., 1990; Bosch et al., 1991; Dentinger et al., 2001; Sánchez et al., 2002), which probably are the result of a highly cohesive capsid conformation mediated through a very accurate folding.

Hepatitis A transmission

Hepatitis A is shed in the feces of infected patients. The viral concentration in such stools is highest (up to 10^{11} genome copies/g of feces) after two weeks of the onset of symptoms and lasts at least four more weeks (Costafreda et al., 2006). HAV infection is mainly propagated via the fecal-oral route as the person-to-person contact is the most common mode of transmission (Mast and Alter, 1993). In fact HAV survival in contaminated fomites, such as sanitary paper, sanitary tile and latex gloves, is very long (Abad et al., 1994a). In consequence, given the high excretion level of HAV, transmission of the infection is facilitated when poor sanitary conditions occur. Nevertheless, transmission through the parental route may occasionally occur (Noble et al., 1984; Sheretz et al., 2005).

Contaminated water and food as critical elements in the route of transmission

Viruses present in the stool of infected patients are discharged into sewage which ultimately may contaminate surface waters and seawater, and consequently be acquired and concentrated by shellfish growing in these waters, or contaminate the vegetables irrigated with the polluted waters. While in approximately 40% of the reported cases of hepatitis A the source of infection cannot be identified, waterborne and foodborne outbreaks of the disease have been reported. Within this latter category, shellfish grown and harvested from waters receiving urban contaminants is a cause of large outbreaks of infectious hepatitis (Halliday et al., 1991; Sánchez et al., 2002). Additionally, large outbreaks associated with the consumption of berry fruits (Reid and Robinson, 1987) and vegetables (Rosemblum et al., 1990; Dentinger et al., 2001) irrigated with contaminated waters may occur. Waterborne outbreaks are less common since the introduction of drinking water treatments. However, reports exist when these measures fail and outbreaks of hepatitis A occur (Bosch et al., 1991).

Detection and quantification of hepatitis A virus

As above stated, HAV may contaminate different types of environmental samples including sewage, surface waters and seawater, and ultimately drinking water and foodstuffs. Thanks to its high stability under very different conditions (Abad et al., 1994b; Bosch et al., 1994), HAV can persist long enough in these edible samples to be transmitted through their ingestion. Thus, although not compulsory, the screening of HAV presence is advisable in specific samples, at least when suspicion of contamination exists.

Cell culture propagation of wild-type strains of HAV is a complex and tedious task, which requires virus adaptation before its effective growth, and even in this case, the virus usually establishes persistent infections resulting in low virus yields (Flehmig, 1980; Daemer et al., 1981; Wang et al., 1986). Thus, infectivity is not nowadays a useful method for primary HAV detection. Alternatively, immunological and, particularly, molecular techniques should be used.

Molecular methods versus immunological methods: a clear choice in water virology

Although fecal excretion of HAV is high, environmental samples usually contain low viral numbers due mainly to the effect of dilution but also to some extent of virus inactivation. Most of the immunological techniques used in clinical diagnosis are aimed at the detection of those antibodies raised upon virus infection, mainly IgM and IgG anti-HAV (Nainan et al., 2006), while antigen detection is uncommon. Additionally, to the lack of immunological kits for antigen detection, their sensitivity would not be high enough to be employed in scenarios of low environmental virus concentration. All this calls for the development of highly sensitive methods for HAV detection, such as those based on nucleic acids amplification. The adoption of these techniques requires selection of the most adequate amplification target. The target region should be highly conserved to increase the chance of detection and with an appropriate structure and length to allow the required sensitivity. Immunological evidence determines the existence of a single serotype of HAV (Lemon and Binn, 1983), although genomic analysis of the virus allows the differentiation of six genotypes (Robertson et al., 1992; Costa-Mattioli et al., 2003). However, all six genotypes are very closely related in the 5'NCR, which is the most conserved region of the genome due to its functional structure in the processes of translation and replication, as above stated, and with a maximum nucleotide divergence of less than 5%. Consequently, the 5'NCR is the region of choice for the design of broad spectrum molecular techniques.

Genetic diversity

In spite of the low antigenic variability of HAV, a degree of nucleotide variability similar to that of other picornavirus in the capsid coding region has been described (Sánchez et al., 2003b). This is explained by a very low number of non-synonymous mutations per non-synonymous site and a number of synonymous mutations per synonymous site similar to that of other picornaviruses. This genetic diversity allows the differentiation of HAV into several genotypes and subgenotypes. Different genomic regions have been used to differentiate the genotypes, including the carboxi-terminus of VP3, the amino-terminus of VP1, the VP1 × 2A junction, the region spanning the carboxi-end of VP1 till the amino-terminus of 2B (VP1/P2B), and finally the entire VP1 region (see the review of Nainan et al., 2006). However, partial genomic sequences never will guarantee the reliability of the complete VP1/2A region. As a matter of fact, the identification of some HAV antigenic variants affecting residues not included in the genotyping regions (Costa-Mattioli et al., 2002a; Sánchez et al., 2002; Gabrieli et al., 2004) could have been elusive in such circumstances. This is the reason why the use of long genomic regions covering at least the entire VP1 including its 2A junction, has recently been recommended (Costa-Mattioli et al., 2002a; European HAV Network, unpublished results) for a more broad molecular typing of HAV. However, the VP1 × 2A junction is still the genomic region most in use worldwide (Robertson et al., 1992). In this region, seven genotypes were initially defined, whose genetic distance was > 15% nucleotide variation. After refining this classification through the addition of more sequences, only six genotypes exist at the present time (Costa-Mattioli et al., 2002a; Lu et al., 2004). Three out of these six genotypes (I, II and III) are of human origin while the other (IV, V and VI) are of simian origin. Genotypes I and II contain subgenotypes (Ia, Ib, IIa and IIb) defined by a nucleotide divergence of 7–7.5%.

Whether the objective is the general broad spectrum detection of HAV or the typing of the isolated strains will determine the use of 5′NCR or VP1 × 2A targets (Sánchez et al., 2004b) or, alternatively, other genotype targets.

The advent of real-time standardized quantitative techniques for the accurate estimation of the HAV titer

Many molecular techniques have been used throughout the years for HAV detection, including hybridization (Bosch et al., 1991; Zhou et al., 1991) and amplification techniques, essentially RT-PCR, usually combined with confirmation tests such as the southern blotting (Calder et al., 2003; Sánchez et al., 2004b), restriction fragment length polimorfism (Goswami et al., 1997) or sequencing (Sánchez et al., 2002). Recently, new approaches have been developed which enable not only the detection but the quantification of the genome copy numbers in a real-time scale. These techniques basically include two different alternatives: the real-time RT-PCR and the nucleic acid sequence based amplification (NASBA). Both techniques take advantage of the use of a combination of enzymes for the amplification of RNA and a reliable method to quantify the final amplified product, usually based on the use of fluorescent probes. In the real-time RT-PCR, the gold standard is the combination of the reverse trascriptase (RT) and the *Taq* polymerase to transform the target ssRNA into the final dsDNA, while the NASBA is again the RT this time with the T7 RNA polymerase and a primer including the T7 promoter that allows the synthesis of great amounts of the final RNA product. Both combinations show a high intrinsic amplification power and although it has been claimed that the NASBA technique allows an even higher amplification than the real-time RT-PCR, more descriptions exist on the use of this latter technique (Costa-Mattioli et al., 2002b; Jothikumar et al., 2005; Costafreda et al., 2006) than with the NASBA (Jean et al., 2001; Abd El Galil et al., 2005) for the quantification of HAV.

Two key issues have to be solved when real-time quantification techniques are developed. Optimal conditions must be found for each particular target rather than using universal settings. It has been claimed that the use of one-step RT-PCR formats reduces the HAV detection sensitivity by up to 1 log unit (Nainan et al., 2006) in comparison to two-step formats. However, when optimum conditions are established for each format, the same level of sensitivity is achieved irrespective of the use of one-step or two-step formats (Costafreda et al., 2006). Not only is the establishment of optimal conditions crucial in the success of a highly sensitive quantification technique, but also standardization of the procedures. Detection of viral RNA by molecular amplification procedures involves several essential steps, the viral RNA extraction and the reverse transcription being the most critical ones. Since no universal reliable nucleic acid extraction method exists, the best option is to control the efficiency of extraction for a particular matrix. The addition of a known concentration of an external virus control to the samples to be quantified has been proposed to assess the efficiency of RNA extraction of enteric viruses in different matrices. Vaccinal poliovirus type 2 (Nishida et al., 2003) and vaccinal Mengo virus (Costafreda et al., 2006) have been proposed as such controls. However, poliovirus presents the inconvenience of being itself a potential contaminant of samples and additionally its use in the context of the poliomyelitis eradication era (see chapter by Hovi et al.) is not advisable. The control of the reverse transcription reaction is more widely performed and most of the real-time techniques described for the quantification of HAV include an RNA standard internal control (Costa-Mattioli et al., 2002b; Jothikumar et al., 2005; Costafreda et al., 2006). Usually, these controls are synthesized by *in vitro* transcription of cloned cDNAs or T7 promoter-containing amplimers corresponding exactly to the targets, and are added to each test tube at known amounts in order to follow up the efficiency of the reaction. Applying these control measures, a fine estimation of the number of HAV genome copies per gram of shellfish was performed for the first time in several clam stocks associated with an outbreak of hepatitis A, and ranged from 1×10^3 to 1×10^5 (Costafreda et al., 2006). These determinations have provided the means for a risk-assessment study linking the level of shellfish contamination and the attack rate of infection (Costafreda et al., unpublished results).

Molecular epidemiology of hepatitis A: the water environment as an overview of the circulating viruses

Environmental surveillance, mostly sewage monitoring, has been used to assess the circulation of different human enteric viruses, notably rotaviruses (Villena et al., 2003; van Zyl et al., 2006), wild-type poliovirus (Deshpande et al., 2003; El Bassioni et al., 2003) and also HAV (Pintó et al., 2006) as well as animal enteric viruses (Jiménez-Clavero et al., 2003, 2005a). The application of molecular typing techniques in this kind of samples allows the molecular and phylogenetic analysis required for the traceability of strains, i.e. in the context of the poliovirus eradication program (see chapter by Hovi et al.). The molecular characterization of viral strains isolated in sewage, i.e. rotavirus, might be very useful in the policy for the design for future vaccines applicable to countries where the clinical surveillance does not cover all the population.

Environmental molecular epidemiology as such should rely not only on the molecular characterization of the isolated strains but also on the epidemiological data regularly collected. When both disciplines are applied together interesting conclusions might be drawn. In the particular case of HAV, a study relating a vaccination campaign, the number of clinical cases and the occurrence of viruses in sewage during a period of 5 years in Barcelona city has been reported (Pintó et al., 2006). Attack rates per 100,000 inhabitants of 9.1, 6.2, 3.3, 1.7 and 8.0 were estimated during the years 1998–2002. While the progressive decline could be clearly associated to a vaccine administration the final increase was attributed to the huge immigration flow, from North Africa, South America and East Asia that the city received (nearly 10% of the total population). The vaccination campaign was, and is, dedicated to children 12 years old, being those < 12 still susceptible to the infection. Immigrant children may act as potential carriers of the infection, and in fact many of the observed cases were school-related, and mostly coincided with the virus incubation period elapsed after the return from the school holiday, and after the immigrant population have returned from visits to their countries of origin. Sewage surveillance data taught us that the similar high attack rates seen at the beginning and final years of the study responded to a different infection pattern. While many asymptomatic cases might occur at the beginning with high levels of virus excreted and with many positive isolations in sewage, at the end only small outbreaks among the non-vaccinated population arose without the massive involvement of asymptomatic carriers and consequently without positivity in sewage. In conclusion, the environmental study reflected that in spite of the increase in clinical cases in the last year of study, the vaccination program was anyway working well.

Finally, it should be stated that molecular environmental surveillance still presents some intrinsic problems, genotyping being one of them. Even when using a short fragment such as the VP1 × 2A region, HAV genotyping is a hard task due to the low concentration of viruses occurring in sewage. Using this genotyping target, a threshold of 10^5 genomes/ml has been estimated. In consequence, only those genotypes, more prevalent in the population, are likely to be detected. However in areas endemic for HAV, such as Egypt (Pintó et al., 2006), environmental surveillance is a powerful tool to complement the clinical epidemiology data, since detection and genotyping as well are possible. In fact, as above stated, this approach has been already implemented in Egypt in the context of the polio eradication program (El Bassioni et al., 2003).

Prevention of hepatitis A

Inactivated HAV vaccines are available since the early 1990s and provide long-lasting immunity against hepatitis A infection (Bell and Feinstone, 2004). The immunity is largely related to the induction of high titers of specific antibodies. Thanks to the existence of a single serotype of HAV, these vaccines are of high efficacy. These vaccines consist of viruses grown in cell culture, purified, inactivated with formalin and adsorbed to an aluminum hydroxide adjuvant, making their economic cost quite high. This is the reason why many discrepancies already exist on their universal use in massive vaccination campaigns. Countries with previous intermediate endemicity of HAV such as Israel or some autonomous communities of Spain such as Catalonia, or some States of United States have performed studies on the impact of child vaccination on the overall incidence of hepatitis A concluding that the immunization is medically (Salleras, 1999; Shouval, 1999; Wasley et al., 2005) and economically (Dagan et al., 2005; http://www.gencat.net/salut/portal/cat/nen.htm; http://http://www.cdc.gov/nip/publications/VIS/vis-hep-a.pdf) justified. In contrast, other countries in a similar situation such as Italy do not recommend at present the implementation of such a measure in terms of cost-benefits (Franco and Vitiello, 2003). In this context is quite evident that high endemic countries that usually have low economic incomes do not regard the vaccination against hepatitis A as a primary policy (Teppakdee et al., 2002).

Although several attenuated vaccine candidates have also been attempted, due to the successful use of inactivated vaccines, its development is hardly plausible.

As a general rule, in low and intermediate endemic regions, where paradoxically the severity of the disease is high, vaccination against hepatitis A should be recommended in high-risk groups, including travelers to high endemic areas, men having sex with men, drug users and patients receiving blood products. In addition, the inclusion of hepatitis A vaccines in mass vaccination programs in those countries receiving high numbers of immigrants from endemic countries is particularly advisable. However, and bearing in mind the quasispecies replication pattern of HAV (Sánchez et al., 2003a) that could lead in populations with continued exposure to the virus to the selection of new antigenic variants escaping immune protection, mass vaccination programs in highly endemic areas is controversial.

Hepatitis E

Natural course and epidemiology of hepatitis E

Hepatitis E, previously known as enterically transmitted non-A, non-B hepatitis, is an infection with clinical and epidemiological features of acute hepatitis. The clinical presentation of hepatitis E is basically similar to that of hepatitis A, but cholestatic jaundice is more common. The clinical course of the infection was first addressed in a human volunteer that ingested a clarified stool preparation from an infected patient (Balayan et al., 1983). The incubation period averages 40 days, with a range between 15 and 60 days. Amino alanine transferase elevation occurs during 30–120 days after infection, and fecal excretion of the virus begins around 1 week before onset of illness and continues for, at least, 2–3 weeks thereafter (Skidmore, 2002). The ecteric phase of the infection is characterized by a flu-like prodrome with epigastric pain, vomiting, fever and discoloration of the urine, but jaundice patients display yellowish skin, scleral icterus, dark urine, and light tan-colored feces; however most infections are asymptomatic and no evidence of chronic disease has been observed (Purcell and Emerson, 2001; Smith, 2001). As a result of viral replication in the liver, the hepatitis E virus (HEV) is found in the bile in large quantities, reaching the intestines by the bile duct and being subsequently shed in the feces. Asymptomatic infected individuals may shed virus and become reservoirs of the virus between epidemics and, therefore, contribute to sporadic infection by person-to-person transmission or by contaminating water and food. In endemic regions, the overall attack rate was estimated to be around 2.5% in adults and around 1.2% in children (Vishwanathan, 1957). The case-fatality rate is usually low (0.2–3%), but in pregnant women during the third trimester of gestation it can be as high as 15–25%, primarily due to fulminant hepatic failure (Purcell and Emerson, 2001; Smith, 2001; Skidmore, 2002).

The significance of the host immune response in the pathogenesis of cell damage is not fully understood. Anti-HEV IgM appears at the time of the onset of symptoms and remains detectable for to 2–3 months. Anti-HEV IgG is detectable shortly after IgM detection, increases during the acute phase and may be present in serum for years after the initial infection (Clayson et al., 1995b).

There is no specific treatment for hepatitis E. Passive immunization with convalescent sera has been accomplished in animal models (Tsarev et al., 1994) but in humans administration of immune globulin obtained from inhabitants of HEV endemic regions was unsuccessful (Khuroo and Dar, 1992); however, it should be noted that this study used unselected plasma and that, even in endemic regions, anti-HEV prevalence and titers are low. Therefore, it cannot be ruled out that the use of selected anti-HEV batches of immunoglobulins with high titers may be useful, particularly for pregnant women and/or during epidemics.

HEV is transmitted primarily by contaminated water, and causes frequent epidemics in areas with inadequate water supplies and poor sanitary conditions (Purcell and Emerson, 2001), being the principal cause of acute, sporadic hepatitis in adults in many areas of Asia, Middle East and Northern Africa (Emerson and Purcell, 2003; Schlauder, 2004). Recently, an increase in the number of cases in regions considered as non-endemic for hepatitis E has been reported (Smith, 2001; Worm et al., 2002a).

Since the main route of HEV transmission is feco-oral, most epidemics can be linked to waterborne outbreaks, particularly in developing countries with warm weather, high population density and poor sanitary conditions. The first documented outbreak of HEV occurred in India in 1955–1956 (Vishwanathan, 1957). The origin of the outbreak, which was initially attributed to hepatitis A and later on confirmed to be hepatitis E, was the contamination by sewage, from 1 to 6 weeks prior to the epidemic, of Jumna River, the source of water for the treatment plant. Alum and chlorine treatment prevented bacterial infections, but 30,000 cases of hepatitis occurred among the population (Wong et al., 1980). One of the highest epidemic areas is China, where al least 11 epidemic outbreaks have been reported to date. The largest one occurred in 1986–1988, with more than 119,000 cases that resulted in more than 700 deaths. Until recently most cases reported in developed countries were attributed to travel to endemic areas, however, as aforementioned, there is an increase in the number of cases of infected patients that had never been abroad (Smith, 2001; Worm et al., 2002a).

In addition, a zoonotic potential for the virus was suggested after detection of HEV infection in wild and domestic animals, and later on confirmed in people who ate HEV infected uncooked deer meat (Meng, 2003; Goens and Purdue, 2004). The risk of a zoonotic spread of the virus, its detection in non-endemic areas, and the continuous occurrence of outbreaks in endemic regions have boosted the interest in the understanding of the biology and life cycle of the virus, and in the improvement of diagnostic tools able to detect the pathogen in polluted waters and other environmental samples.

Seroprevalence

The overall anti-HEV prevalence reported in endemic countries is quite variable, but lower than expected, 3–27% (Purcell 1994; Purcell and Emerson, 2001; Worm et al., 2002b). In contrast to other enteric viruses such as poliovirus or HAV, the prevalence of anti-HEV IgG is lower in children than in adults (Arankalle et al., 1995, Meng et al., 2002). A possible explanation for this could be that HEV immunity acquired with subclinical infection during childhood wanes with time. In non-endemic areas with good sanitary conditions and control of water supplies, there is a low but constant increase in the number of HEV sporadic cases non-related to travel, and the anti-HEV antibody prevalence among the healthy population is relatively high, even higher than that reported in endemic areas (Meng et al., 2002; Worm et al., 2002a; Meng, 2003).

After the first description of HEV in swine (Balayan et al., 1990; Clayson et al., 1995a; Meng et al., 1997), an initial epidemiological survey in North America reported a higher prevalence of anti-HEV antibodies among swine veterinarians (26%) than among blood donors (18%) (Meng et al., 2002). A further study described a 35, 11 and 2.5% seroprevalence among swine, swine workers and non-swine workers, respectively (Whiters et al., 2002). Recently, a Chinese study has reported that swine workers have a 74% higher risk of HEV infection than people engaged in other occupations (Zheng et al., 2006). In contrast, no statistical difference in anti-HEV antibodies prevalence has been noted between pig farmers (13.0%) and control subjects (9.3%) in Sweden (Olsen et al., 2006). In any case, it should be noted that differences in study design (population features, health status, demographical variables, etc...) make difficult the comparison of the reported data, and that there are contradictory results about the reliability of the anti-HEV detection test used in the different studies (Worm et al., 2002a; Emerson and Purcell, 2003; Schlauder, 2004). Hence, analyses of well-selected population with standardized reagents are needed to have a more clear understanding of the actual incidence of HEV infection.

General features of the hepatitis E virus (HEV) with a special emphasis to the genomic organization

The identity of the causative agent of hepatitis E was first described in 1990 (Reyes et al., 1990). One year latter, the entire sequence of the viral genome was published (Tam et al., 1991). HEV was provisionally classified as a member of the

Caliciviridae family, but it is now ascribed to a separate family, *Hepeviridae*, in the prototypic genus *Hepevirus* (Mayo and Ball, 2006). HEV is a spherical, non-enveloped viral particle of around 32–34 nm in diameter. The genome is a ssRNA molecule of positive polarity of approximately 7.2 Kb containing 3 overlapping ORF and a 3'poly (A) tail (Worm et al., 2002b; Emerson and Purcell, 2003; Schlauder, 2004).

In vitro analysis suggested that HEV RNA is capped at the 5'end (Kabrane-Lazizi et al., 1999a). After a non-coding region of 27–35 nucleotides (nt), ORF-1 encodes about 1693 aa encompassing non-structural proteins with enzymatic activity that are involved in viral replication, transcription and protein processing, including the viral replicase (Emerson and Purcell, 2003). ORF-2 extends 1980 nt, terminating 65 nt upstream of the poly-A tail, and renders a 660 aa protein likely representing the structural capsid protein(s) (Tam et al., 1991). In vitro experiments suggested that ORF-2 protein is synthesized as a large glycoprotein precursor of around 88 kDa, which is cleaved into the mature protein (Jameel et al., 1996, Zafrullah et al., 1999). ORF-2 protein contains epitopes that induce neutralizing antibodies and are mainly located near the carboxi-end (Tam et al., 1991). ORF-3 overlaps the 5'end of ORF-1 by only 1 nt and ORF-2 by 328 nt. It encodes a 123 aa protein which is post-translationally modified by phosphorylation giving a mature protein of around 13.5 kDa of unknown function (Emerson and Purcell, 2003). This phosphoprotein is associated with the hepatocellular cytoskeleton (Zafrullah et al., 1997) and form a complex with capsid protein of ORF-2 and, thus, it is believed to be involved in the assembly of the viral particle (Jameel et al., 1996). However, it has been recently shown that, in contrast to its requirement in vivo, ORF-3 protein is not required for infection of Huh-7 cells or production of infectious virus in vitro (Emerson et al., 2006). ORF-3 may also have regulatory functions implicated in modulation of cell signaling (Emerson and Purcell, 2003). In addition, ORF-3 protein also beards neutralizing epitopes near its 3'end (Tam et al., 1991). In any case, it should be noted that the lack of a suitable and efficient cell culture system for replication of HEV has hampered the study of the viral life cycle (Emerson and Purcell, 2003).

Genetic variation

The genome sequence of HEV is quite stable (Arankalle et al., 1999). A high genomic homology is found among isolates from the same outbreak, and serial passages in animal models did not result in genetic drift (Worm et al., 2002b; Schlauder, 2004). However, data supporting a quasispecies organization of HEV genome during epidemics have also been reported (Grandadam et al., 2004). Additionally, isolates from different geographical regions are relatively diverse. Based on this genomic heterogeneity, HEV has been classified into four different genotypes (Worm et al., 2002b; Meng, 2003; Schlauder, 2004). Genotype I is mainly presented in endemic areas from Asia and Africa. Genotype II includes the Mexican isolates and some Nigerian strains. Isolates from regions considered as non-endemics (USA, Spain, Italy, Greece, etc.) represent a more diverse cluster of sequences and are grouped into genotype III. Finally, genotype IV includes isolates from China. Besides this genotypic diversity, no evidence of serological heterogeneity has been reported and, therefore, it seems that there is only one HEV serotype.

HEV transmision

Waterborne transmission

Epidemics of hepatitis E in endemic areas are usually due to fecally contaminated water (Aggarwal and Naik, 1994) and most outbreaks can be traced back to contaminated water sources (Smith, 2001). Adequate circumstances for HEV epidemics arise when raw sewage enters in contact with water reservoirs during heavy rain sessions, floods, monsoons, etc. For instance, heavy and flooding rains preceded the Indian epidemic of the 1950s (Vishwanathan, 1957; Khuroo and Kamili, 1994). In most instances, people affected by HEV outbreaks lives near rivers with inadequate sanitary conditions (Bile et al., 1994) and a high incidence of HEV seropositivity has been correlated with the use of non-boiled river water for drinking, cooking and washing. Likewise, refugees and people living in urban crowded slums and camps are at increased risk for fecal-oral transmitted diseases, including HEV infection (Khuroo and Kamili, 1994; Mast et al., 1994), as it has been recently demonstrated in the displaced population from Darfur (Sudan) where, in 6 months, 2.621 hepatitis E cases were recorded (attack rate 3.3%). The case-fatality rate was 1.7%, with 45 deaths, including 19 pregnant women (Guthmann et al., 2006).

HEV was detected in all sewage influent samples and in 67–89% of effluent samples from sewage treatment plants in Madras, India, showing that treatment was not as effective as it should be. Viral particles have also been detected in sewage from industrialized countries (Pina et al., 2000). HEV RNA was detected in a pretreated sewage sample collected in Washington, DC, and it showed a very high homology with human and swine isolates from the US (Clemente-Casares et al., 2003). In contrast, no HEV RNA has been detected in drinking and surface waters collected from pig farms where HEV was present (Kasorndorkbua et al., 2005).

Person-to-person transmission

Person-to-person transmission seems to be low (Aggarwal and Naik, 1994; Bile et al., 1994; Mast et al., 1994). Secondary cases among household members of patients with documented HEV infection occurs in 1–2% (Aggarwal and Naik, 1994). Person-to-person transmission in hospital settings has been described, although results about the incidence of HEV infection in hospitalized patients and in people that received contaminated blood are inconclusive (Smith, 2001). Data on mother-to-child transmission rates of HEV are quite variable, ranging between 30 and 100% (Khuroo et al., 1995; Kumar et al., 2004). Additionally, it has been reported that up to 2/3rd of pregnant HEV-infected women may have preterm delivery (Kumar et al., 2004). HEV RNA has also been detected in the blood of newborns at a time when no virus was detectable in the mother (Khuroo et al., 1995).

Foodborne transmission

Washing, irrigating and processing of food with HEV-contaminated water could lead to HEV outbreaks if the food is eaten uncooked. Food manipulation by an HEV-infected person may also transmit the disease. Acute hepatitis E in Sicily (Italy) was attributed to contaminated shellfish consumption (Cacopardo et al., 1997), and a case of hepatitis E after ingestion of Chinese medicinal herbs has also been reported (Ishikawa et al., 1995). Likewise, sporadic acute or fulminant hepatitis has been linked to uncooked pig liver and wild boar meat consumption in Japan (Yazaki et al., 2003; Li et al., 2005). Finally, a clear demonstration of acute HEV infection after consumption of HEV-infected uncooked deer meat has been reported (Tei et al., 2003). In this latter study, 4 out of 5 individuals who ate the infected meat presented hepatitis, while the 3 other members of the families who did not eat it were not infected. One child who was not infected claimed to have eaten a very small portion of deer meat, suggesting that HEV infection is dose-dependant. Sequence analysis of HEV RNA from patients and from frozen leftover deer meat showed a 100% similarity (Tei et al., 2003).

Zoonotical transmission

Presence of anti-HEV antibodies in pigs and characterization of swine HEV were first described in the 1990s (Balayan et al., 1990; Clayson et al., 1995a; Meng et al., 1997). Later on, experimental infection of pigs with either swine or human HEV isolates was achieved, and showed that infected animals presented viremia and shed virus in feces, although no clinical or biochemical signs of disease were observed (Balayan et al., 1990; Meng et al., 1998; Halbur et al., 2001). After that, several evidences have raised the hypothesis of a zoonotic potential for HEV and its possible risk in xenotransplantation (Meng, 2003). For instance, HEV has been detected in sewage polluted with pig feces (Pina et al., 2000), people drinking water from downstream of pig farms seem to have a higher risk of HEV infection (Zheng et al., 2006) as do workers engaged in occupations related to swine farming (Meng, 2003; Zheng et al., 2006), and anti-HEV antibodies have been found in swine herds from endemic and non-endemic areas (Emerson and Purcell, 2003; Meng, 2003; Goens and Purdue, 2004). Furthermore, in general, swine isolates are genetically more closely related to human HEV strains of the same geographical region than to swine strains of other parts of the world (Meng et al., 1997; Meng et al., 2002; Meng, 2003). Finally, as aforementioned, HEV infection in humans after ingestion of HEV-infected raw deer meat has been demonstrated (Tei et al., 2003).

Besides pigs, specific anti-HEV antibodies and HEV strains have also been detected in rodents (Clayson et al., 1995a; Kabrane-Lazizi et al., 1999b), wild boar (Matsuda et al., 2003), donkeys (Guthmann et al., 2006), chickens, cattle and dogs (Meng, 2003; Goens and Purdue, 2004). More recently, an avian HEV has been described (Haqshenas et al., 2001) and, although it is genetically less related to human HEV than swine isolates, it shares antigenic epitopes with both of them (Haqshenas et al., 2002). All these observations have strengthened the zoonotic potential of HEV, but the assessment of a zoonotical transmission to humans through animal waste still needs further evaluation.

Diagnosis

Serological diagnosis

Enzyme-linked immunosorbent assay (EIA) is the main diagnostic tool to detect anti-HEV IgG and IgM (Worm et al., 2002b). In general, a positive result for anti-HEV IgM indicates acute disease, however, to avoid false negative results, testing should be done in the acute phase of the infection. Although detection of anti-HEV IgG is not conclusive of HEV infection, a high IgG titer or increasing titers in consecutive samples, support the diagnosis of acute hepatitis E.

Several antigenic domains have been identified in the three ORFs of HEV. Based on this information, different synthetic peptides and recombinant proteins derived from the carboxi-end of ORF-2 and/or ORF-3 have been assayed for specific antibody detection (Worm et al., 2002a). Studies carried out with these tests have reported a relatively high seroprevalence in non-endemic countries, raising concerns about the possible detection of non-specific cross-reactive antibodies (Mast et al., 1998; Worm et al., 2002a). However, a blind comparison of a test based on a recombinant ORF-2 protein showed that it was 98% specific for anti-HEV (Mast et al., 1998). When this approach was applied to sera from different regions of the world, it confirmed the previous results obtained with other assays (Emerson and Purcell, 2003). At present, the few commercially available EIAs kits are based on the Mexican and/or the Burmese prototypes, and although a recent comparison of different tests using outbreak samples has shown that they can be highly specific and sensitive (Myint et al., 2006), their reliability still needs to be fully confirmed worldwide.

Molecular detection

Detection of HEV by RT-PCR is indicative of active infection. The availability of an increased number of HEV sequences from different sources and geographical regions has enabled the design of specific oligonucleotide primers that match conserved regions of the HEV genome and allows the detection of HEV in acute phase sera, stools and contaminated water and sewage (Schlauder, 2004). Several conventional "in-house" RT-PCR assays have been published for detection of HEV in serum, feces and bile of infected individuals (Smith, 2001). For water analysis, efficient concentration procedures and highly sensitive detection methods are required for viral detection (Jiménez-Clavero et al., 2005b; and chapter by Wyn-Jones). Recently, several real-time RT-PCR detection methods have been described (Mansuy et al., 2004; Orrù et al., 2004; Enouf et al., 2006; Jothikumar et al., 2006). Using an internal control and spiked water samples, detection of as few as 4 genome equivalent copies of HEV plasmid DNA and of 0.12 pig ID_{50} of swine HEV has been achieved (Jothikumar et al., 2006). Development of a quantitative, broadly reactive, quick, easy and reproducible HEV detection method would be of special interest for testing water and environmental samples, and may allow tracking of the polluting sources. Recently, as few as 100 fM of an ORF-2 amplicon were detected using an HEV specific microarray (Liu et al., 2006). Ideally, development of a microarray assay able to detect as much waterborne pathogens as possible in a single reaction would greatly improved our current capacity for detection of water pollutants representing human and/or animal health risk.

Prevention of hepatitis E

The feasibility of HEV vaccines is based on several evidences: (i) specific antibodies are raised after HEV infection; (ii) HEV infected people are usually protected following epidemics; and (iii) animal experimentation has shown that passive immune prophylaxis induces humoral immunity (Emerson and Purcell, 2001; Wang and Zhuang, 2004). Additionally, only one HEV serotype has been described, thus, production of a broadly cross-reactive vaccine should be possible. Such a vaccine would be useful in protection against HEV infection, mainly in pregnant women and in people from endemic regions and travelers to these areas. Nevertheless, the lack of a susceptible cell culture system has hampered the development of live attenuated or killed vaccines (Wang and Zhuang, 2004) and hence no commercial vaccines against HEV are available.

   To date, most research on HEV vaccines is focused on ORF-2-derived proteins or peptides that contain neutralizing epitopes common to different genotypes (Emerson and Purcell, 2001; Meng et al., 2001; Worm et al., 2002a; Wang and Zhuang, 2004). Several ORF-2 vaccine candidate products have been expressed in insect, prokaryotic, yeast, animal and plant cells (Emerson and Purcell, 2001; Wang and Zhuang, 2004). Animal experimentation has shown that administration of some of these ORF-2 recombinant proteins protected against homologous and heterologous challenge (Purdy et al., 1993; Tsarev et al., 1994; Tsarev et al., 1997; Im et al., 2001). DNA immunization of mice with an HEV-cDNA elicits high titers of specific anti-HEV antibodies (He et al., 1997), immunologic memory (He et al., 2001) and protection in cynomolgus macaques (Kamili et al., 2004). Truncated ORF-2 protein expressed in baculovirus spontaneously assembles into viral-like particles (VLPs) and are also good immunogens (Li et al., 2001). A recombinant

HEV baculovirus vaccine candidate that protects against intravenous administration of heterologous HEV strains has entered into preclinical trials (Stevenson, 2000; Zhang et al., 2002; Purcell et al., 2003).

Concluding remarks

Enterically transmitted hepatitis represent by large the most common manifestation of acute hepatitis worldwide. Regarding hepatitis A, although the increase in living standards and public health sanitation are greatly contributing to a decrease in its global incidence, the total number of cases per year is still extremely high with estimations up to the scale of millions. Overall the situation encourages policies of immunization, but in spite of the availability of vaccines, their use in highly endemic regions may be limited on the basis of their cost and even on the basis of medical and molecular aspects. Thus in these areas prevention should, at least, rely on the implementation of effective control measures such as water sanitation and virus monitoring.

In the case of hepatitis E, the disease that was restricted to endemic areas is now been increasingly reported also in regions considered as non-endemic. However, it remains unclear whether this increase is related to the emergence of the pathogen or to the new epidemiological and public health interest in this virus and the availability of diagnostic procedures. This concern derives from the associated mortality in pregnant women, as well as the morbidity and disability in the general population, mainly in endemic areas. Since HAV and HEV are transmitted by contaminated waters, improvement of water quality by proper sewage disposal and water treatment is very important in preventing spread of these infections.

Although it may appear unlikely that infectious hepatitis outbreaks occur in industrialized countries with properly treated waters and sewages and good sanitation and hygienic conditions, potential risk derived form consumption of vegetables, fruits and other products imported from endemic regions exist. Bivalve mollusks grown and harvested in polluted waters represent a particular potential threat due to their capacity to filter large volumes of water and accumulate the viruses in their edible tissues.

The possible zoonotic transmission of HEV may also contribute to the continuous spread of the virus, despite improved sanitation, and calls for a more deep knowledge of HEV prevalence in animals in order to prevent hepatitis E transmission.

References

Abad FX, Pintó RM, Bosch A. Survival of enteric viruses on environmental fomites. Appl Environ Microbiol 1994a; 60: 3704–3710.

Abad FX, Pintó RM, Diez JM, Bosch A. Disinfection of human enteric viruses in water by copper and silver in combination with low levels of chlorine. Appl Environ Microbiol 1994b; 60: 2377–2383.

Abd El Galil KH, El Sokkary MA, Kheira SM, Salazar AM, Yates MV, Chen W, Mulchandani A. Real-time nucleic acid sequence-based amplification assay for detection of hepatitis A virus. Appl Environ Microbiol 2005; 71: 7113–7116.

Adzhubei AA, Adzhubei IA, Krasheninnikov IA, Neidle S. Non-random usage of "degenerate" codons is related to protein three-dimensional structure. FEBS Lett 1996; 399: 78–82.

Aggarwal R, Naik SR. Hepatitis E: intrafamiliar transmission versus waterborne spread. J Hepatol 1994; 21: 718–723.

Agol VA. Picornavirus genome: an overview. In: Molecular Biology of Picornaviruses (Semler BL, Wimmer E, editors). Washington, DC: ASM Press; 2002; pp. 127–148.

Akriviadis EA, Redeker AG. Fulminant hepatitis A in intravenous drug users with chronic liver disease. Ann Intern Med 1989; 110: 838–839.

Arankalle VA, Paranjape S, Emerson SU, Purcell RH, Walimbe AM. Phylogenetic analysis of hepatitis E virus isolates from India (1976–1993). J Gen Virol 1999; 80: 691–1700.

Arankalle VA, Tsarev SA, Chadha MS, Alling DW, Emerson S. Age-specific prevalence of antibodies to hepatitis A and E viruses in Pune, India, 1982 and 1992. J Infect Dis 1995; 171: 447–450.

Balayan MS, Andjaparidze AG, Savisnkaya SS, Kelitadze ES, Braginsky DM, Savinov AP, Poleschuk VF. Evidence for a virus in non-A, non-B hepatitis transmitted via the fecaloral route. Intervirology 1983; 20: 23–31.

Balayan MS, Usmanov RK, Zamyatina NA, Djumalieva DI, Karas FR. Brief report: experimental hepatitis E infection in domestic pigs. J Med Virol 1990; 32: 58–59.

Baroudy BM, Ticehurst JR, Miele TA, Maizel JV, Purcell RH, Feinstone SM. Sequence analysis of hepatitis A virus cDNA coding for capsid proteins and RNA polymerase. Proc Natl Acad Sci 1985; 82: 2143–2147.

Bell BP, Feinstone SM. Hepatitis A vaccine. In: Vaccine (Plotkin SA, Orenstein WA, Offit PA, editors). Philadelphia, PA: Saunders; 2004; pp. 269–297.

Bile K, Isse A, Mohamud O, Allebeck P, Nilsson L, Norder H, Mushahwar IK, Magnus LO. Contrasting roles of rivers and wells as sources of drinking water on attack and fatality rates in a hepatitis E epidemic in Somalia. Am J Trop Med Hyg 1994; 51: 466–474.

Borman AM, Kirchweger R, Ziegler E, Rhoads RE, Skern T, Kean KM. eIF4G and its proteolytic cleavage products: effect on initiation of protein synthesis from capped, uncapped, and IRES-containig mRNAs. RNA 1997; 3: 186–196.

Bosch A, Abad FX, Gajardo R, Pintó RM. Should shellfish be purified before public consumption? Lancet 1994; 344: 1024–1025.

Bosch A, Lucena F, Díez JM, Gajardo R, Blasi M, Jofre J. Human enteric viruses and indicator microorganisms in a water supply associated with an outbreak of infectious hepatitis. J Am Water Works Assoc 1991; 83: 80–83.

Bower WA, Nainan OV, Han X, Margolis HS. Duration of viremia in hepatitis A virus infection. J Infect Dis 2000; 182: 12–17.

Braun EA, Zajac AJ, Lemon SM. In vitro characterization of an internal ribosome entry site (IRES) present within the 5′ nontranslated region of hepatitis A virus RNA: comparison with the IRES of encephalomyocarditis virus. J Virol 1994; 68: 1066–1074.

Cacopardo B, Russo R, Preiser W, Benanti F, Brancati G, Nunnari A. Acute hepatitis E in Catania (eastern Sicily) 1980–1994. The role of hepatitis E virus. Infection 1997; 25: 313–316.

Calder L, Simmons G, Thornley C, Taylor P, Pritchard K, Greening G, Bishop J. An outbreak of hepatitis A associated with consumption of raw blueberries. Epidemiol Infect 2003; 131: 745–751.

Chou T, Lakatos G. Clustered bottlenecks in mRNA translation and protein synthesis. Phys Rev Lett 2004; 93: 198101–198104.

Cianciara J. Hepatitis A shifting epidemiology in Poland and Eastern Europe. Vaccine 2000; 18: S68–S70.

Clayson ET, Innis BL, Myint KS, Narupiti S, Vaughn DW, Giri S, Ranabhat P, Shrestha MP. Detection of hepatitis E virus infections among domestic swine in the Kathmandu Valley of Nepal. Am J Trop Med Hyg 1995a; 53: 228–232.

Clayson ET, Myint KS, Snitbhan R, Vaughn DW, Innis BL, Chan L, Cheung P, Shrestha MP. Viremia, fecal shedding, and IgM and IgG responses in patients with hepatitis E. J Infect Dis 1995b; 172: 927–933.

Clemente-Casares P, Pina S, Buti M, Jardi R, Martin M, Bofill-Mas S, Gironés R. Hepatitis E virus epidemiology in industrialized countries. Emerg Infect Dis 2003; 9: 448–454.

Cohen JI, Ticehurst, Purcell RH, Buckler-White A, Baroudy BM. Complete nucleotide sequence of wild-type hepatitis A virus: comparison with different strains of hepatitis A and other picornaviruses. J Virol 1987; 61: 50–59.

Costafreda MI, Bosch A, Pintó RM. Development, evaluation, and standardization of a real-time TaqMan reverse transcription-PCR assay for quantification of hepatitis A virus in clinical and shellfish samples. Appl Environ Microbiol 2006; 72: 3846–3855.

Costa-Mattioli M, Cristina J, Romero H, Pérez-Bercoff R, Casane D, Colina R, García L, Vega I, Glikman G, Romanovsky V, Castello A, Nicand E, Gassin M, Billaudel S, Ferre V. Molecular evolution of hepatitis A virus: a new classification based on the complete VP1 protein. J Virol 2002a; 76: 9516–9525.

Costa-Mattioli M, Monpoeho S, Nicand E, Aleman MH, Billaudel S, Ferre V. Quantification and duration of viraemia during hepatitis A infection as determined by real-time RT-PCR. J Viral Hepat 2002b; 9: 101–106.

Costa-Mattioli M, Napoli AD, Ferre V, Billaudel S, Perez-Bercoff R, Cristina J. Genetic variability of hepatitis A virus. J Gen Virol 2003; 84: 3191–3201.

Daemer RJ, Feinstone SM, Gust ID, Purcell RH. Propagation of human hepatitis hepatitis A virus in African green monkey kidney cell culture: primary isolation and serial passage. Infect Immun 1981; 32: 388–393.

Dagan R, Leventhal A, Anis E, Ashur Y, Shuval D. Incidence of hepatitis A in Israel following universal immunization of toodlers. J Am Med Assoc 2005; 294: 202–210.

Dentinger CM, Bower WA, Nainan OV, Cotter SM, Myers G, Dubusky LM, Fowler S, Salehi ED, Bell BP. An outbreak of hepatitis A associated with green onions. J Infect Dis 2001; 183: 1273–1276.

Deshpande JM, Shetty SJ, Siddiqui ZA. Environmental surveillance system to track wild poliovirus transmission. Appl Environ Microbiol 2003; 69: 2919–2927.

Dotzauer A, Gebhardt U, Bieback K, Göttke U, Kracke A, Mages J, Lemon SM, Vallbracht A. Hepatitis A virus-specific immunoglobulin A mediates infection of hepatocytes with hepatitis A virus via the asialoglycoprotein receptor. J Virol 2000; 74: 10950–10957.

Ehrenfeld E, Teterina NL. Initiation of translation of picornavirus RNAs: structure and function of the Internal Ribosome Entry Site. In: Molecular Biology of Picornaviruses (Semler BL, Wimmer E, editors). Washington, DC: ASM Press; 2002; pp. 159–170.

El Bassioni L, Barakat I, Nasr E, De Gourville EM, Hovi T, Blomqvist S, Burns C, Stenvik M, Gary H, Kew OM, Pallansch MA, Wahdan MH. Prolonged detection of indigenous wild polioviruses in sewage from communities in Egypt. Am J Epidemiol 2003; 158: 807–815.

Emerson SU, Nguyen H, Torian U, Purcell RH. ORF3 protein of hepatitis E virus is not required for replication, virion assembly or infection of hepatoma cellas *in vitro*. J Virol 2006; 80: 10457–10464.

Emerson SU, Purcell RH. Recombinant vaccines for hepatitis E. Trends Microbiol 2001; 7: 462–466.

Emerson SU, Purcell RH. Hepatits E virus. Rev Med Virol 2003; 13: 145–154.

Enouf V, Dos Reis G, Guthman JP, Guerin JP, Caron M, Nizou JY, Andraghetti R. Validation of single real-time TaqMan PCR assay for the detection and quantitation of four major genotypes of hepatitis E virus in clinical specimens. J Virol Meth 2006; 78: 1076–1082.

Evans MS, Clarke TF, Clark PL. Conformations of co-translational folding intermediates. Protein Pept Lett 2005; 12: 189–195.

Fauquet C, Mayo MA, Maniloff J, Desselberger U, Ball LA. Virus Taxonomy: Eighth Report of the International Committee on Taxonomy of Viruses. Amsterdam: Elsevier Academic Press; 2005.

Feigelstock D, Thompson P, Mattoo P, Zhang Y, Kaplan GG. The human homolog of HAVcr-1 codes for a hepatitis A virus cellular receptor. J Virol 1998; 72: 6621–6628.

Flehmig B. Hepatitis A-virus in cell culture. I. Propagation of different hepatitis A-virus isolates in a fetal rhesus monkey kidney cell line (FrhK-4). Med Microbiol Immunol 1980; 168: 239–248.

Franco E, Vitiello G. Vaccination strategies against hepatitis A in Southern Europe. Vaccine 2003; 21: 696–697.

Gabrieli R, Sánchez G, Macaluso A, Cenko F, Bino S, Palombi L, Buonuomo E, Pintó RM, Bosch A, Divizia M. Hepatitis in Albanian children: molecular analysis of hepatitis A virus isolates. J Med Virol 2004; 72: 533–537.

Gagneux P, Varki A. Evolutionary considerations in relating oligossaccharide diversity to biological function. Glycobiology 1999; 9: 747–755.

Gavrilin GV, Cherkasova EA, Lipskaya GY, Kew OM, Agol VA. Evolution of circulating wild poliovirus and of vaccine-derived poliovirus in an immunodeficient patient: a unifying model. J Virol 2000; 74: 7381–7390.

Germinario C, Lopalco PL. Mass hepatitis A vaccination considered for Puglia, Italy. Viral Hepat 1999; 8: 7.

Glikson M, Galun E, Oren R, Tur-Kaspa R, Shouval D. Relapsing hepatitis A: review of 14 cases and literature survey. Medicine (Baltimore) 1992; 71: 14–23.

Goens SD, Purdue ML. Hepatitis E virus in humans and animals. Anim Health Res Rev 2004; 5: 145–156.

Goswami BB, Burkhardt Jr. W, Cebula TA. Identification of genetic variants of hepatitis A virus. J Virol Meth 1997; 65: 95–103.

Graff J, Richards OC, Swiderek KM, Davis MT, Rusnak F, Harmon SA, Jia X-Y, Summers DF, Ehrenfeld E. Hepatitis A virus capsid protein VP1 has a heterogeneous C terminus. J Virol 1999; 73: 6015–6023.

Grandadam M, Tebbal S, Caron M, Siriwardana M, Larouze B, Koeck JL, Buisson Y, Enouf V, Nicand E. Evidence for hepatitis E virus quasispecies. J Gen Virol 2004; 85: 3189–3194.

Gust ID. Epidemiological patterns of hepatitis A in different parts of the world. Vaccine 1992; 10(Suppl. 1): S56–S58.

Guthmann JP, Klovstad H, Boccia D, Hamid N, Pinoges L, Nizou JY, Tatay M, Diaz F, Moren A, Grais RF, Ciglenecki I, Nicand E, Guerin PJ. A large outbreak of hepatitis E among displaced population in Dafur, Sudan, 2004: the role of water treatment methods. Clin Infect Dis 2006; 42: 1685–1691.

Halbur PG, Kosorndorkbuam C, Gilbert C, Guenette D, Potters MB, Purcell RH, Emerson SU, Toth TE, Meng XJ. Comparative pathogenesis of infection of pigs with hepatitis E virus recovered from a pig and a human. J Clin Microbiol 2001; 39: 918–923.

Halliday ML, Kang L-Y, Zhou T-Z, Hu M-D, Pan Q-C, Fu T-Y, Huang YS, Hu S-L. An epidemic of hepatitis A attributable to the ingestion of raw clams in Shanghai, China. J Infect Dis 1991; 164: 852–859.

Haqshenas G, Huang FF, Fenaux M, Guenette DK, Pierson FW, Larsen CT, Shivaprasad HL, Toth TE, Meng XJ. The putative capsid protein of the newly identified avian hepatitis E virus shares antigenic epitopes with that of swine and human hepatitis E virus and chicken big liver and spleen disease virus. J Gen Virol 2002; 83: 2201–2209.

Haqshenas G, Shivaprasad HL, Woolcock PR, Read DH, Meng XJ. Genetic identification and characterization of a novel virus related to human hepatitis E virus from chickens with hepatitis-splenomegaly syndrome in the United States. J Gen Virol 2001; 82: 2449–2462.

He J, Hayes CG, Binn LN, Seriwatana J, Vaughn DW, Kuschner RA, Innis BL. Hepatitis E virus DNA vaccine elicits immunological memory in mice. J Biomed Sci 2001; 8: 223–226.

He J, Hoffman SL, Hayes CG. DNA inoculation with a plasmid vector carrying the hepatitis E virus structural protein gene induces immune response in mice. Vaccine 1997; 15: 357–362.

Hollinger FB, Emerson SU. Hepatitis A virus. In: Fields Virology (Knipe DM, Howley PM, editors). Vol. 1. Philadelphia, PA: Lippincott Williams and Wilkins; 2001; pp. 799–840.

Im SW, Zhang JZ, Zhuang H, Che XY, Zhu WF, Xu GM, Li K, Xia NS, Ng MH. A bacterial expressed peptide prevents experimental infection of primates by hepatitis E virus. Vaccine 2001; 19: 3726–3732.

Ishikawa K, Matsui K, Madarame T, Sato S, Oikawa K, Uchida T. Hepatitis E probably contracted via a Chinese herbal medicine, demonstrated by nucleotide sequencing. J Gastroenterol 1995; 30: 534–538.

Jameel S, Zafrullah M, Ozdener MH, Panda SK. Expression in animal cells and characterization of the hepatitis E virus structural proteins. J Virol 1996; 70: 207–216.

Jackson RJ. Proteins involved in the function of picornavirus Internal Ribosome Entry sites. In: Molecular Biology of Picornaviruses (Semler BL, Wimmer E, editors). Washington, DC: ASM Press; 2002; pp. 171–186.

Jean J, Blais B, Darveau A, Fliss I. Detection of hepatitis A virus by the nucleic acid sequence-based amplification technique and comparison with reverse transcription-PCR. Appl Environ Microbiol 2001; 67: 5593–5600.

Jiménez-Clavero MA, Escribano-Romero E, Mansilla C, Gómez N, Córdoba L, Roblas N, Ponz F, Ley V, Saiz JC. Survey of bovine enterovirus in biological and environmental samples by a highly sensitive real-time reverse transcrption-PCR. Appl Environ Microbiol 2005a; 71: 3536–33543.

Jiménez-Clavero MA, Fernández C, Ortiz JA, Pro J, Carbonell G, Tarazona JV, Robles N, Ley V. Teschovirus as indicators of porcine fecal contamination of surface water. Appl Environ Microbiol 2003; 69: 6311–6315.

Jiménez-Clavero MA, Ley V, Gómez N, Saiz JC. Detection of enterovirus. In: Methods in Biotechnology, Food-Borne Pathogens: Methods and protocols (Adley CC, editor). Totowa, NJ: Human Press Inc; 2005b; pp. 153–169.

Jothikumar N, Cromeans TL, Robertson BH, Meng XJ, Hill VR. A broadly reactive one-step real-time RT-PCR assay for rapid and sensitive detection of hepatitis E virus. J Virol Meth 2006; 131: 65–71.

Jothikumar N, Cromeans TL, Sobsey MD, Robertson BH. Development and evaluation of a broadly reactive TaqMan assay for rapid detection of hepatitis A virus. Appl Environ Microbiol 2005; 71: 3359–3363.

Kabrane-Lazizi Y, Fine JB, Elm J, Glass GE, Higa H, Diwan A, Gibbs Jr. CJ, Meng XJ, Emerson SU, Purcell RH. Evidence for widespread infection of wild rats with hepatitis E virus in the United States. Am J Trop Med Hyg 1999b; 61: 331–335.

Kabrane-Lazizi Y, Meng XJ, Purcell RH. Evidence that the genomic RNA of hepatitis E virus is capped. J Virol 1999a; 73: 8848–8850.

Kamili S, Spelbring J, Carson D, Krawczynski K. Protective efficacy of hepatitis E virus DNA vaccine administered by gene gun in the cynomologus macaque model of infection. J Infect Dis 2004; 189: 258–264.

Kasorndorkbua C, Opriessing T, Huang FF, Guenette DK, Thomas PJ, Meng XJ, Hallbur PG. Infectious swine hepatitis E virus is present in pig manure storage facilities on United States farms, but evidence of water contamination is lacking. Appl Environ Microbiol 2005; 71: 7831–7837.

Khuroo MS, Dar MY. Hepatitis E: evidence for person-to-person transmission and inability of low dose immune serum globulin from an Indian source to prevent it. Indian J Gastroenterol 1992; 11: 113–116.

Khuroo MS, Kamili S. Hepatitis E: from hypothesis to reality. Indian J Gastroenterol 1994; 13: 39–43.

Khuroo MS, Kamili S, Jameel S. Vertical transmission of hepatitis E virus. Lancet 1995; 345: 1025–1026.

Kuechler E, Seipelt J, Liebig H-D, Sommergruber W. Picornavirus proteinase-mediated shutoff of host cell translation: direct cleavage of a cellular initiation factor. In: Molecular Biology of Picornaviruses (Semler BL, Wimmer E, editors). Washington, DC: ASM Press; 2002; pp. 301–312.

Kumar A, Beniwal M, Kar P, Sharma JB, Murthy NS. Hepatitis E in pregnancy. J Gynaecol Obstet 2004; 85: 240–244.

Lemon SM, Binn LN. Antigenic relatedness of two strains of hepatitis A virus determined by cross neutralization. Infect Immun 1983; 42: 418–420.

Leong LEC, Cornell CT, Semler BL. Processing determinants and functions of cleavage products of picornavirus. In: Molecular Biology of Picornaviruses (Semler BL, Wimmer E, editors). Washington, DC: ASM Press; 2002; pp. 187–198.

Li T, Takeda N, Miyamura T. Oral administration of hepatitis E virus-like particles induces a systemic and mucosal immune response in mice. Vaccine 2001; 19: 3476–3484.

Li TC, Chijiwa K, Sera N, Ishibashi T, Etoh Y, Shinohara Y, Kurata Y, Ishida M, Sakamoto S, Takeda N, Miyamura T. Hepatitis E virus transmission from wild boar meat. Emerg Infect Dis 2005; 11: 1958–1960.

Liu HH, Cao X, Yang Y, Liu MG, Wang YF. Array-based nano-amplification technique was applied in detection of hepatitis E virus. J Biochem Mol Biol 2006; 31: 247–252.

Lu L, Ching KZ, de Paula VS, Nakano T, Siegl G, Weitz M, Robertson BH. Characterization of the complete genomic sequence of genotype II hepatitis A virus (CF53/Berne isolate). J Gen Virol 2004; 85: 2943–2952.

Mansuy JM, Peron JM, Abarbanel F, Poirson IL, Dubois M, Miedouge M, Vischi F, Alrie L, Vinel JP, Izopet J. Hepatitis E in the south west of France in individuals who have never visited an endemic area. J Med Virol 2004; 74: 419–424.

Martin A, Benichou D, Chao SF, Cohen LM, Lemon SM. Maturation of the hepatitis A virus capsid protein VP1 is not dependent on processing by the 3Cpro proteinase. J Virol 1999; 73: 6220–6227.

Martin A, Lemon SM. Hepatitis A virus: from discovery to vaccines. Hepatology 2006; 43: S164–S172.

Mast EE, Alter MJ. Epidemiology of viral hepatitis: an overview. Semin Virol 1993; 4: 273–283.

Mast EE, Alter MJ, Holland PV, Purcell RH. Evaluation of assays for antibody to hepatitis E virus by a serum panel. Hepatitis E virus serum panel evaluation group. Hepatology 1998; 27: 857–861.

Mast EE, Polish LB, Favorov MO, Khudyakova NS, Collins C, Tukei PM, Kopitch D, Khudyakov YE, Fields HA, Margolis HS. The Somali Refugee Medical Team. Hepatitis E among refugees in Kenya: minimal apparent person-to-person transmission, evidence of ageñdependent disease expression, and new serological assays. In: Viral Hepatitis and Liver Disease (Nishioka N, Suzuki H, Mishiro S, Oda T, editors). Tokyo: Springer-Verlag; 1994; pp. 375–378.

Matsuda H, Okada K, Takahashi K, Mishiro S. Severe hepatitis E virus infection after ingestion of uncooked liver from a wild boar. J Infect Dis 2003; 188: 944.

Mayo MA, Ball LA. ICTV in San Francisco: a report from the Plenary Session. Arch Virol 2006; 151: 413–422.

Meng J, Dai X, Chang JC, Lopareva E, Pillot J, Fields HA, Khudyakov YE. Identification and characterization of the neutralization epitope(s) of the hepatitis E virus. Virology 2001; 288: 203–211.

Meng XJ. Swine hepatitis E virus: cross-species infection and risk in xenotransplantation. Curr Top Microbiol Immunol 2003; 278: 85–216.

Meng XJ, Halbur PG, Shapiro MS, Govindarajan S, Bruna JD, Mushahwar IK, Purcell RH, Emerson SU. Genetic and experimental evidence for cross-species infection by the swine hepatitis E virus. J Virol 1998; 72: 9714–9721.

Meng XJ, Purcell RH, Halbur PG, Lehman JR, Webb DM, Tsareva TS, Haynes JS, Thacker BJ, Emerson SU. A novel virus in swine is closely related to the human hepatitis E virus. Proc Natl Acad Sci Unit States Am 1997; 94: 9860–9865.

Meng XJ, Wiseman B, Elvinger F, Guenette DK, Toth TE, Engle RE, Emerson SU, Purcell RH. Prevalence of antibodies to hepatitis E virus in veterinarians working with swine and in normal blood donors in the United States and other countries. J Clin Microbiol 2002; 40: 117–122.

Myint KS, Endy TP, Gibbons RV, Laras K, Mamenn Jr. MP, Sedyaningsih ER, Seeriwatana J, Glass JS, Narupeti S, Corwin AL. Evaluation of diagnostic assays for hepatitis E virus outbreaks settings. J Clin Microbiol 2006; 44: 1581–1583.

Nainan OV, Xia G, Vaughan G, Margolis HS. Diagnosis of hepatitis A virus infection: a molecular approach. Clin Microbiol Rev 2006; 19: 63–79.

Nishida T, Kimura H, Saitoh M, Shinohara M, Kato M, Fukuda S, Munemura T, Mikami T, Kawamoto A, Akiyama M, Kato Y, Nishi K, Kozawa K, Nishio O. Detection, quantitation, and phylogenetic analysis of noroviruses in Japanese oysters. Appl Environ Microbiol 2003; 69: 5782–5786.

Noble RC, Kane EM, Reeves SA, Roeckle I. Posttransfusional hepatitis A in a neonatal intensive care unit. J Am Med Assoc 1984; 252: 2711–2715.

Olsen B, Axelsson-Olsson D, Thelin A, Weiland O. Unexpected high prevalence of IgG-antibodies to hepatitis E virus in Swedish pig farmers and controls. Scand J Infect Dis 2006; 38: 55–58.

Orrù G, Masia G, Orrù G, Romanò L, Piras V, Coppola RC. Detection and quantitation of hepatitis E virus in human faeces by real-time quantitative PCR. J Virol Meth 2004; 118: 77–82.

Pina S, Buti M, Cortina M, Piella J, Girones R. HEV identified in serum from humans with acute hepatitis and in sewage of animal origin in Spain. J Hepatol 2000; 33: 826–833.

Pintó RM, Alegre D, Domínguez A, El-Senousy WM, Sánchez G, Villena C, Costafreda MI, Aragonès Ll, Bosch A. Hepatitis A virus in urban sewage from two Mediterranean countries. Epidemiol Infect 2006; 134: 1–4.

Previsani N, Lavanchy D, Siegl G. Hepatitis A. In: Viral Hepatitis Molecular Biology, Diagnosis, Epidemiology and Control (Mushahwar IS, editor). Amsterdam: Elsevier; 2004; pp. 1–30.

Probst C, Jecht M, Gauss-Muller V. Intrinsic signals for the assembly of hepatitis A virus particles: role of structural proteins VP4 and 2A. J Biol Chem 1999; 274: 4527–4531.

Purcell RH. Hepatitis viruses: changing patterns of human disease. Proc Natl Acad Sci Unit States Am 1994; 91: 2401–2406.

Purcell RH, Emerson SU. Hepatitis E virus. In: Fields Virology (Fields BN, Knipe DM, Howley PM, editors). Philadelphia, PA: Lippincott-Raven Publishers; 2001; pp. 2831–2843.

Purcell RH, Nguyen H, Shapiro M, Engle RE, Govindarajan S, Blackwelderm WC, Wong DC, Prieels JP, Emerson SU. Pre-clinical immunogenicity and efficacy trial of a recombinant hepatitis E vaccine. Vaccine 2003; 21: 2607–2615.

Purdy MA, McCaustland KA, Krawczynski K, Spelbring J, Reyes GR, Bradley DW. Preliminary evidence that a trpE-HEV fusin protein protects cynomolgus macaques against challenge with wild-type hepatitis E virus (HEV). J Med Virol 1993; 41: 90–94.

Reid TMS, Robinson HG. Frozen raspberries and hepatitis A. Epidemiol Infect 1987; 98: 109–112.

Reyes GR, Purdy MA, Kim JP, Luk KC, Young LM, Fry KE, Bradley DW. Isolation of a cDNA from the virus responsible for enterically transmitted non-A, non-B hepatitis. Science 1990; 247: 1335–1339.

Rieder E, Wimmer E. Cellular receptors of picornaviruses: an overview. In: Molecular Biology of Picornaviruses (Semler BL, Wimmer E, editors). Washington, DC: ASM Press; 2002; pp. 61–70.

Robertson BH, Jansen RW, Khanna B, Totsuka A, Nainan OV, Siegl G, Widell A, Margolis HS, Isomura S, Ito K, Ishizu T, Moritsugu Y, Lemon SM. Genetic relatedness of hepatitis A virus strains recovered from different geographical regions. J Gen Virol 1992; 73: 1365–1377.

Robinson M, Lilley R, Little S, Emtage JS, Yarranton G, Stephens P, Millican A, Eaton M, Humphreys G. Codon usage can affect efficiency of translation of genes in *Escherichia coli*. Nucleic Acids Res 1984; 12: 6663–6671.

Rosemblum LS, Mirkin IR, Allen DT, Safford S, Hadler SC. A multifocal outbreak of hepatitis A traced to commercially distributed lettuce. Am J Publ Health 1990; 80: 1075–1079.

Salleras LL. Catalonia, Spain introduces mass hepatitis A vaccination programme. Viral Hepat 1999; 8: 3–4.

Sánchez G, Aragonès L, Costafreda MI, Ribes E, Bosch A, Pintó RM. Capsid region involved in the hepatitis A virus binding to the glycophorin A of the erythrocyte membrane. J Virol 2004a; 78: 9807–9813.

Sánchez G, Bosch A, Gómez-Mariano G, Domingo E, Pinó RM. Evidence for quasispecies distributions in the human hepatitis A virus genome. Virology 2003a; 315: 34–42.

Sánchez G, Bosch A, Pintó RM. Genome variability and capsid structural constraints of hepatitis A virus. J Virol 2003b; 77: 452–459.

Sánchez G, Pintó RM, Vanaclocha H, Bosch A. Molecular characterization of hepatitis A virus isolates from a transcontinental shellfish-borne outbreak. J Clin Microbiol 2002; 40: 4148–4155.

Sánchez G, Villena C, Bosch A, Pintó RM. Molecular detection and typing of hepatitis A virus. In: Methods in Molecular Biology (Spencer JFT, Ragout de Spencer AL, editors). Vol. 268. Totowa, NJ: Humana Press; 2004b; pp. 103–114.

Schlauder GG. Viral hepatitis: Molecular biology, diagnosis, epidemiology and control. In: Perspective in Medial Virology (Zuckerman AJ, Mushahwar IK, editors). Elsevier; 2004; pp. 199–222.

Schultheiss T, Kusov YY, Gauss.Muller V. Proteinase 3C of hepatitis A virus (HAV) cleaves the HAV polyprotein P2-P3 at all sites including VP1/2A and 2A/2B. Virology 1994; 198: 275–281.

Schultheiss T, Sommergruber W, Kusov Y, Gauss-Müller V. Cleavage specificity of purified recombinant hepatitis A virus 3C proteinase on natural substrates. J Virol 1995; 69: 1727–1733.

Sheretz RJ, Russell BA, Reunman PD. Transmission of hepatitis A by transfusion of blood products. Arch Intern Med 2005; 1441: 1579–1580.

Shouval D. Israel implements universal hepatitis A immunization. Viral Hepat 1999; 8: 2.

Silberstein E, Xing L, van de Beek W, Lu J, Cheng H, Kaplan GG. Alteration of hepatitis A virus (HAV) particles by a soluble form of HAV cellular receptor 1 containing the immunoglobulin- and mucin-like regions. J Virol 2003; 77: 8765–8774.

Sjogren MH, Tanno H, Fay O, Sileoni S, Cohen BD, Burke DS, Feighny RJ. Hepatitis A virus in stool during clinical relapse. Ann Intern Med 1987; 106: 221–226.

Skidmore SJ. Overview of hepatitis E virus. Curr Infect Dis Rep 2002; 4: 118–123.

Smith JL. A review of hepatitis E virus. J Food Protect 2001; 64: 572–586.

Sørensen MA, Kurland CG, Pedersen S. Codon usage determines translation rate in *E. coli*. J Mol Biol 1989; 207: 365–377.

Stevenson P. Nepal calls the shots in hepatitis E virus vaccine trial. Lancet 2000; 355: 1623.

Tallo T, Norder H, Tefanova V, Ott K, Ustina V, Prukk T, Solomonova O, Schmidt J, Zilmer K, Priimägi L, Krispin T, Magnius LO. Sequential changes in hepatitis A virus genotype distribution in Estonia during 1994 to 2001. J Med Virol 2003; 70: 187–193.

Tam AW, Smith MM, Guerra ME, Huang CC, Bradley DW, Fry KE, Reyes GR. Hepatitis E virus (HEV): molecular cloning and sequencing of the full-length viral genome. Virology 1991; 185: 120–131.

Tei S, Kitajima N, Takahashi K, Mishiro S. Zoonotic transmission of hepatitis E virus from deer to human beings. Lancet 2003; 362: 371–373.

Teppakdee A, Tangwitoon A, Khemasuwan D, Tangdhanakanond K, Suramaethakul N, Sriratanaban J, Poovorawan Y. Cost-benefit analysis of hepatitis a vaccination in Thailand. Southeast Asian J Trop Med Pub Health 2002; 33: 118–127.

Tsarev SA, Tsareva TS, Emerson SU, Govindarajan S, Shapiro M, Gertin JL, Purcell RH. Successful passive and active immunization of cynomolgus monkeys against hepatitis E. Proc Natl Acad Sci Unit States Am 1994; 91: 10198–10202.

Tsarev SA, Tsareva TS, Emerson SU, Govindarajan S, Shapiro M, Gertin JL, Purcell RH. Recombinant vaccine against hepatitis E virus: dose response and protection against heterologous challenge. Vaccine 1997; 15: 1834–1838.

Van Damme P, Hallauer J, Papaevangelou G, Bonanni P, de la Torre J, Grob P. Viral Hepat 1999; 8: 12–13.

van Zyl WB, Page NA, Grabow WOK, Steele AD, Taylor MB. Molecular epidemiology of group A rotaviruses in water sources and selected raw vegetables in Southern Africa. Appl Environ Microbiol 2006; 72: 4554–4560.

Villena C, El-Senousy WM, Abad FX, Pintó RM, Bosch A. Group A rotavirus in sewage samples from Barcelona and Cairo: emergence of unusual genotypes. Appl Environ Microbiol 2003; 69: 3919–3923.

Vishwanathan R. Infectious hepatitis in Delhi (1955–1956): a critical study: epidemiology. Indian J Med Res 1957; 45: 1–29.

Wang ZQ, Ma L, Gao H, Dong XG. Propagation of hepatitis A virus in human diploid fibroblast cells. Acta Virol 1986; 30: 463–467.

Wang L, Zhuang H. Hepatitis E: an overview and recent advances in vaccine research. World J Gatroenterol 2004; 10: 2157–2162.

Wasley A, Samandari T, Bell BP. Incidence of hepatitis A in the United States in the era of vaccination. J Am Med Assoc 2005; 294: 194–201.

Weitz M, Baroudy BM, Maloy WL, Ticehurst JR, Purcell RH. Detection of a genome-linked protein (VPg) of hepatitis A virus and its comparison with other picornaviral VPgs. J Virol 1986; 60: 124–130.

Whetter LE, Day SP, Elroy-Stein O, Brown EA, Lemon SM. Low efficiency of the 5′ non-translated region of hepatitis A virus RNA in directing cap-independent translation in permissive monkey kidney cells. J Virol 1994; 68: 5253–5263.

Whiters MR, Correa MT, Morrow M, Stebbins ME, Seriwatana J, Webster WD, Boak MB, Vaughn DW. Antibody levels to hepatitis E virus in North Carolina swine workers, non-swine workers, swine, and murids. Am J Trop Med Hyg 2002; 66: 384–388.

Wong DC, Purcell RH, Sreenivasan MA, Prasad SR, Pavri KM. Epidemic and endemic hepatitis in India: evidence for a non-A, non-B hepatitis virus aetiology. Lancet 1980; 2: 876–879.

Worm HC, Schlauder GG, Brandstätter G. Hepatitis E and its emergence in non-endemic areas. Wien Klin Wochenschr 2002a; 114: 663–670.

Worm HC, van der Poel WHM, Brandstätter G. Hepatitis E: an overview. Microb Infect 2002b; 4: 657–666.

Yazaki Y, Mizuo H, Takahashi M, Nishizawa T, Sasaki N, Gotanda Y, Okamoto H. Sporadic acute or fulminant hepatitis E in Hokkaido, Japan, may be food-borne, as suggested by the presence of hepatitis E virus in pig liver as food. J Gen Virol 2003; 84: 2351–2357.

Zafrullah M, Ozdener MH, Kumar R, Panda SK, Jameel S. Mutational analysis of glycosilation, membrane translocation, and cell surface expression of the hepatitis E virus ORF-2 protein. J Virol 1999; 73: 4074–4082.

Zafrullah M, Ozdener MH, Panda SK, Jameel S. The PRF-3 protein of hepatitis E virus is a phosphoprotein that associates with the cytoskeleton. J Virol 1997; 71: 9045–9053.

Zhang M, Emerson SU, Nguyen H, Engle RE, Govindarajan S, Blackwelder WC, Gertin JL, Purcell RH. Recombinant vaccine against hepatitis E virus: duration of protective immunity in rhesus macaques. Vaccine 2002; 20: 3285–3291.

Zheng Y, Ge S, Zhang J, Guo Q, Ng MH, Wang F, Xia N, Jiang Q. Swine as a principal reservoir of hepatitis E virus that infects humans in eastern China. J Infect Dis 2006; 15: 1643–1649.

Zhou YJ, Estes MK, Jiang X, Metcalf TG. Concentration and detection of hepatitis A virus and rotavirus from shellfish by hybridization tests. Appl Environ Microbiol 1991; 57: 2963–2968.

© 2007 Elsevier B.V. All rights reserved
DOI 10.1016/S0168-7069(07)17004-X

Chapter 4

Enteroviruses with Special Reference to Poliovirus and Poliomyelitis Eradication

Tapani Hovi, Merja Roivainen, Soile Blomqvist

*Enterovirus Laboratory, National Public Health Institute (KTL), WHO
Collaborating Centre for Poliovirus Surveillance and Enterovirus Research,
Mannerheimintie 166, 00300 Helsinki, Finland*

Human enteroviruses (HEVs) can be found in natural and man-made environmental
water bodies only as a consequence of contamination by human excreta. Detection
of enteroviruses (EVs) in sources of drinking water or in recreational water bodies
can, hence, be used as a marker of a possible failure of the sanitation systems and as
an indicator of the potential role of water in disease outbreaks. Survival of EVs in
water depends on several factors but in general, examination of samples of water
bodies for EVs can also be used as one approach in studies on EV epidemiology,
defined as environmental surveillance. This is best exploited in the monitoring of the
progress of the WHO-coordinated poliomyelitis eradication initiative (PEI). This
review summarises some key observations in this field and discusses possibilities and
limitations of environmental surveillance of HEV infections.

Background

General features of human enteroviruses

Human enteroviruses (HEVs) exist as four species (HEV-A through HEV-D) in the
genus *Enterovirus* of the *Picornaviridae* family (Stanway et al., 2005). Each
species contains several serotypes still referred to by their traditional names: polioviruses, coxsackieviruses of subtypes A and B, echoviruses and the newer enteroviruses (EVs) given numbers 68 (EV68) and so forth (Table 1). EVs are small
RNA viruses with about 7.5 kilobase messenger-sense genome and an icosahedral
non-enveloped capsid composed of 60 copies of each of the four different capsid

Table 1

Classification of human enterovirus (HEV) serotypes in the old sub-groups of the genus and in the current species

Sub-group	Species			
	HEV-A	HEV-B	HEV-C	HEV-D
Poliovirus			PV1, PV2, PV3	
Coxsackie-virus A (CVA)[a]	CVA2–8	CVA9	CVA1, 11, 13, 17	
	CVA10, 12, 14, 16		CVA19–22, 24	
Coxsackie-virus B (CVB)		CVB1–6		
Echovirus (E)[b]		E1–7, 9		
		E11–21, 24–27		
		E29–33		
Newer enterovirus (EV) serotypes[c]	EV71 **EV76, 89–91**	EV69 **EV73–75,77–79** **EV81–83, 85–86** **EV97, 100**	**EV96**	EV68, 70 **EV94**

[a]Coxsackievirus serotypes A15 and A18 have been reclassified as strains of CAV11 and CAV13, respectively (Brown et al., 2003). Coxsackievirus A23 has been found to be of the same serotype as echovirus E9.
[b]Echovirus E8 has been found to be a strain of E1; E10 has been reclassified as Reovirus 1; E22 and 23 have been reclassified in the new *Parechovirus* genus; E28 has been reclassified as human rhinovirus 1A.
[c]New types identified by the molecular methods only are in bold phase.

proteins referred to as VP1 through VP4. Point mutations and recombinations occur frequently (rate for both is about 10^{-4}) during replication; and probably mainly due to bottleneck selection during transmission, broad genetic diversification of virus strains within any given serotype has developed. For many serotypes, distinct sub-serotypic clusters, referred to as genotypes (Rico-Hesse et al., 1987), are known to co-circulate. Interestingly, echovirus E-30 appears to be one exception to this "rule" with a single prominent genotype found among European strains through almost three decades (Savolainen et al., 2001). While most of the diversification appears to be based on fixation of neutral mutations, antigenic drift is also known to occur, sometimes resulting in difficulties in serotyping of recent virus isolates with antisera made against old strains (Pallansch and Roos, 2001).

For entry into the host cells, EVs use a wide range of different receptor molecules, several of them belonging to the immunoglobulin superfamily of proteins or to integrins as reviewed by Pietiainen et al. (2005). Polioviruses use exclusively an Ig-superfamily protein (CD155), referred to as the poliovirus receptor (PVR), a fact exploited in isolation of poliovirus from a mixture of different EVs with the help of recombinant murine cell lines expressing the human PVR (Pipkin et al., 1993; Hovi and Stenvik, 1994). However, possible susceptibility of the parental murine cell to some non-polio HEVs must be remembered as a possible confounding factor (Nadkarni and Deshpande, 2003).

Natural course of human enterovirus infection

Most scientific evidence for the "typical" faeco-orally transmitted infection resulting in a biphasic course of disease symptoms is derived from studies on poliovirus infection, and it is obvious that these observations must not be interpreted to automatically hold true for each and every HEV serotype. Indeed, most HEV-C serotypes and EV68 are mainly respiratory pathogens causing common cold-like symptoms while EV70 and some strains of coxsackievirus A24 may directly infect the conjunctiva and cause haemorrhagic conjunctivitis. For more detailed information on pathogenesis of EV infection, see Pallansch and Roos (2001).

In the classical poliovirus model the virus primarily infects the (naso)pharyngeal mucosa—or is swallowed and may directly start infection in the intestines. Initial replication may result in transient non-specific symptoms. From the mucosae, the virus may enter the lymphatics and eventually blood circulation. After a brief viraemic phase, secondary target tissues in different organs may be infected resulting in the second phase of disease with infected organ-specific symptoms. In the case of poliovirus, the target tissues of the secondary phase include the motor neurons of the central nervous system (CNS) with the infection-induced cytopathology resulting in typical paralytic symptoms. This is, however, an exception; more than 99% of poliovirus infections proceed without paralysis. Similarly, other EV infections sometimes, but not regularly, result in aseptic meningitis, myocarditis, generalized neonatal infection, etc.; serotype-specific variation in the proportion of asymptomatic infections has been reported (Cherry, 1987).

Irrespective of the location and intensity of symptoms of infection, EV replication is considered to continue in the sub-mucosal lymphatic tissue for several weeks, up to a couple of months. The virus is excreted into the faeces and translocated to the environment. The amounts excreted in the case of poliovirus are known to vary with maximal amounts reaching 10^6 infectious doses per gram (Dowdle et al., 2002). For other EV serotypes, very little data are available. Poliovirus is also known to be excreted in the nasopharyngeal mucosa during first 2 weeks of infection. This may have a role in virus transmission between close contacts but is not likely to contribute to the load of virus in the environmental water bodies. EVs are relatively stabile in the natural environment (see article by C.P. Gerba in this volume) but again, differences between species and serotypes may exist.

Detection and identification of enteroviruses

Goal-driven selection of approach

Traditionally, EV detection in water samples as well as in clinical specimens, after suitable processing of the sample, is based on inoculation of cell cultures and subsequent serotyping of cytopathogenic agents with pools of monotypic antisera (Lennette and Smith, 1999, pp. 703–719). This is challenging as no single cell line is known to support productive replication of all HEV serotypes and the neutralization assay with pools of antisera is a tedious procedure. Poliovirus and coxsackievirus B strains readily replicate in several human or monkey cell lines while echoviruses prefer human fibroblastoid cell strains or certain human tumour cell lines. Some coxsackie A-virus serotypes were originally isolated in suckling mouse brain and cannot in practice be isolated in cell culture (Pallansch and Roos, 2001).

Optimal choice of cell culture for EV detection depends on the goals of the study. If one is monitoring potential faecal contamination in sources of household water, one or two "broad-spectrum" cell lines such as the buffalo green monkey (BGM) kidney cell line, human rhabdomyosarcoma (RD) or the colon tumour cell CaCo-2, is likely to be sufficient. A proposal for an international standard for safety monitoring of natural water bodies and swimming pools includes a search for EVs using a plaque assay in BGM (prEN 14486, European Committee for Standardization, water quality—detection of HEVs by monolayer plaque assay).

On the other hand, if one is interested in the entire EV serotype spectrum in the environment, human embryonic lung or kidney cells, primary monkey kidney cells or a monkey kidney cell line Vero, as well as an ICAM-1 overexpressing HeLa cell line, might further broaden the spectrum. Replication pattern of a virus in a set of multiple cell lines has also been used as a tentative approach to identify the subgroup of EVs (Sedmak et al., 2003).

Reverse transcriptase polymerase chain reaction (RT-PCR) can also be used for detection of EV genome in water samples. Most often, primers targeted to conserved sequences in the 5'-non-coding part of the genome (Fig. 1) are used (Hyypia et al., 1989; Halonen et al., 1995; Gregory et al., 2006). Species-specific detection can be accomplished by amplification of the 3'-non-coding region of the genome (Oberste et al., 2006). In interpretation of the obtained results it is important to remember that enteroviral RNA may be stabile inside dead particles and, especially in quantitative assays, more than 100 particles may be needed for an infectious unit (Pallansch and Roos, 2001). Therefore, infectivity titres and RNA concentrations do not necessarily correlate (Pusch et al., 2005).

Old and new identification methods

Faeces-contaminated waters, especially community sewage, usually contain a mixture of viruses. Inoculation of a "bulk" cell culture will usually reveal only the most common virus or viruses present in the examined sample. For a better coverage of the variety of viruses, plaque assays or multiple parallel wells have to be used. Even then it is clear that minority components present in concentrations lower than a majority serotype may never be revealed without specific enrichment techniques. Identification of the isolate is traditionally carried out by neutralization with pools of monotypic antisera. This is tedious and the availability of thoroughly tested antiserum pools is limited. Sets of monoclonal antibodies for EV identification are also commercially

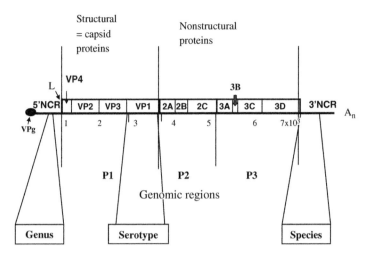

Fig. 1 A sketch view of EV genome indicating the regions most commonly used for molecular identification of the genus, species and serotypes.

available and typically used in immunofluorescence assays. Again, validation of the monoclonals may be limited as one could imagine that they are even more sensitive than polyclonal antisera to antigenic drift of the virus strains.

Recently, molecular methods for identification of the serotype of an EV isolate have been developed (Oberste et al., 1999). Partial sequence of the VP1-coding part of the genome appears to be a highly reliable marker of serotype identity. Sequence-based identification of EV isolates has also revealed the existence of much larger number of different EV serotypes, or more accurately "types" than previously recognized (Oberste et al., 2004; Oberste et al., 2005). The running number of EV types, registered by the Picornavirus Study Group of the International Committee for Taxonomy of Viruses, is well beyond the "classical" number 71 and has exceeded 100 (Nick Knowles, personal communication). For sequence-based typing, separation of the components of a putative mixture in advance is also necessary.

Sewage derived from a water closet-using human population almost always contains EVs and occasionally even in concentrations allowing detection without concentration. However, usually concentration is necessary. Techniques used for virus concentration are described elsewhere in this volume (see article by P. Wyn-Jones in this volume). Especially when using RT-PCR for virus detection, it is important to evaluate possible enrichment of putative interfering component in the samples.

Incongruous distribution of serotypes as compared to clinical isolates

Although numerous reports describe detection of HEVs in sewage and also in natural water bodies, relatively few articles have combined prospective monitoring of the environment and concurrent results from analysis of clinical specimens. Results in some reports suggest that the serotype distribution detected in sewage appears to well reflect that it concomitantly causes a typical disease in the patient population (Sellwood et al., 1981; Sedmak et al., 2003). However, longer-term monitoring programmes have revealed definite inconsistencies of the coincidence. Several years ago we published a 20-year follow-up of EV infections in Finland, based on results from analysis of clinical specimens and on systematic analysis of sewage specimens (Hovi et al., 1996). Some major outbreaks caused by a given EV serotype were found to coincide in both categories of samples but in general, the annual top lists appeared to differ remarkably. Yet, for the total 20-year period, as many as 8 out of the top 10, and 11 out of the top 14 serotypes were shared by the two materials (Table 2). One confounding factor was the fact that the environmental samples had been examined in a single laboratory while results from the clinical specimens were a combination of several laboratories using different cell lines for virus isolation (Hovi et al., 1996).

A more recent study from USA, reporting serotype distribution of 1068 environmental and 694 clinical EV isolates revealed a similar situation in spite of the fact that in this case both environmental specimens and the clinical specimens had been analysed in the same laboratory using the same set of cell lines (Sedmak et al., 2003). Some major outbreaks were easily detected in both categories of samples while some serotypes frequently encountered in the environment very rarely were isolated from patients and vice versa. Some EV serotypes, notably

Table 2

Prevalence of the most common enterovirus (EV) serotypes in clinical samples and in sewage in Finland in 1971–1992

Serotype	Clinical	Environment
CVA9	**8.6**	0.6
CVA16	2.0	<0.1
CVB1	2.7	**3.5**
CVB2	4.6	**11.1**
CVB3	**7.3**	**7.8**
CVB4	6.2	**18.7**
CVB5	**14.9**	**17.3**
E3	1.3	**2.6**
E6	**4.5**	**15.2**
E9	3.9	0.4
E11	**12.1**	**16.6**
E18	3.2	<0.1
E25	1.3	**0.9**
E30	**6.7**	**1.3**

Source: Modified from Hovi et al., 1996.
Numbers are percentages of a total of 1681 clinical and 1161 sewage samples, respectively. Ten most common serotypes in the given sample category are shown in bold.

echovirus E6, were found in both cited prospective studies (Hovi et al., 1996; Sedmak et al., 2003) to be highly prevalent in the environment and relatively rare in the clinical specimens, suggesting that E6 probably is sub-clinical or is causing only minor or non-specific symptoms not resulting in attempts to identify the causative agent of the symptoms. However, even in the case of E6, the ratio of clinical cases to sewage isolates varied a lot from year to year. On the other hand, rarity of some serotypes in the sewage in spite of frequent occurrence in the clinical samples (e.g. coxsackievirus A9 and echovirus E-22, now classified as human parechovirus type 1) could result from poor stability of the virus in the environment, or alternatively, from a putative small quantity or short duration of virus excretion.

We have recently added yet another dimension to this problem by examining collections of human sera for prevalence of neutralizing antibodies to a large number of different EV serotypes. While the studies confirmed the high prevalence of infections caused by some serotypes common in either the clinical material or in the environment, other serotypes were rarely detected in either material. Furthermore, some of the only recently described new types were already found to bewidely circulating for decades (Smura et al., 2007). In conclusion, while detection of a given EV serotype in sewage or other environmental specimens naturally is an indication of circulation of the virus in the respective human population, we should be cautious in interpreting such a finding—without additional studies—as an indication of a causative role of the virus in a concurrently occurring disease in the population. Yet, sometimes this coincidence may be true, and environmental surveillance might be useful in, for instance, revealing the geographical extent of circulation of an epidemic virus strain as well documented for poliovirus (see below).

Monitoring the progress of the poliomyelitis eradication initiative

Long history and short future of paralytic poliomyelitis

Poliomyelitis is a devastating neurological disease caused by three serotypes of "wild" poliovirus (WPV1 through WPV3). It has been occurring among humans for several thousands of years but was described as a definite clinical entity only in the 19th century. During the first half of the 20th century it caused severe epidemics in several industrialized countries. The switch to an epidemic pattern of virus transmission was based on improved hygiene and sanitary conditions, and hence partial inhibition of the faeco-oral transmission of the virus (Nathanson and Martin, 1979). As a consequence, successive cohorts of newborn children were saved from infection and remained susceptible to the virus, and when the virus eventually arrived into the population, it was able to cause an epidemic.

Two types of vaccine were developed against poliomyelitis in the 1950s and systematic use of either of them eliminated WPV transmission in industrialized world by mid-1970s. A small number of countries succeeded in the elimination by using the inactivated poliovirus vaccine (IPV) (Salk, 1981) only, while most of the world switched to the use of the somewhat later introduced live oral poliovirus vaccine (OPV) (Sabin, 1985). However, in developing countries with high birth rates and poor sanitation, the infection pressure was high enough to make frequent breakthroughs in spite of the use of OPV and most of the world still had poliovirus circulation going through the 1980s. In 1988, the World Health Assembly accepted the resolution targeted to global eradication of poliomyelitis, later defined as poliomyelitis eradication initiative (PEI). WHO has been coordinating this exceptional international collaboration ranging all over the world from the western industrialized countries to equatorial Africa, Northern Korea and the smallest Pacific islands. Major cooperating organizations are UNICEF, Rotary International and CDC USA supplemented with both donations and operational support from a number of other countries and organizations.

Success and challenges of the eradication programme

The routine newborn immunizations defined in the Expanded Programme of Immunizations (EPI) in 1970s never reached sufficiently high coverage in developing countries to stop virus circulation. Therefore, different modes of supplementary immunization activities (SIA) were designed in the PEI. National and sub-national immunization days (NIDs, sNIDs) have been critical for the success and include at least two rounds of annual immunisations targeted to all and everyone under a certain age limit, usually 4 or 5 years. Possible defects in the coverage of NIDs or sNIDs are corrected in the mop-up rounds with vaccinators going from home to home or from hut to hut to reach every child in the target group. Although the initially planned deadline of eradication, year 2000, is well in the past, the success of the programme is spectacular. The number of new cases has been reduced by more than 99% and endemic circulation of indigenous poliovirus strains has been limited to four countries only: Afghanistan, India, Nigeria, and Pakistan. Furthermore, one of the three serotypes, WPV2, has been entirely eradicated from the mankind (Centers for Disease Control and Prevention (CDC), 2001). On the other hand, recent reemergence of poliomyelitis in more than 20 previously polio-free countries (Centers for Disease Control and Prevention (CDC), 2006) demonstrates that the herd immunity in apparently polio-free developing countries may be very fragile indeed and even temporary deficits in vaccinations may rapidly worsen the situation again. The PEI still seems to have several years to go.

Outline of standard poliovirus analysis in the PEI

The gold standard approach of WHO in monitoring the progress of PEI is the surveillance of cases of acute flaccid paralysis (AFP) connected to a standardized virological analysis of two faecal samples of the patients, and/or sometimes those from contacts (Birmingham et al., 1997). While the starting point in the virus hunt thus is case-driven, i.e. a clinical suspicion of a potential case of poliomyelitis, development of a special network of polio laboratories (Polio Labnet) has been instrumental to the success of PEI. Sample analysis starts in more than 100 National Polio Laboratories (NPL) distributed all over the world. In larger countries there may also be a network of provincial or other sub-national laboratories responsible for the primary virus isolation. Poliovirus isolates made at NPLs are being sent to more than 20 Regional Reference Laboratories (RRLs) for a procedure called intratypic differentiation (ITD), i.e. a pair of antigenic and molecular screening tests revealing whether the strain is a WPV or derived from the OPV (van der Avoort et al., 1995a). To avoid unnecessary transport times between different laboratories, some of the NPLs have been recently "upgraded" to carry out ITD tests. In this way it is possible to get the answers faster and allow more time to proper programmatic responses. All WPV strains and all isolates presenting with discordant results in the ITD tests are being further referred to the "highest level" of the Polio Labnet, to one of the seven Global Specialized Laboratories (GSL). Here, the sequence of the VP1-coding part of the genome will be determined and using computer-assisted phylogenetic analysis, the genetic origin of the strains will be established. Again, some RRLs and even some NPLs now have access to sequence analysis, and strain characterizations made locally may naturally facilitate the overall process and potential onset of necessary interventions.

Vaccine-derived polioviruses

After eradication of the WPV infections from mankind, a remaining obstacle will be: how to get safely rid of the OPV. OPV is well known to induce, very rarely, vaccine-associated paralytic poliomyelitis (VAPP) in the vaccinee or in his or her immediate contact. One feature of OPV is the capability of the virus to spread from the vaccinee to nearby contacts. This was initially considered to be an advantage, as the observed small-scale spreading of the vaccine virus was thought to improve the herd immunity. The latter may well be true on one side but the other, a shadowy side of the coin has been discovered more recently: In the absence of WPV of the same serotype and in the presence of sub-optimal coverage of immunizations, any serotype of OPV seems to be able to establish circulation in human population. The attenuated phenotype is rapidly lost in circulation and circulating vaccine-derived polioviruses (cVDPV) have caused several outbreaks of paralytic poliomyelitis (Kew et al., 2002). So far, these outbreaks have been contained by rigorous OPV campaigns, but this is not necessarily the wisest thing to do after stopping the overall vaccinations to poliomyelitis. If cVDPV can cause outbreaks in partially immunized population, what would the result be in totally unimmunized cohorts of newborns after the planned post-eradication stopping of the vaccinations? Even in high OPV-coverage populations, some genetic drifting of the Sabin strains can be observed (Fig. 2) and it is easy to imagine that there is a risk that some of these initiated chains of transmission will continue in the presence of sufficient amounts of fully susceptible individuals.

Circulating VDPV strains associated with outbreaks of paralytic poliomyelitis have shown a moderate degree of genetic drifting, not usually exceeding 3% in the

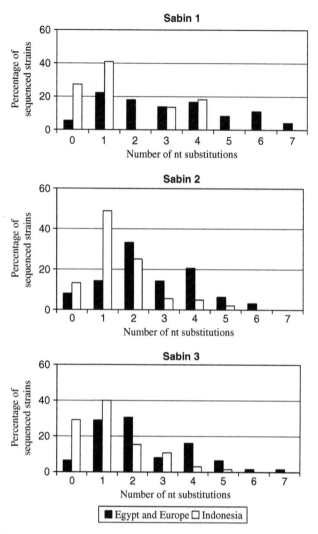

Fig. 2 Distribution of nucleotide substitutions in the VP1-coding part of genome in poliovirus isolates collected from populations with high-coverage vaccination with oral poliovirus vaccine (OPV). Strains isolated in Europe and Egypt were selected on the basis of an aberrant result in one or both of the intratypic differentiation (ITD) tests. Indonesian strains were all Sabin-like in both tests.

VP1-coding part of the genome. Often but not exclusively, cVDPV strains have a recombinant genome with the non-structural P2P3 region originating from one or more unknown HEV-C—specific genetic lineage(s) (Kew et al., 2002). In contrast, VDPV strains isolated from immunodeficient (ID) individuals (iVDPV) may be much more drifted (up to 15%) from Sabin, but the genome—in spite of also being a recombinant—is apparently always totally of poliovirus origin (Martin, 2006).

ID patients occasionally come down with paralytic poliomyelitis only after several years' excretion of the virus and may spread the virus to close contacts. However, no secondary cases have been reported for iVDPV patients so far. Again, we can only imagine what could happen after stopping polio immunizations with increasing amounts of totally unimmunized newborns and young children in the surroundings.

Yet another category of VDPV strains is that isolated from clinical specimens or environment without any epidemiological linkage to AFP cases. Genetic drift in these isolates range from strains marginally exceeding the 1% classification border to environmental isolates resembling the most vastly differentiated iVDPV strains (Blomqvist et al., 2004). They have been referred to as unclassified VDPVs, other VDPVs or ambiguous VDPVs (aVDPV). The last name seems the most appropriate as it reflects the ambiguous origin of the strain. The least drifted ones may represent relatively brief and perhaps still eventually abortive chains of transmission, while the strongly drifted ones could reflect presence of an unidentified ID patient in the region, whose excreta only happened to be in the harvested environmental sample "by accident". Most of the characterized aVDPVs are, indeed, sporadic strains well fitting into these explanations. Even if presence of a VDPV strain in a population can be confirmed by repeated isolation, the origin may be difficult to find out. The most frustrating episode in this sense was experienced in the Slovakian Republic recently (see below). More than 100 VDPV strains were isolated from sewage samples within a 2-year follow-up time but no cases of poliomyelitis were detected and the origin of the virus remained unresolved (unpublished). Whatever the origin, the accumulation of aVDPV isolations is adding to the general worry of increasing emergence of potentially paralytic OPV-derived strains whose real significance may be uncovered only after the designated stopping of polio immunizations.

Environmental surveillance of poliovirus circulation

Occurrence of poliovirus in environmental specimens

Poliovirus strains have been frequently isolated from sewage, from contaminated natural water bodies and also occasionally from drinking water sources and recreational water as long as these types of samples have been examined for cytopathic viruses anywhere in the world. On the other hand, countries that have eliminated indigenous WPV circulation by using the IPV usually also got rid of poliovirus in the environment. This was systematically demonstrated in Finland by studies in the 1970s through 1980s (Lapinleimu, 1984). However, importation of both OPV-derived and WPVs might still take place and was documented both in Sweden (Böttiger and Herrström, 1992) and in The Netherlands (Mulders et al., 1997).

Various techniques have been used for sample collection and processing. It is obvious that they may differ in sensitivity but as no systematic comparative studies have been published, they will not be commented here. In the following, we will briefly discuss the use of environmental surveillance for poliovirus in attempts to answer selected specific epidemiological questions.

Studies aiming at assessing the geographical extent of polio outbreaks

Two major studies in this category will be discussed, corresponding to outbreaks in Finland (1984), and The Netherlands (1992). Finland had been free of WPV for 20 years, and had only recently stopped routine environmental surveillance, as the outbreak of type 3 WPV took the country by surprise in late 1984. This event has been previously dealt with in several publications, e.g. (Hovi et al., 1986; Hovi, 1989). The serotype causing the outbreak was not a surprise because seroepidemiological studies published earlier the same year had revealed that in spite of relatively high coverage of the childhood vaccination programme involving several IPV doses, the prevalence of neutralizing antibodies to poliovirus type 3 was alarmingly poor up to 15 years of age (Lapinleimu, 1984). Several contacts of the index case were found to excrete the same WPV3, and environmental surveillance rapidly reinstituted also revealed the virus in the community and elsewhere in the Greater Helsinki region. Environmental surveillance was subsequently extended to about 40 locations all over the country, usually only one sample was collected per location. More than half of the specimens contained the epidemic WPV3 strain (Poyry et al., 1988), indicating a very widespread circulation of the virus, and reassuring that the decision to combat the outbreak by launching a very intensive vaccination campaign was well motivated. As an *ad hoc* active surveillance had revealed only nine paralytic cases altogether in the country, this observation meant that WPV3 was being detected in many parts of the country not showing the virus infection-associated cases of AFP at all. In other words, under these conditions environmental surveillance was definitely more sensitive than AFP surveillance in detecting WPV3 circulation.

In The Netherlands in 1992, an outbreak of poliomyelitis was once again detected among members of a minority religious sect not accepting any type of vaccinations. The main population of the country is immunized with a high coverage IPV programme. During the outbreak, OPV vaccinations were actively offered to the unvaccinated ones but no campaign was organized for the main population. No cases were diagnosed among the only IPV-immunized population. The extent of virus spreading in the country was examined, among other things, by environmental surveillance (van der Avoort et al., 1995b). The virus was frequently found in sewage derived from villages mainly inhabited by the unvaccinated people, but not in regions where the regularly immunized Dutch people were living, a result in accordance with a large stool survey carried out during the outbreak (Conyn-van Spaendonck et al., 1996).

Investigation of putative WPV reservoirs

Environmental surveillance has been used as a supplementary tool in some instances to evaluate possible WPV circulation at a low level. Mathematical modelling suggests that the sensitivity of AFP surveillance decreases when the intensity of WPV circulation decreases (Gary et al., 1997). In Egypt, new cases of poliomyelitis occurred in the second half of 1990s sporadically or in small clusters in spite of the fact that the coverage of immunizations was reportedly good and that AFP surveillance usually fulfilled the quality criteria of WHO. Phylogenetic analysis of WPV strains showed "jumps" in some branches of the tree suggesting that the surveillance had temporarily missed some transmission chains. One proposed interpretation was that, somewhere in Egypt, there was an unrecognized reservoir of WPV circulation occasionally fuelling the main population with virus transmission. Environmental surveillance was instituted to uncover this reservoir. Raw sewage samples initially collected from inlets of sewage plants in two provinces (governorates) were concentrated using the two-phase separation technique (Pöyry et al., 1988) and inoculated in cell cultures in two laboratories in parallel (VACSERA, Cairo and KTL, Helsinki). Later on the number of sampling sites was increased to cover most provinces of the country. Rather than uncovering a presumed reservoir region, environmental surveillance rapidly showed that WPV1 was circulating in many regions of the country, not only in those provinces which had shown paralytic cases in recent years but in several others as well including some distant ones (El Bassioni et al., 2003). These observations resulted in intensified immunization campaigns and improved AFP surveillance. No cases have been reported since 2004 and the continuing intensive environmental surveillance has not detected WPV1 after January 2005. (unpublished).

This study has also been used to assess some technical problems of environmental surveillance. These are largely based on the fact that the samples always contain a mixture of different viruses. Increasing the number of parallel vials inoculated with a given sample usually increased the number of different virus strains isolated, i.e. improved the sample sensitivity. At the same time, it limited the number of samples that could be analysed with the same resources, i.e. possibly decreased the population sensitivity of detecting virus circulation. In most situations, parallel culture vials inoculated with identical aliquots of the sample could reveal surprisingly variable results. An example of the variability is given in Table 3 (Hovi et al., 2005). This observation emphasizes the difficulty in interpretation of a negative result: "absence of evidence is not evidence of absence".

In India, WPV circulation has continued intensively in the northern states of Uttar Pradesh and Bihar until recently while, for instance, in Mumbai only sporadic cases occurred in the slum areas. In this area, no sewage networks exist and wastewater samples had to be collected from the open canals containing waste from the lodging area. Polyethylene glycol precipitation was used for virus concentration (Shieh et al., 1995). WPV strains were frequently isolated from samples even in the absence of concurrent paralytic cases in the corresponding population. Phylogenetic analysis later showed that virus strains detected in these samples were rather diverse as regards their origin with representatives from different parts of India where endemic circulation still took place. Rather than representing a local reservoir of PV circulation, repeated isolation of WPV in the studied regions of

Table 3

Variation of virus isolation results in five parallel L20B cell culture inoculations of sewage samples collected in Egypt

Sample code	Isolation results for individual L20B cell flasks				
	Flask 1	Flask 2	Flask 3	Flask 4	Flask 5
EGY/02/12	SL3	SL2+SL3	-	-	SL3
EGY/02/49	-	SL1	SL1+SL3	DR1	NPEV
EGY/02/52	-	SL1	SL2	-	-
EGY/02/122	SL2+SL3	SL3	SL3	-	SL3
EGY/02/124	-	SL2	-	SL1	-
EGY/02/125	SL3	SL3	SL1	SL3	SL3
EGY/04/201	-	SL3	-	DR2	-
EGY/04/203	SL3	-	-	PV1W	-
EGY/04/204	-	PV1W	SL3	SL1+SL3	SL3

PV1W, wild type poliovirus of serotype 1; SL, Sabin-like poliovirus strain of the indicated serotype; -, no cytopathic effect; DR, a poliovirus showing double reactivity in the enzyme immunoassay intratypic differentiation (ITD) test; NPEV, non-poliovirus enterovirus.

Mumbai with poor sanitation thus appeared to be caused by repeated introductions of the strains along with people moving from the endemic regions to Mumbai (Deshpande et al., 2003). Along with the success of control of poliomyelitis in India, the number of virus strains isolated from the sewage in Mumbai has also decreased (J. Deshpande, personal communication).

Evaluation of the persistence of OPV-derived viruses in the population after an OPV/IPV switch

It is generally thought that possible transmission chains after OPV administration are very short in OPV-immunized populations with good herd immunity. Recent occurrence of cVDPV outbreaks has, however, shown that the OPV-derived strains have under certain situation a potential to cause serious problems. This is especially worrying considering the desire to eventually stop polio immunizations after guaranteed WPV eradication as discussed above. A traditionally held view is that IPV immunization does not provide the population with good herd immunity, and therefore a question has been asked, is it really safe to switch from only-OPV immunization to only-IPV immunization. Many western industrialized countries have now switched from OPV to IPV without any reported detection of circulating PV strains. In Finland the 1984–1985 WPV3 outbreak was contained by a campaign including a 5-week period with OPV administration throughout the country. Subsequent survey of environmental samples readily revealed OPV-derived viruses during the following 2–3 months but only exceptionally after that. The very last PV strain was isolated about 6 months after stopping the OPV campaign (Pöyry et al., 1988).

New Zealand switched from OPV to IPV immunization of newborns from February 2002 onwards. Environmental surveillance, among other approaches, was used to monitor possible establishment of OPV-derived virus transmission. Three locations were followed for 18 months. After the initially high detection rate, the occurrence of polioviruses rapidly declined in the sewage and from May 2002 onwards, poliovirus was recorded only on three occasions in a small proportion of specimens (Huang et al., 2005). Thus, so far all data available support the view that it is fully safe to switch from OPV to IPV. However, some people think that the situation might be different in developing countries with high population density and poor sanitation. To evaluate this question, a study was designed to monitor the range of genetic drift of PV strains detected in the environment before and after a switch from OPV to IPV. A province on the Java isle, Indonesia, was selected for the site but the recent spread of WPV1 to Indonesia from Africa has postponed the switch (Centers for Disease Control and Prevention (CDC), 2006).

Monitoring potential reemergence of WPV or onset of VDPV circulation

WPV strains may be imported from the countries with remaining circulation almost to anywhere in the world. Persistent monitoring of reemergence of WPV circulation is therefore necessary in all countries as long as WPV circulation exists in any part of the world. In addition, in countries with sub-optimal vaccination coverage, a danger of onset of VDPV circulation is eminent. While AFP surveillance remains the gold standard of WHO in this context, environmental surveillance is a useful supplementary approach in situations where there is a suspicion that the sensitivity or coverage of AFP surveillance is not optimal (WHO Guideline for environmental surveillance of poliovirus (http://www.who.int/vaccines-documents/DocsPDF03/www737.pdf). In addition, some industrialized countries do not have a working AFP surveillance at all, and then, environmental surveillance is even more important. Unlike AFP surveillance, environmental surveillance can never be targeted to the entire population in a country for one or both of the following reasons: Firstly, all the population is not served by converging sewage networks necessary for collecting samples that represent groups of people larger than a household, and secondly, the number of samples to be examined very quickly grows to levels that are impossible to deal with the available laboratory capacity. According to WHO guidelines environmental surveillance should therefore be limited to carefully selected situations. The guidelines also contain principles for selecting the sampling sites and proposes methods for sample processing (WHO guideline for environmental surveillance of poliovirus (http://www.who.int/vaccines-documents/DocsPDF03/www737.pdf).

The power of environmental surveillance in detecting importations and/or reemergence of PV circulation in the population is well documented in Israel and the adjacent Palestinian Authority. A systematic monitoring of environmental samples in 25–30 locations was established after the 1988 outbreak (Slater et al.,

1990). As the vaccination programme comprises a combination of IPV and OPV, the virus isolation procedure includes a selective step (growth at 40°C) aiming at exclusion of regular vaccine viruses from the strains taken for detailed character- ization (Manor et al., 1999). Between 1989 and 1997 environmental surveillance detected four "silent outbreaks" comprising 17 WPV strains mostly from sites in the Gaza strip. During the same period, high-quality AFP surveillance revealed no cases with associated WPV infection (Manor et al., 1999). Environmental surveil- lance in Israel has also revealed several aVDPV strains, some of them vastly drifted from the vaccine strains (Shulman et al., 2000).

Finland has used environmental surveillance as the main approach in poliovirus surveillance for decades. About 60 samples are being collected annually and the catchment areas of the sampling sites jointly cover about 20% of the population. Not a single poliovirus was isolated from Finnish sewage samples from autumn 1985 to end of 2005 (only IPV was used in polio immunizations) in spite of the fact that annually millions of passengers crossed the border to the neighbouring Estonia or Russia both of which were still using OPV.

The proportion of annual samples found positive for non-polio EVs has varied from year to year but is usually 50% or more. We have also tried to analyse the population sensitivity of the system both by experimentally and by modelling. In two series of experiments a given amount of attenuated vaccine virus was flushed down from a toilet in the Institute and the flow of the virus in the sewage network was monitored by daily sampling from the inlet of the main sewage treatment plant in the Greater Helsinki region. The sampling site is located about 20 km down- stream the sewage network. One observation was that while the virus reached the site within 24 h it continued to be detectable there for 4 days indicating that a bolus introduction of the virus into the network will not remain as a bolus but part of it is temporarily retained in the complexity of the network, though eventually arriving at the sewage plant. This not only means that the momentarily introduced virus was diluted to a larger volume than a daily output of the network but also indicates that, at this distance of introduction, the timing of the grab sampling may not be important and, for instance, there is no need to correlate the sampling time with the likely peak hours of toilet use. We calculated that under optimal conditions, anal- ysis of a single sewage sample would reveal poliovirus circulation in the Helsinki region if there were about altogether 100 infected individuals somewhere in the catchment area of the network (Hovi et al., 2001). Using mathematical modelling we also showed that repeated sampling will rapidly increase the population sen- sitivity and that using this approach, an emerging poliovirus circulation would be detected at least as quickly as by using optimal AFP surveillance (Ranta et al., 2001).

Systematic environmental surveillance of poliovirus is also being used in some other European countries including Slovakia. Slovakia was one of the first coun- tries to eliminate indigenous WPV circulation in early 1960s. The immunization policy comprised OPV given to newborns in annual spring campaigns rather than individually at a given age. The four-dose coverage at the age of 2 years in 2002 was more than 97%. AFP surveillance was established but the sensitivity coefficients were sub-optimal through 1990s. Therefore, the country had also introduced environmental surveillance with 47 sites all over the country sampled usually at 2-month intervals. Any poliovirus isolates from environment or from patient samples were regularly sent to KTL for ITD. A sewage sample collected in Bratislava in April 2003 revealed a type 2 poliovirus strain with aberrant ITD results (Cernakova et al., 2005). Partial sequencing in the VP1-coding region showed that the virus, if vaccine derived, had drifted extensively. Comparison of the complete VP1-coding sequence showed that it was almost as far from the Sabin 2 sequence as the closest WPV2 sequences in the GenBank. Slovakia was not a likely place for cVDPV circulation and this extent of sequence variation was more typical of iVDPV strains than cVDPVs. Therefore, this result prompted an intensified search for the virus not only in patients with acute paralytic or other neurological symptoms, but also in ID individuals known to the health-care system. However, no PV was isolated from these samples. In Autumn 2003 a similar strain was again isolated not only from Bratislava but also from another location in Slovakia, a small town of Skalica located some 60 km north from the capital. The vastly drifted aVDPV was then repeatedly isolated through a 2-year period from the sewage of this town. A retrospective "walk" of sampling sites along the bifurcations of the sewage network (Hovi, 2006) was successfully used to "corner" the virus excretors in a given village but in spite of much effort the source of the strain could not be identified before the occurrence of the virus in the sewage samples (so far) completely stopped (unpublished). Through all this time, the virus was found neither in clinical specimens nor in sewage in other locations than Skalica (and initially in 2003 in Bratislava). For the time being, we might thus conclude that the VDPV excretion was highly local perhaps due to one single infected individual with possibly a few contacts occasionally.

Concluding remarks

HEV infections are very common all over the world and especially under poor hygienic conditions and high population densities, a significant percentage of young children are excreting one or more EV strains at any sampling time. A large proportion of HEV infections are sub-clinical or cause only mild atypical symptoms that may pass without recognition. Infected persons excrete relatively large amounts of the virus into the faeces through several weeks irrespective of the clinical symptoms. Therefore, it is obvious that examination of possible presence of EVs in household and recreational waters can be used as a test for potential contamination of the water sources by faecal pathogens. Identification of the strains isolated from sewage and other environmental water bodies can also be used as a tool to study HEV epidemiology in the corresponding human population but several confounding factors have to be remembered before drawing conclusions about possible cause–effect relationships. On the other hand, sewage may be a good starting material for identifying new EV types that do not cause severe clinical symptoms and thus may remain undetected through clinical sampling. Systematic search for polioviruses in sewage samples, environmental surveillance for poliovirus circulation, is being used in several countries as a major approach to monitor potential reemergence of WPV circulation or to detect potential VDPV circulation. It also has a recognized role in the WHO strategy in the end-game of poliomyelitis eradication. With the conventional techniques, the approach is rather labour-intensive and another significant limiting factor is the fact that most people in the developing countries are not living in houses connected to a converging sewage network. Therefore, it is not possible to collect sewage samples representing larger groups of people. More simple techniques and high-capacity turnover equipment are desired in order to increase the availability of this approach more widely.

Acknowledgements

Our own original work cited in this review has been supported by WHO and the Academy of Finland. Studies carried out in Egypt and Slovakia were in collaboration with the corresponding NPLs and epidemiologists as well as with staff of the Eastern Mediterranean and European regional offices of WHO, respectively.

References

Birmingham ME, Linkins RW, Hull BP, Hull HF. Poliomyelitis surveillance: the compass for eradication. J Infect Dis 1997; 175(Suppl 1): S146–S150.

Blomqvist S, Savolainen C, Laine P, Hirttio P, Lamminsalo E, Penttila E, Joks S, Roivainen M, Hovi T. Characterization of a highly evolved vaccine-derived poliovirus type 3 isolated from sewage in Estonia. J Virol 2004; 78: 4876–4883.

Böttiger M, Herrström E. Isolation of polioviruses from sewage and their characteristics: experience over two decades in Sweden. Scand J Infect Dis 1992; 24: 151–155.

Brown B, Oberste MS, Maher K, Pallansch MA. Complete genomic sequencing shows that polioviruses and members of human enterovirus species C are closely related in the noncapsid coding region. J Virol 2003; 77: 8973–8984.

Centers for Disease Control and Prevention (CDC). Apparent global interruption of wild poliovirus type 2 transmission. MMWR Morb Mortal Wkly Rep 2001; 50: 222–224.

Centers for Disease Control and Prevention (CDC). Resurgence of wild poliovirus type 1 transmission and consequences of importation—21 countries, 2002–2005. MMWR Morb Mortal Wkly Rep 2006; 55: 145–150.

Cernakova B, Sobotova Z, Rovny I, Blahova S, Roivainen M, Hovi T. Isolation of vaccine-derived polioviruses in the Slovak Republic. Eur J Clin Microbiol Infect Dis 2005; 24: 438–439.

Cherry JD. Enteroviruses: poliviruses (poliomyelitis), coxsackieviruses, echoviruses, and enteroviruses. In: Textbook of Paediatric Infectious Diseases (Feigin RD, Cherry JD, editors). 2nd ed. Philadelphia: Saunders; 1987; pp. 1729–1790.

Conyn-Van Spaendonck MA, Oostvogel PM, Van Loon AM, Van Wijngaarden JK, Kromhout D. Circulation of poliovirus during the poliomyelitis outbreak in The Netherlands in 1992–1993. Am J Epidemiol 1996; 143: 929–935.

Deshpande JM, Shetty SJ, Siddiqui ZA. Environmental surveillance system to track wild poliovirus transmission. Appl Environ Microbiol 2003; 69: 2919–2927.

Dowdle WR, Gary HE, Sanders R, Van Loon AM. Can post-eradication laboratory containment of wild polioviruses be achieved? Bull World Health Organ 2002; 80: 311–316.

El Bassioni L, Barakat I, Nasr E, De Gourville EM, Hovi T, Blomqvist S, Burns C, Stenvik M, Gary H, Kew OM, Pallansch MA, Wahdan MH. Prolonged detection of indigenous wild polioviruses in sewage from communities in Egypt. Am J Epidemiol 2003; 158: 807–815.

Gary Jr. HE, Sanders R, Pallansch MA. A theoretical framework for evaluating the sensitivity of surveillance for detecting wild poliovirus: I. Factors affecting detection sensitivity in a person with acute flaccid paralysis. J Infect Dis 1997; 175(Suppl 1): S135–S140.

Gregory JB, Litaker RW, Noble RT. Rapid one-step quantitative reverse transcriptase PCR assay with competitive internal positive control for detection of enteroviruses in environmental samples. Appl Environ Microbiol 2006; 72: 3960–3967.

Halonen P, Rocha E, Hierholzer J, Holloway B, Hyypia T, Hurskainen P, Pallansch M. Detection of enteroviruses and rhinoviruses in clinical specimens by PCR and liquid-phase hybridization. J Clin Microbiol 1995; 33: 648–653.

Hovi T. The outbreak of poliomyelitis in Finland in 1984–1985: significance of antigenic variation of type 3 polioviruses and site specificity of antibody responses in antipolio immunizations. Adv Virus Res 1989; 37: 243–275.

Hovi T. Surveillance for polioviruses. Biologicals 2006; 34: 123–126.

Hovi T, Blomqvist S, Nasr E, Burns CC, Sarjakoski T, Ahmed N, Savolainen C, Roivainen M, Stenvik M, Laine P, Barakat I, Wahdan MH, Kamel FA, Asghar H, Pallansch MA, Kew OM, Gary Jr. HE, Degourville EM, El Bassioni L. Environmental surveillance of wild poliovirus circulation in Egypt—balancing between detection sensitivity and workload. J Virol Methods 2005; 126: 127–134.

Hovi T, Cantell K, Huovilainen A, Kinnunen E, Kuronen T, Lapinleimu K, Poyry T, Roivainen M, Salama N, Stenvik M, et al. Outbreak of paralytic poliomyelitis in Finland: widespread circulation of antigenically altered poliovirus type 3 in a vaccinated population. Lancet 1986; 1: 1427–1432.

Hovi T, Stenvik M. Selective isolation of poliovirus in recombinant murine cell line expressing the human poliovirus receptor gene. J Clin Microbiol 1994; 32: 1366–1368.

Hovi T, Stenvik M, Partanen H, Kangas A. Poliovirus surveillance by examining sewage specimens. Quantitative recovery of virus after introduction into sewerage at remote upstream location. Epidemiol Infect 2001; 127: 101–106.

Hovi T, Stenvik M, Rosenlew M. Relative abundance of enterovirus serotypes in sewage differs from that in patients: clinical and epidemiological implications. Epidemiol Infect 1996; 116: 91–97.

Huang QS, Greening G, Baker MG, Grimwood K, Hewitt J, Hulston D, Van Duin L, Fitzsimons A, Garrett N, Graham D, Lennon D, Shimizu H, Miyamura T, Pallansch MA. Persistence of oral polio vaccine virus after its removal from the immunisation schedule in New Zealand. Lancet 2005; 366: 394–396.

Hyypia T, Auvinen P, Maaronen M. Polymerase chain reaction for human picornaviruses. J Gen Virol 1989; 70(Pt 12): 3261–3268.

Kew O, Morris-Glasgow V, Landaverde M, Burns C, Shaw J, Garib Z, Andre J, Blackman E, Freeman CJ, Jorba J, Sutter R, Tambini G, Venczel L, Pedreira C, Laender F, Shimizu H, Yoneyama T, Miyamura T, Van Der Avoort H, Oberste MS, Kilpatrick D, Cochi S, Pallansch M, De Quadros C. Outbreak of poliomyelitis in Hispaniola associated with circulating type 1 vaccine-derived poliovirus. Science 2002; 296: 356–359.

Lapinleimu K. Elimination of poliomyelitis in Finland. Rev Infect Dis 1984; 6(Suppl 2): S457–S460.

Lennette EH, Smith TF. Laboratory diagnosis of viral infections. New York: Dekker; 1999.

Manor Y, Handsher R, Halmut T, Neuman M, Bobrov A, Rudich H, Vonsover A, Shulman L, Kew O, Mendelson E. Detection of poliovirus circulation by environmental surveillance in the absence of clinical cases in Israel and the Palestinian authority. J Clin Microbiol 1999; 37: 1670–1675.

Martin J. Vaccine-derived poliovirus from long term excretors and the end game of polio eradication. Biologicals 2006; 34: 117–122.

Mulders MN, Reimerink JH, Koopmans MP, van Loon AM, van der Avoort HG. Genetic analysis of wild-type poliovirus importation into The Netherlands (1979–1995). J Infect Dis 1997; 176: 617–624.

Nadkarni SS, Deshpande JM. Recombinant murine L20B cell line supports multiplication of group A coxsackieviruses. J Med Virol 2003; 70: 81–85.

Nathanson N, Martin JR. The epidemiology of poliomyelitis: enigmas surrounding its appearance, epidemicity, and disappearance. Am J Epidemiol 1979; 110: 672–692.

Oberste MS, Maher K, Kilpatrick DR, Flemister MR, Brown BA, Pallansch MA. Typing of human enteroviruses by partial sequencing of VP1. J Clin Microbiol 1999; 37: 1288–1293.

Oberste MS, Maher K, Michele SM, Belliot G, Uddin M, Pallansch MA. Enteroviruses 76, 89, 90 and 91 represent a novel group within the species Human enterovirus A. J Gen Virol 2005; 86: 445–451.

Oberste MS, Maher K, Williams AJ, Dybdahl-Sissoko N, Brown BA, Gookin MS, Penaranda S, Mishrik N, Uddin M, Pallansch MA. Species-specific RT-PCR amplification of human enteroviruses: a tool for rapid species identification of uncharacterized enteroviruses. J Gen Virol 2006; 87: 119–128.

Oberste MS, Michele SM, Maher K, Schnurr D, Cisterna D, Junttila N, Uddin M, Chomel JJ, Lau CS, Ridha W, Al-Busaidy S, Norder H, Magnius LO, Pallansch MA. Molecular identification and characterization of two proposed new enterovirus serotypes, EV74 and EV75. J Gen Virol 2004; 85: 3205–3212.

Pallansch MA, Roos RP. Enteroviruses: polioviruses, coxsackieviruses, echoviruses and new enteroviruses. In: Fields' Virology (Knipe DM, Howley P, Griffin DE, editors). 4th ed. Philadelphia: Lippincott Williams & Wilkins; 2001; pp. 723–735.

Pietiainen VM, Marjomaki V, Heino J, Hyypia T. Viral entry, lipid rafts and caveosomes. Ann Med 2005; 37: 394–403.

Pipkin PA, Wood DJ, Racaniello VR, Minor PD. Characterisation of L cells expressing the human poliovirus receptor for the specific detection of polioviruses in vitro. J Virol Methods 1993; 41: 333–340.

Poyry T, Stenvik M, Hovi T. Viruses in sewage waters during and after a poliomyelitis outbreak and subsequent nationwide oral poliovirus vaccination campaign in Finland. Appl Environ Microbiol 1988; 54: 371–374.

Pusch D, Ihle S, Lebuhn M, Graeber I, Lopez-Pila JM. Quantitative detection of entero-viruses in activated sludge by cell culture and real-time RT-PCR using paramagnetic capturing. J Water Health 2005; 3: 313–324.

Ranta J, Hovi T, Arjas E. Poliovirus surveillance by examining sewage water specimens: studies on detection probability using simulation models. Risk Anal 2001; 21: 1087–1096.

Rico-Hesse R, Pallansch MA, Nottay BK, Kew OM. Geographic distribution of wild poliovirus type 1 genotypes. Virology 1987; 160: 311–322.

Sabin AB. Oral poliovirus vaccine: history of its development and use and current challenge to eliminate poliomyelitis from the world. J Infect Dis 1985; 151: 420–436.

Salk D. Herd effect and virus eradication with use of killed poliovirus vaccine. Dev Biol Stand 1981; 47: 247–255.

Savolainen C, Hovi T, Mulders MN. Molecular epidemiology of echovirus 30 in Europe: succession of dominant sublineages within a single major genotype. Arch Virol 2001; 146: 521–537.

Sedmak G, Bina D, Macdonald J. Assessment of an enterovirus sewage surveillance system by comparison of clinical isolates with sewage isolates from Milwaukee, Wisconsin, col-lected August 1994 to December 2002. Appl Environ Microbiol 2003; 69: 7181–7187.

Sellwood J, Dadswell JV, Slade JS. Viruses in sewage as an indicator of their presence in the community. J Hyg (Lond) 1981; 86: 217–225.

Shieh YS, Wait D, Tai L, Sobsey MD. Methods to remove inhibitors in sewage and other fecal wastes for enterovirus detection by the polymerase chain reaction. J Virol Methods 1995; 54: 51–66.

Shulman LM, Manor Y, Handsher R, Delpeyroux F, Mcdonough MJ, Halmut T, Silber-stein I, Alfandari J, Quay J, Fisher T, Robinov J, Kew OM, Crainic R, Mendelson E. Molecular and antigenic characterization of a highly evolved derivative of the type 2 oral poliovaccine strain isolated from sewage in Israel. J Clin Microbiol 2000; 38: 3729–3734.

Slater PE, Orenstein WA, Morag A, Avni A, Handsher R, Green MS, Costin C, Yarrow A, Rishpon S, Havkin O, et al. Poliomyelitis outbreak in Israel in 1988: a report with two commentaries. Lancet 1990; 335: 1192–1195; Discussion, 1196–1198.

Smura TP, Junttila N, Blomqvist S, Norder H, Kaijalainen S, Paananen A, Magnius LO, Hovi T, Roivainen M. Enterovirus 94, a proposed new serotype in human enterovirus species D. J Gen Virol 2007; 88: 849–858.

Stanway G, Brown F, Christian P, Hovi T, Hyypiä T, King AMO, Knowles NJ, Lemon SM, Minor PD, Pallansch MA, Palmenberg AC, Skern T. Family Picornaviridae. Virus taxonomy. In: Virus Taxonomy—Eighth Report of the International Committee on Taxonomy Viruses (Fauquet CM, Mayo MA, Maniloff J, Desselberger U, Bull LA, editors). New York: Elsevier Academic Press; 2005.

Van Der Avoort HG, Hull BP, Hovi T, Pallansch MA, Kew OM, Crainic R, Wood DJ, Mulders MN, Van Loon AM. Comparative study of five methods for intratypic differ-entiation of polioviruses. J Clin Microbiol 1995a; 33: 2562–2566.

Van Der Avoort HG, Reimerink JH, Ras A, Mulders MN, Van Loon AM. Isolation of epidemic poliovirus from sewage during the 1992–3 type 3 outbreak in The Netherlands. Epidemiol Infect 1995b; 114: 481–491.

Human Viruses in Water
Albert Bosch (Editor)
© 2007 Elsevier B.V. All rights reserved
DOI 10.1016/S0168-7069(07)17005-1

Chapter 5

Virus Occurrence and Survival in the Environmental Waters

Charles P. Gerba
Department of Soil, Water and Environmental Science, University of Arizona, Tucson, AZ

Background

Critical to the assessment of the risks from waterborne transmission of viruses is determining their occurrence and survival in the environment. Sensitivity analysis has shown that the greatest amount of uncertainty in quantitative microbial risk assessment comes from assessment of exposure (Haas et al., 1999). The more accurately we can predict exposure, the more precise we can be in our assessment of risk. Quantitative methods for the detection of viruses in large volumes of water did not become available until the 1970s, and until the development of the polymerase chain reaction (PCR) for virus detection in water, most studies focused on the occurrence of enteroviruses. The development of PCR has allowed for detection of any virus in water. Almost any virus, which infects man, can be found in sewage. This includes viruses often considered blood or insect borne, such as AIDS virus or arboviruses (Ansari et al., 1992; Mathur et al., 1995). However, these viruses may be excreted inactivated or do not persist long in the environment. A characteristic of enteric viruses is their prolonged survival in the environment. This is probably a result of their adaptation to the waterborne route of transmission.

Factors controlling virus survival

Temperature

Factors controlling virus survival in aquatic environments are listed in Table 1. The most significant factor controlling virus survival is temperature (Yates et al., 1985).

Table 1

Factors affecting enteric virus survival in water

Temperature	Probably the most important factor; longer survival at lower temperatures
Light	UV light in sunlight can damage the nucleic acid causing dimmers to form
pH	Most viruses are stable at pH values of most natural waters
Salts	Some viruses are protected against heat inactivation by the presence of certain cations
Organic matter	The presence of sewage usually results in longer survival
Suspended solids or sediments	Association with solids prolongs survival
Air–water interfaces	Viruses that have greater hydrophobicity are more likely to be attracted to air water interfaces where denaturization of capsid proteins may occur
Biological factors	Aquatic microflora is usually antagonistic

Table 2

Mean inactivation rates for viruses in freshwater sources

Water type	Mean inactivation rate ($\log_{10}$/day)	Reported temperature range tested (°C)
Tap water	0.576	4–22.5
Polluted surface water	0.325	5–28.0
Unpolluted surface water	0.250	4–37.0
Groundwater	0.174	4–30.5

Data from Kutz and Gerba, 1988.

The survival of viruses in most environments can be predicted by temperature. The lower the temperature, the longer viruses persist. Temperature affects the rate at which protein and nucleic acid denaturization occurs as well as chemical reactions in general that can degrade the viral capsid (e.g. enzymes). At freezing or near-freezing temperatures, viruses may survive for many months (Goyal, 1984). Kutz and Gerba (1988) analyzed all the data on virus survival in surface waters and found that viruses survive longer in ground than surface waters held in the laboratory (Table 2). At 8°C in groundwater and 4°C in surface waters, virus inactivation approached less than 0.01 $\log_{10}$ per day.

Hepatitis A, adenoviruses, and parvoviruses are among the most thermal resistant of the enteric viruses (Crance et al., 1998; Bruniger et al., 2000) (Table 3). The composition of the water also influences the thermal inactivation of viruses. Factors such as, pH, salts, organics, and the presence of particular matter have been shown to affect the thermal stability of viruses (Pohjanpelto, 1961; Wallis et al.,

Table 3

Inactivation rates of viruses in groundwater

Virus	Temperature (°C)	Mean inactivation rate ($\log_{10}$/day)
Polio	0–10	0.075
	11–15	0.0868
	16–20	0.108
	21–25	0.289
	26–30	1.03
Hepatitis A	0–10	0.0055
	20–25	0.0557
	26–30	0.0375
Echo	11–15	0.107
	16–20	0.121
	21–25	0.179
Coxsackie	3–15	0.19
Rota	3–15	0.36
PRD-1	0–10	0.019
	21–25	0.324

Data from John and Rose, 2005.

1965; Salo and Cliver, 1976). Some cations enhance the effects of temperature, while others have the opposite effect. Rotavirus SA-11 is more rapidly inactivated by heating at 50°C in a 1 M $MgCl_2$, while reovirus type 1 infectivity is stabilized (Estes et al., 1979). Poliovirus type 1 is more stable in seawater than in distilled water (Liew and Gerba, 1980) and it has been shown to be thermostabilized by association with marine sediments (Liew and Gerba, 1980).

Virus inactivation at elevated temperatures probably results primarily from protein denaturization. Poliovirus RNA becomes more sensitive to ribonuclease after it has been exposed to high temperatures, indicating structural alterations in the viral capsid (Dimmock, 1967). At lower temperatures, inactivation due to damage of the RNA may also be important.

pH

At the pH of most natural waters (pH 5–9), enteric viruses are very stable. Most enteric viruses are more stable at a pH between 3 and 5 than at pH 9 and 12. Enteroviruses can survive at a pH of 11.0–11.5 and 1.0–2.0 for short periods of time; however, adenoviruses and rotaviruses are sensitive to inactivation at a pH of 10.0 or greater (Gerba and Goyal, 1982). The sensitivity of enteric viruses to pH may be strain dependent.

Light

The ultraviolet (UV) light in sunlight can inactivate viruses by causing cross-linking among the nucleotides. Fujioka and Yoneyama (2002) found that several enteroviruses in seawater were inactivated much more rapidly in the presence of sunlight than in the dark. The rate of inactivation was much less in the winter than in the summer (Fattal et al., 1983). The double-stranded DNA viruses (e.g. adenoviruses) are significantly much more resistant to UV light inactivation than enteroviruses because they can use host cell repair enzymes to repair the UV light damage (Gerba et al., 2002).

Visible light may also affect the survival of viruses by a process called photodynamic inactivation. In this process, the virus is "sensitized" to photooxidation by interaction of the viral genome with certain substances (Wallis and Melnick, 1965). Numerous synthetic and natural substances, such as ligins, fulvic acids, humic acids, and vitamins, may act as photosensitizers. These substances adsorb radiation and selectively transfer that energy to dissolved oxygen, which is excited to its highly oxidized state. The length of light exposure and its intensity also influence the rate of virus inactivation. A photodynamic substance has been demonstrated in algae cells, which is antiviral (Melnick and Gerba, 1980).

Salts and metals

Viruses almost always survive longer in freshwater than in seawater (Table 2). This appears to be largely due to the presence of antagonistic microorganisms rather than the increased salt concentration (Kapuscinski and Mitchell, 1980). Certain heavy metals such as copper and silver are known to have antiviral properties; their concentrations in water are usually too low to have an effect on viruses in natural waters (Thurman and Gerba, 1988).

Interfaces

Certain coliphages (MS2 and PRD1) have been shown to be inactivated at dynamic air–water–solid interfaces (Thompson and Yates, 1999). This is probably due to denaturization of the viral protein capsid. The effect has been shown to increase as the ionic strength of a solution increases. The more hydrophobic the viral capsid, the more pronounced the effect.

Virus adsorption to solids plays a major role in their survival in natural waters. Enteric viruses readily associate with particulate matter in water and sediments. Thus, viruses are usually found in greater numbers in sediments than in the overlaying water (Goyal et al., 1978). Adsorption may act to prolong the survival of the virus, but such associations may also enhance virus inactivation (Murray and Laband, 1979). Studies have shown that viruses readily adsorb to sand, pure clays, bacterial cells, particulate organic matter, silts, etc. Adsorption of enteric viruses to sediments has been demonstrated to prolong the survival of enteroviruses (Smith et al., 1978). The protective effect of clays on virus survival may be due to several factors, including the adsorption of enzymes or other substances that inactivate viruses, increased stability of the viral capsid, prevention of aggregate formation, and interference with the action of virucidal substances. Benonite clays have been shown to protect viruses against inactivation by ribonuclease by the adsorption of this enzyme to the clay (Singer and Fraenkel-Conrat, 1961). Marine sediments and clays also protect viruses against thermal inactivation in seawater (Liew and Gerba, 1980).

Microflora

The significance of bacteria and higher forms of living organisms on the survival of viruses in any body of water will vary spatially and temporally depending on the types and numbers of antagonistic microflora present. Failure to demonstrate the involvement of bacteria in the inactivation of viruses in a given water may simply mean that an organism affecting virus survival was not collected in that particular sample. Evidence for involvement of microflora has been demonstrated for surface freshwaters, groundwater, and seawater. Most of the evidence comes from studies done with seawater. Laboratory studies have indicated that bacteria of natural waters can play a significant role on virus survival. How important a role they play in natural water on virus survival where factors involved in survival may be present in much lower concentrations is uncertain. For example, conducting survival experiments in water samples placed in containers in the laboratory can allow for the build-up of organic acids, resulting in a drop in pH which in itself will affect virus survival (Gerba, 2006).

Treating natural waters to eliminate or kill bacteria or fungi by autoclaving, filtration, addition of antibiotics, or treatment with UV light increases survival of viruses (Melnick and Gerba, 1980; Gerba, 2006). The greater antiviral activity in seawater compared to freshwater appears to be related to the presence of bacteria indigenous to seawater (Fujioka et al., 1980). Studies in freshwaters have suggested that enteroviruses are degraded by the indigenous microflora (Cliver and Hermann, 1972). The longer survival of enteric virus in water compared to other viruses may be due to their greater resistance to proteolytic enzymes (Chang, 1971). Polioviruses and coxsackie viruses have been shown to be resistant to a wide range of photolytic enzymes (Cliver and Hermann, 1972). However, this may vary with the serotype of enterovirus. Other substances may also be produced that inactivate viruses by processes other than enzymatic. *Pseudomonas aeruginosa* produces substances with a molecular weight below 500, which appears to result in the virus being dismantled. Substances of this low molecular weight cannot act enzymatically. Substances are also produced that react with the viron to prevent its adsorption to host cells (Fujsaki et al., 1978).

Others

High hydrostatic pressures can inactivate enteric viruses. A 7 $\log_{10}$ drop in hepatitis A virus occurred when exposed to 450 MPa for 5 min (Kingsley et al., 2002). However, poliovirus was unaffected by a 5 min exposure to 600 MPa. In the presence of seawater, the resistance of hepatitis A virus increased. Exposure of the virus to RNase indicated that inactivation was due to alteration of the viral capsid proteins.

96 C.P. Gerba

Table 4

Detection of non-enteric viruses in feces or urine

Virus	Remarks	Reference
SARS (a cornavirus)	Detected by PCR	Chan-Yeung and Xu (2003)
JC	Detected by PCR	Bofill-Mass et al. (2000)
TT	Detected by PCR	Vaidya et al. (2002)
HIV	Not excreted in infectious form	Ansari et al. (1992)
Japanese encephalitis	Excreted in the urine in infectious form	Mathur et al. (1995)
Smallpox	Excreted in the urine in infectious form in concentrations as high as 10^5/ml	Sarker et al., 1973

Occurrence of human viruses in water

There are probably hundreds of viruses that can infect man. Most, if not all, are released into the environment by excretion or secretion of bodily fluids or skin and hair. The viruses that infect the gastrointestinal tract are the most suited for transmission through the aquatic environment because they are released in large numbers in the stools of infected persons (up to $10^{11}\,g^{-1}$) and have a prolonged survival in water. However, many respiratory viruses are excreted in the feces and many viruses are excreted in the urine. Respiratory adenoviruses have been shown to be transmitted by recreational waters; thus, there is the potential for viruses that may not primarily replicate in the intestinal tract to be transmitted by water. A list of non-enteric viruses excreted in feces or urine is shown in Table 4. Most are thought not to be excreted in viable form or only remain infectious for short periods of time in the environment; for these reasons no studies have been conducted on their occurrence in natural waters.

Methods for virus detection in water

Virus detection in natural waters involves a concentration step followed by assay in cell culture or by a molecular method, PCR being the most common. Viruses are usually concentrated from volumes ranging from 10 to 1000 l or more (Gerba et al., 1978; Gajardo et al., 1991). Most methods result in a final sample volume of 20–30 ml, which is easily assayed in cell culture. Enteroviruses are the easiest to detect and can produce visible cell destruction or cytopathogenic effects (CPE) within a few days. However, other viruses such as adenoviruses require up to 2 weeks and multiple passages on new cell monolayers to produce visible CPE. In addition, serum neutralization assays in cell culture were required to identify the virus. Completion of these procedures usually required several months to complete.

Before the advent of PCR, most studies on viruses in water focused on the enteroviruses. PCR allows for the detection of almost any virus in water, but suffers from a number of limitations including inability to assess viability, small assay volumes, and the presence of interfering substances. Combining PCR with cell culture has led to the ability to detect infectious viruses that do not produce CPE (Pinto et al., 1994; Reynolds et al., 1996; Chapron et al., 2000), reducing the time to detect the virus in cell culture (e.g. you do not have to wait for the production of CPE). Thus, overall costs and time for the laboratory detection of viruses in natural waters have been significantly reduced.

Cell lines used to detect viruses usually are only permissive to a few groups or types of human enteric viruses. For example, coxsackie viruses B will grow very easily in the Buffalo green monkey (BGM) cell line, but coxsackie viruses A will not grow at all. For this reason, some investigators have used multiple cell lines. In recent years some cell lines, such as CaCo-2 (Pinto et al., 1994) and PLC/PLF/5 (Grabow et al., 1992), have been found to be permissive to a wide range of enteric viruses, although CPE may not always be produced. Another aspect that must be taken into account is that cell-culture assay methods are not 100% efficient in detecting all of the viruses despite being capable of supporting growth. The viruses must come into contact with the cells for infection to take place. Thus, techniques such as rocking the cells, including additives that enhance attachment of the virus to the cells (Benton and Hurst, 1986), suspended cell techniques (Morris and Waite, 1980), and multiple passages, have been shown to increase the number of viruses detected in a sample. Standard cell-culture monolayer methods appear to only detect about 10% of the viruses of a given type in a sample (Morris and Waite, 1980).

Another factor to consider when reviewing the data on virus concentration/ occurrence in natural waters is that the method of concentration is not 100% efficient. Most methods are capable of concentrating enteroviruses and adenoviruses with a 30–50% efficiency (Gerba et al., 1978; Gajardo et al., 1991). However, the efficiency will vary with the type and strain of virus as well as water quality factors such as turbidity and soluble organic matter.

Considering the factors it is not surprising that the numbers of viruses detected by PCR methods is much greater than that detected by cell culture (Hot et al., 2003). When reviewing studies on the concentration of viruses in natural waters, it is important to realize that the true number of viruses is much greater, likely somewhere from 10 to 100 times greater (Morris and Waite, 1980).

Sources of viruses in natural waters

Sewage discharges in surface and groundwaters are the major source of human viruses in the environment. The concentration of viruses in sewage discharges is controlled by a number of factors shown in Table 5. The incidence of infection in a population (Hejkal et al., 1984), not necessarily disease, is the most important factor followed by the type of sewage treatment and disinfection. Virus occurrence

Table 5

Factors controlling the occurrence of viruses in surface waters

Factor	Remarks
Sewage treatment	Fewer numbers of viruses are found in surface waters in the United States than Europe because disinfection is required
Type of disinfectant	Because of the greater resistance of adenoviruses to UV light disinfection—greater numbers occur than in chlorine-disinfected sewage
Time of year	Greater concentrations of enteroviruses in the summer than in winter in temperate climates. Longer survival of viruses in winter water.
Incidence of infection	Greater concentrations in lower social–economic groups
Rainfall	During periods of high rainfall sewage treatment plants may bypass or reduce treatment times

and concentration in surface waters is usually greater in Europe and locations outside the United States because sewage is usually not disinfected before discharge. As disinfection became common practice in the United States, virus concentrations declined in surface waters. The amount of per capita water use may also affect the concentration of viruses in sewage.

Rainfall also affects the occurrence of viruses in surface and groundwaters. During major storm events sewage treatment plants may not be able to treat all of the incoming sewage because their collection systems also collect storm water runoff. Thus, incompletely treated or untreated sewage may be discharged into surface waters. Rainfall also results in the greater percolation of water into the soil, resulting in higher groundwater levels reducing the unsaturated soil distances between septic tank drains and the groundwater. Leaching from unlined landfills may result in greater migration of viruses through the subsurface. These factors also probably account why the majority of documented outbreaks of waterborne disease outbreaks increases after periods of above normal rainfall (Curriero et al., 2001).

Other sources of viruses in surface waters include swimmers and resuspension of contaminated sediments. Bathers in streams or lakes may release significant numbers of enteric viruses to the water body (Rose et al., 1987; Gerba, 2000).

Examination of Tables 6–9 indicates that almost every enteric virus has been detected in surface waters. Seawater probably has received the greatest amount of attention because of concern with the contamination of shellfish, which are often consumed raw or lightly cooked and concerns with transmission to bathers (Bosch et al., 2006). Since fresh surface waters are usually treated and disinfected before use as drinking water there has been less interest in virus occurrence in these waters. Only in recent years has groundwater received much attention. Recent surveys in the United States suggest that virus occurrence is not uncommon, probably originating largely from septic tanks and leaking sewer lines (Table 10).

Table 6

Recent studies on the occurrence of enteric viruses in surface freshwaters

Location	Virus concentration/l or % positive	Remarks	Reference
United States (Arizona)	0–0.75 PFU enteroviruses 0–0.25 IFF rotaviruses	Recreational river	Rose et al. (1987)
United States	27.57% astroviruses ICC-PCR 27.5% enteroviruses ICC-PCR 37.9% adenoviruses ICC-PCR	Drinking water treatment plant intakes from across the United States	Chapron et al. (2000)
United States (Florida)	0.002–0.014 MPN/CPE	Lakes	Betancourt and Rose (2005)
Spain	15.5–16.2MPNCU	Ripoll and Besos Rivers	Bosch et al. (1986)
Italy	Positive by cell culture in 3 of 5 samples	Tiber	Divizia et al. (1989)
Germany	0.5–56 MPNCU	Two rivers receiving poorly treated sewage	Walter et al. (1989)
Canada	0.1–29 MPNCU	St. Lawrence River	Payment et al. (2000)
Korea	33.3% enterovirus ICC-PCR 30.4% adenovirus ICC-PCR	Han River	Lee et al. (2005)
Netherlands	0.3–2 enteroviruses PFU 2–10 reovirus PFU 57–5386 rotavirus PCR 4–4900 PCR	Maal and Waal Rivers	Lodder and de Roda Husman (2005)
South Africa	13% rotavirus	Dam water	Van Zyl et al. (2004)
France	3% enteroviruses cell culture 88% enterovirus PCR 1.5% HAV PCR 1.5% norovirus group I PCR 0% norovirus group II PCR 3% astrovirus PCR 0% rotavirus PCR	Various rivers	Hot et al. (2003)
South Africa	35.3% HAV PCR (River) 37.3% HAV PCR Dam water)	River and dam water	Taylor et al. (2001)
Thailand	15% HAV PCR	Canals	Kittigul et al. (2000)

CPE, cytopathogenic effects; MPNCU, most probable number culturable unit; ICC-PCR, integrated cell-culture polymerase chain reaction; PFU, plaque forming unit.

Table 7

Occurrence of adenoviruses in surface waters

Adenovirus type(s)	Water source	Concentration/ frequency	Location	Reference
Ad2, Ad3, Ad5, Ad6	River	0–25 PFU/l	Japan	Tani et al. (1995)
Ad40, Ad41	Surface	49–88%	South Africa	Genthe et al. (1995)
Ad$_{NS}$	River	100%		
Ad$_{NS}$	River	74%	Spain	Pina et al. (1998)
Ad$_{NS}$	River	100%	Spain	Girones et al. (1995)
Ad 40	Surface water	48% 0–2.11 MPN	United States	Chapron et al. (2000)
Ad	Surface water	880–7500 genomes/l	California	Jiang et al. (2001)
Ad	River	50%	California	Jiang and Chu (2004)
Ad	Surface water	66.7%	South Korea	Lee et al. (2005)
Ad	Surface water	12.7%	South Africa	Van Heerden et al. (2003)
Ad 2,40,41	Surface water	22.2%	South Africa	Van Heerden et al. (2005)
Ad	River receiving sewage discharge	20%	Germany	Pusch et al. (2005)
Ad	Seawater		France	Hugues et al. (1980)

Ad$_{NS}$, Adenovirus, exact type not specified; IU, infectious unit; PFU, plaque forming unit; INS, Not Specified.

Adenoviruses have been receiving more attention in recent years because they appear to be more common, at least in sewage, and are more resistant to UV disinfection, which is becoming more popular for both sewage discharges and drinking water disinfection.

Data gaps and the future

Information on the concentration of viruses in natural waters is critical to understanding our risk of infection and the effectiveness of controls to limit our exposure. Unfortunately, the amount of data we have is very limited, largely because of the cost and time needed for detection of viruses in water. Faster and lower cost methods are needed for the collection and concentration of viruses. Costs and time for detection of viruses have been significantly reduced by the application of molecular methods.

Our knowledge on the occurrence of viruses in surface and groundwaters is largely limited to the enteroviruses. Application of molecular methods should allow a better understanding of the types and concentrations of almost any virus that can be found in water. Development of molecular methods to differentiate infectious from noninfectious virons would also greatly reduce the cost and effort of studying viruses in environmental waters.

Table 8

Occurrence of enteroviruses in seawater

Location	Virus concentration/l or % positive	Reference
USA (Texas)	0.04–0.04 PFU	Goyal et al. (1978)
USA (New York)	0–2.1 PFU	Vaughn et al. (1979)
USA (Florida)	0.05–0.14 PFU	Schaiberger et al. (1982)
USA (Texas)	0.06–0.0.026 PFU	Rao et al. (1984)
USA (Texas)	0.007–2.6 PFU	Rao et al. (1986)
USA (Hawaii)	0–0.37 MPN	Reynolds et al. (1998)
USA (Florida Keys)	79% PCR	Griffin et al. (1999); Griffin et al. (2003)
USA (Florida Keys)	0.01 MPN 93% by PCR	Lipp et al. (2001a)
USA (Sarasota, FL)	1.7–7.7 MPN 91% PCR	Lipp et al. (2001b)
USA (Charlotte Harbor, FL)	10–20 MPN	Lipp et al. (2001c)
USA (Florida Keys)	60 by PCR	Donaldson et al. (2002)
USA (Florida Keys)	0.113–0.0126 PFU	Wetz et al. (2004)
USA (San Monica Bay, CA)	32% PCR	Noble and Fuhrman (2001)
Italy	0.4–16 $TCID_{50}$	De Flora et al. (1975)
Italy (Pesaro)	32.6% by cell culture	Pianetti et al. (2000)
Italy (Venice)	3–1614 genome copies by PCR	Rose et al. (2006)
France	0.05–6.5 MPNCU	Hugues et al. (1980)
Spain	0.12–1.72 MPNCU	Finance et al. (1982)
Spain	0.12–0.15 MPNCU	Lucena et al. (1986)
Spain (Barcelona)	44% PCR	Pina et al. (1998)
Greece (Patras)	83.4% PCR	Vantarakis and Papapetropoulou (1998)
Israel	1–6 PFU	Fattal et al. (1983)
Japan (Tokyo Bay)	10–50% PCR	Katayama et al. (2004)

PFU, plaque forming unit.
MPNCU, most probable number culturable unit.
$TCID_{50}$, tissue culture infectious dose 50%.

Increases in higher intensity rainfalls because of climate change may act to increase concentrations of viruses in surface waters because of the need to bypass sewage treatment and the raising groundwater causing greater transport of viruses in the subsurface. The tendency towards the use of UV light over chlorination of sewage discharges may increase the occurrence of adenoviruses and other viruses more resistant to UV light in surface waters. Information on long-term trends of

Table 9

Virus occurrence in groundwater

Location	Virus concentration/l or % positive	Remark	Reference
Israel	20%	Cell culture; 99 samples	Marzouk et al. (1979)
United States	30.1% enterovirus PCR 8.6% HAV PCR 8.6% rotavirus PCR 8.7% CPE	139 samples of water supply wells used by utilities from across the United States	Abbaszadegan et al. (1999)
United States (Wisconsin)	50% for enteroviruses, rotavirus, HAV and norovirus	48 samples from municipal drinking water supply wells	Borchardt et al. (2004)
United States	72% overall 62% reovirus PCR 38% enterovirus PCR 0% rotavirus PCR 4% HAV PCR 21% norovirus PCR	21 water supply wells; 321 samples	Fout et al. (2003)
United States	4.8% CPE 31.5% overall PCR 15.2% enterovirus 13.8% rotavirus PCR 6.9% HAV PCR 0.9% norovirus	448 sites from 35 states	Abbaszadegan et al. (2003)
United States (Wisconsin)	8% overall PCR 6% HAV PCR 2% rotavirus PCR 2% enterovirus PCR 2% norovirus PCR	50 individual home owner wells	Borchardt et al. (2003)
United States (Florida)	0.0002–0.0011 MPN	5 wells	Betancourt & Rose (2005)
Italy	6% CPE	35 samples	Carducci et al. (2003)
United Kingdom	40% norovirus, enterovirus PCR, and CPE	5 wells	Powell et al. (2003)
South Africa	0% rotavirus PCR	30 samples	Van Zyl et al. (2004)

CPE, cytopathogenic effects.

viruses in natural waters would also be helpful to document the impacts of development and waste control strategies in watersheds. Without this we really do not have an idea how our exposure to viruses via water has changed over the years since the first methods became available for the detection of viruses in water.

Table 10

Sources of viruses in groundwater

Source	Remarks
Septic tanks	Concentration is affected by the density of septic tanks, depth to groundwater, and rainfall
Leaking sewer lines	All sewer lines leak. May be the largest source of viruses detected in groundwater
Unlined landfills	Disposal diapers are believed to be a major source in unlined landfills
Wastewater irrigation or land application of wastewater	If viruses are present in land applied wastewater they may reach the groundwater
Subsurface injection of wastewater	In some areas wastewater is injected via wells as a method of disposal or aquifer recharge

References

Abbaszadegan M, LeChevallier M, Gerba CP. Occurrence of viruses in U.S. groundwaters. J Am Water Works Assoc 2003; 95(9): 107–120.

Abbaszadegan M, Stewart P, LeChevallier M. A strategy for detection of viruses in groundwater by PCR. Appl Environ Microbiol 1999; 65: 444–449.

Ansari SA, Farrah SR, Chaudhry GR. Presence of human immunodeficiency virus nucleic acids in wastewater and their detection by polymerase chain reaction. Appl Environ Microbiol 1992; 58: 3984–3990.

Benton WH, Hurst CJ. Evaluation of mixed cell types and 5-iodo-2′-deoxyuridine treatment upon plaque assay titers of human enteric viruses. Appl Environ Microbiol 1986; 51: 1036–1040.

Betancourt WQ, Rose JB. Microbiological assessment of ambient waters and proposed water sources for restoration of a Florida wetland. J Water Health 2005; 3: 89–100.

Bofill-Mass S, Pina S, Girones R. Documenting the epidemiologic patterns of polymaviruses in human populations by studying their presence in urban sewage. Appl Environ Microbiol 2000; 66: 238–245.

Borchardt MA, Bertz PD, Spencer SK, Battigelli DA. Incidence of enteric viruses in groundwater from household wells in Wisconsin. Appl Environ Microbiol 2003; 69: 1172–1180.

Borchardt MA, Haas NL, Hunt RJ. Vulnerability of drinking-water wells in La Crosse, Wisconsin, to enteric-virus contamination from surface water contributions. Appl Environ Microbiol 2004; 70: 5937–5946.

Bosch A, Lucena F, Girones R, Cofre J. Survey of viral pollution in Besos River. J Water Pollut Control Fed 1986; 58: 87–91.

Bosch A, Pinto RM, Abad FX. Survival and transport of enteric viruses in the environment. In: Viruses in Foods (Goyal SM, editor). New York, NY: Springer; 2006; pp. 151–187.

Bruniger S, Peters J, Borchers U, Kao M. Further studies on the thermal resistance of bovine parvovirus against moist and dry heat. Int J Hyg Environ Health 2000; 203: 71–75.

Carducci A, Casni B, Bani A, Rovini E, Verani M, Masón F, Giuntin A. Virological control of groundwater quality using bimolecular tests. Water Sci Technol 2003; 47: 261–266.

Chan-Yeung M, Xu RH. SARS: epidemiology. Respirology 2003; 8(Suppl.): S9–S14.

Chang SL. Interactions between animal viruses and higher forms of microbes. J Sanit. Eng Div Am Soc Civil Eng 1971; 96: 151–161.

Chapron CD, Ballester NA, Fontaine JH, Frades CN, Margolin AB. Detection of astroviruses, enteroviruses, and adenovirus types 40 and 41 in surface water collected and evaluated by the information collection rule and an integrated cell culture-nested PCR procedure. Appl Environ Microbiol 2000; 66: 2525–2529.

Cliver DO, Hermann JE. Proteolytic and microbial inactivation of enteroviruses. Water Res 1972; 6: 797–805.

Crance JM, Gantzer C, Schwartzbrod, Deloince R. Effect of temperature on the survival of hepatitis A virus and its capsidal antigen in synthetic seawater. Environ Toxicol Water Qual 1998; 13: 89–92.

Curriero FC, Patz JA, Rose JB, Leile S. The association between extreme precipitation and waterborne disease outbreaks in the United States, 1948–1994. Am J Public Health 2001; 91: 1194–1199.

De Flora S, De Renzi G, Badolati G. Detection of animal viruses in coastal seawater and sediments. Appl Environ Microbiol 1975; 30: 472–475.

Dimmock NJ. Differences between the thermal inactivation of picornaviruses at "high" and "low" temperatures. Virology 1967; 31: 338–353.

Divizia M, De Filippi P, Di Napoli A, Venuti A, Perez-Bercoff R, Pana A. Isolation of wild-type hepatitis A virus from the environment. Water Res 1989; 23: 1155–1160.

Donaldson KA, Griffin DW, Paul JH. Detection, quantitation and identification of enteroviruses from surface waters and sponge tissue from the Florida Keys using real time PCR. Water Res 2002; 36: 2414–2505.

Estes MK, Graham DY, Smith EM, Gerba CP. Rotavirus stability and inactivation. J Gen Virol 1979; 43: 403–409.

Fattal B, Vasl RJ, Katzenelson E, Shuval HI. Survival of bacterial indicator organisms and enteric viruses in the Mediterranean coastal waters of Tel Aviv. Water Res 1983; 17: 397–402.

Finance C, Briguad M, Lucena F, Aymard M, Bosch A, Schwartzbrod L. Viral pollution of seawater of Barcelona. Zentralbl Bakteriol Mikrobiol Hyg 1982; B176: 530–536.

Fout GS, Martinson BC, Moyer MWN, Dahling DR. A multiplex reverse transcription-PCR method for detection of human enteric viruses in groundwater. Appl Environ Microbiol 2003; 69: 3158–3164.

Fujioka RS, Loh PC, Lau LS. Survival of human enteroviruses in the Hawaiian ocean environment: evidence for virus-inactivating microorganisms. Appl Environ Microbiol 1980; 39: 1105–1110.

Fujioka RS, Yoneyama BS. Sunlight inactivation of human enteric viruses and fecal bacteria. Water Sci Technol 2002; 46: 291–295.

Fujsaki M, Uchida S, Kojima M, Hota S, Kuroda H, Hamada K. Studies inhibiting substances of bacterial origin. III. Antiviral substances extractable from *Streptococcus faecalis*: its in vitro activities and some biological characteristics. Kobe J Med Sci 1978; 24: 99–114.

Gajardo R, Diez JM, Cofre J, Bosch A. Adsorption-elution with negatively and positively-charged glass powder for the concentration of hepatitis A virus from water. J Virol Methods 1991; 31: 345–351.

Genthe B, Gericke M, Bateman B, Mjoli N, Kfir R. Detection of enteric adenoviruses in South African water using gene probes. Water Sci Tech 1995; 31: 345–350.

Gerba CP. Assessment of the enteric pathogens shedding by bathers during recreational activity and its impact on water quality. Quant Microbiol 2000; 2: 55–68.

Gerba CP. Survival of viruses in the marine environment. In: Ocean and Health (Belkin S, Colwell RR, editors). New York, NY: Springer; 2006; pp. 133–142.

Gerba CP, Farrah SR, Goyal SM, Wallis C, Melnick JL. Concentration of enteroviruses from large volumes of tap water, treated sewage, and sweater. Appl Environ Microbiol 1978; 35: 540–548.

Gerba CP, Goyal SM. Methods in Environmental Virology. New York, NY: Marcel Dekker; 1982.

Gerba CP, Gramos DM, Nwachuku. Comparative inactivation of enteroviruses and adenovirus 2 by UV Light. Appl Environ Microbiol 2002; 68: 5167–5169.

Girones R, Puig M, Allard A, Lucena F, Wadekk G, Jofre J. Detection of adenoviruses and enteroviruses by amplification in polluted waters. Waer Sci Technol 1995; 31: 351–357.

Goyal SM, Gerba CP, Melnick JL. Human enteroviruses in oysters and their overlaying waters. Appl Environ Microbiol 1978; 37: 572–581.

Goyal SM. Viral pollution of the marine environment. Crit Rev Environ Contr 1984; 14: 1–32.

Grabow WO, Puttergill DL, Bosch A. Propagation of adenovirus types 40 and 41 in the PLC/PLF/5 primary liver cacinonma cell line. J Virol Meth 1992; 37: 201–207.

Griffin D, Gibson III CJ, Lipp EK, Riley K, Paul JH, Rose JB. Detection of viral pathogens by reverse transcriptase PCR and of microbial indicators by standard methods in the canals of the Florida Keys. Appl Environ Microbiol 1999; 65: 4118–4125.

Griffin DW, Donaldson KA, Paul JH, Rose JB. Pathogenic human viruses in coastal waters. Clin Microbiol Rev 2003; 16: 129–143.

Haas CN, Rose JB, Gerba CP. Quantitative Microbial Risk Assessment. New York, NY: Wiley; 1999.

Hejkal TW, Smith EM, Gerba CP. Seasonal occurrence of rotavirus in sewage. Appl Environ Microbiol 1984; 47: 588–590.

Hot D, Legeay D, Jacques J, Gantzer C, Caudrelier Y, Guyard K, Lange M, Andreoletti L. Detection of somatic phages, infectious enteroviruses and enterovirus genomes as indicators of human enteric viral pollution in surface water. Water Res 2003; 37: 4703–4710.

Hugues B, Cini A, Plissier M, Lefebre JR. Resherche des virus dans le marina a parit'echantilons de volumes differents. Eau Ouectc 1980; 13: 199–203.

Jiang S, Chu W. PCR detection of pathogenic viruses in southern California urban rivers. J Appl Microbiol 2004; 97: 17–28.

Jiang S, Noble R, Chu W. Human adenoviruses and coliphages in urban runoff-impacted coastal waters of Southern California. Appl Environ Microbiol 2001; 67: 179–184.

John DE, Rose JB. A review of factors affecting microbial survival in ground water. Environ Sci Technol 2005; 39: 7345–7356.

Kapuscinski RB, Mitchell R. Processes controlling virus inactivation in coastal waters. Water Res 1980; 14: 363–371.

Katayama H, Okuma K, Furumai H, Ohgaki S. Series of surveys for enteric viruses and indicator organisms in Tokyo Bay after an event of combined sewer over flows. Water Sci Technol 2004; 50: 259–362.

Kingsley R, Hoover DG, Papafragjou W, Richards GP. Inactivation of hepatitis A virus and a calicivirus. J Food Protect 2002; 65: 1605–1609.

Kittigul L, Raengsakulrach B, Sirtanikorn S, Kanyok R, Utraachkij F, Diraphat P, Thirawuth V, Sirpanichgon K, Pungchitton S, Chitpirom K, Chaichantanakit N, Vethanophas K. Detection of poliovirus, hepatitis A virus and rotavirus from sewage and water samples. Southeast Asian J Trop Med Public Health 2000; 31: 41–46.

Kutz SM, Gerba CP. Comparison of virus survival in freshwater sources. Water Sci Technol 1988; 20: 467–471.

Lee SH, Lee C, Lee KW, Cho HB, Kim SJ. The simultaneous detection of both enteroviruses and adenoviruses in environmental water samples including tap water with an integrated cell culture- multiplex-nested PCR procedure. J Appl Microbiol 2005; 98: 1020–1029.

Liew PF, Gerba CP. Thermostabilization of enteroviruses by estuarine sediment. Appl Environ Microbiol 1980; 40: 249–305.

Lipp EK, Farrah SR, Rose JB. Assessment and impact of microbial fecal contamination and human enteric pathogens in a coastal community. Mar Pollut Bull 2001a; 42: 258–286.

Lipp EK, Jarrell JL, Griffin DW, Lukasik J, Jacukiewicz J, Rose JB. Preliminary evidence for human fecal contamination in corals of the Florida Keys. Mar Pollut Bull 2001b; 44: 666–670.

Lipp EK, Kurz R, Vincent R, Rodriguez-Palacios C, Farrah SR, Rose JB. The effects of seasonal variability and weather on microbial fecal pollution and enteric pathogens in a subtropical estuary. Estuaries 2001c; 24: 238–258.

Lodder WJ, de Roda Husman AM. Presence of noroviruses and other enteric viruses in sewage and surface waters in the Netherlands. Appl Environ Microbiol 2005; 71: 1453–1461.

Lucena F, Schwartzbrod L, Bosch A. The effect of mass poliomyelitis vaccination program on the occurrence of enterovirus in seawater. Zentralbl Bakteriol Mikrobiol Hyg 1986; B183: 67–69.

Marzouk Y, Goyal SM, Gerba CP. Prevalence of enteroviruses in ground water of Israel 1979; 17: 487–491.

Mathur AK, Nivedita K, Rejesh SC, Maitra, Chaturvedi UC. Viruria during acute Japanese encephalitis virus infection. Int J Exp Path 1995; 76: 103–109.

Melnick JL, Gerba CP. The ecology of enteroviruses in natural waters. Crit Rev Environ Contr 1980; 10: 65–93.

Morris R, Waite WM. Evaluation of procedures for recovery of viruses from water-II detection systems. Water Res 1980; 14: 795–798.

Murray JP, Laband JJ. Degradation of poliovirus adsorption on inorganic surfaces. Appl Environ Microbiol 1979; 37: 480–486.

Noble RT, Fuhrman JA. Enteroviruses detected by reverse transcripase polymerase chain reaction from coastal waters of Santa Monica Bay, California: low correlation to bacterial indicator levels. Hydrobiologia 2001; 460: 175–184.

Payment P, Berte A, Prevost M, Menard B, Barbeau B. Occurrence of pathogenic micro-organisms in the Saint-Lawrence River (Canada) and comparison of health risks for populations using it as their source of drinking water. Can J Microbiol 2000; 46: 565–576.

Pianetti A, Baffone B, Citterio A, Casaroli F, Bruscolini F, Salvaggio L. Presence of enteroviruses and reoviruses in waters of the Italian coast of the Adriatic Sea. Epidemiol Infect 2000; 125: 455–462.

Pina S, Puig M, Lucena F, Cofre J, Girones R. Viral pollution in the environment and in shellfish: human adenovirus detection by PCR as an index of human viruses. Appl Environ Microbiol 1998; 64: 3376–3382.

Pinto RM, Diez JM, Bosch A. Use of the colonia carcinoma cell line CaCo-2 for in vivo amplification and detection of enteric viruses. J Med Virol 1994; 44: 310–315.

Pohjanpelto P. Response of enteroviruses to cystine. Virology 1961; 15: 225–230.

Powell KL, Taylor RG, Cronin AA, Barrett MH, Pedley S, Sellwood J, Trowsdale SA, Lerner DN. Microbial contamination of two urban sandstone aquifers in the UK. Water Res 2003; 37: 339–352.

Pusch D, Oh DY, Wolf S, Dumke R, Schroter-Bobsin U, Hohne M, Roske I, Schreier E. Detection of enteric viruses and bacterial indicators in German environmental waters. Arch Virol 2005; 150: 929–947.

Rao VC, Metcalf TC, Melnick JL. Development of a method for concentration of rotavirus and its application to recovery of rotaviruses in estuarine waters. Appl Environ Microbiol 1986; 52: 484–488.

Rao VC, Seidel KN, Goyal SM, Metcalf TC, Melnick JL. Isolation of enteroviruses from water, suspended solids and sediments from Galveston bay; survival of poliovirus and rotavirus adsorbed to sediments. Appl Environ Microbiol 1984; 48: 404–409.

Reynolds KA, Gerba CP, Pepper IL. Detection of infectious enteroviruses by an integrated cell culture-PCR procedure. Appl Environ Microbiol 1996; 62: 1424–1427.

Reynolds KA, Roll K, Fujioka RS, Gerba CP, Pepper IL. Incidence of enteroviruses in Mamala Bay, Hawaii using cell culture and direct polymerase chain reaction methodologies. Can J Microbiol 1998; 44: 598–604.

Rose JB, Mullinax RL, Singh SN, Yates MV, Gerba CP. Occurrence of rotaviruses and enteroviruses in recreational waters of Oak Creek, Arizona. Water Res 1987; 21: 1375–1381.

Rose MA, Dhar AK, Brooks HA, Zecchini F, Gersberg RM. Quanitation of hepatitis A virus and enterovirus levels in the lagoon canals and Lido beach of Venice, Italy, using real-time RT-PCR. Water Res 2006; 40: 2387–2396.

Salo RJ, Cliver DO. Effect of acid, salts, and temperature on the infectivity and physical integrity of enteroviruses. Arch Virol 1976; 52: 269–282.

Sarker JK, Mitra AC, Mukherjee MK, De SK. Virus excretion in smallpox. Bull World Organ 1973; 48: 523–527.

Schaiberger GE, Edmond TD, Gerba CP. Distribution of enteroviruses in sediments contiguous with a deep marine sewage outfall. Water Res 1982; 16: 1425–1428.

Singer B, Fraenkel-Conrat H. Effects of benonite and stability of TMV-RNA. Virology 1961; 14: 59–65.

Smith EM, Gerba CP, Melnick JL. Role of sediment in the persistence of enteroviruses in the estuarine environment. Appl Environ Microbiol 1978; 35: 685–689.

Tani N, Dohi Y, Kurumatani N, Yonemasu K. Seasonal distribution of adenoviruses, enteroviruses and reoviruses in urban river water. Microbiol Immunol 1995; 39: 577–580.

Taylor MB, Cox N, Very MA, Grabow WO. The occurrence of hepatitis A and astroviruses in selected river and dam waters in South Africa. Water Res 2001; 35: 2653–2660.

Thompson SS, Yates MV. Bacteriophages inactivation at the air-water-solid interface in dynamic batch systems. Appl Environ Microbiol 1999; 65: 1186–1190.

Thurman RB, Gerba CP. Molecular mechanisms of viral inactivation by water disinfectants. Adv Appl Microbiol 1988; 33: 75–105.

Vaidya SR, Chitambar SD, Arankalle VA. Polymerase chain reaction-based prevalence of hepatitis A, hepatitis E and TT viruses in sewage from an endemic area. J Hepatol 2002; 37: 131–136.

Van Heerden J, Ehlers MM, Hiem A, Grabow WOK. Prevalence, quantification and typing of adenoviruses detected in river and treated drinking water in South Africa 2005; 99: 234–242.

Van Heerden J, Ehlers MM, van Zyl WV, Grabow WOK. Incidence of adenoviruses in raw and treated water. Water Res 2003; 37: 3704–3708.

Van Zyl WB, Williams PJ, Grabow WO, Taylor MB. Application of a molecular method for the detection of group A rotavirus in raw and treated water. Water Sci Technol 2004; 50: 223–228.

Vantarakis AC, Papapetropoulou M. Detection of enteroviruses and adenoviruses in coastal waters in SW Greece by nested polymerase chain reaction. Water Res 1998; 32: 2356–2372.

Vaughn JM, Landry EF, Thomas MZ, Vicale TJ, Penello WF. Survey of human enterovirus occurrence in fresh and marine waters on Long-Island. Appl Environ Microbiol 1979; 38: 290–296.

Wallis C, Melnick JL. Photodynamic inactivation of animal viruses: a review. Photochem Photobiol 1965; 4: 159–170.

Wallis C, Melnick JL, Rapp F. Different effects of $MgCl_2$ and $MgSO_4$ on the thermostability of viruses. Virology 1965; 26: 694–695.

Walter R, Macht W, Durkrop J, Hecht R, Hornig U, Schulze P. Virus levels in river water. Water Res 1989; 23: 133–138.

Wetz JJ, Lipp EK, Griffin DW, Lukasik J, Wait D, Sobsey MD, Scott TM, Rose JB. Presence, infectivity and stability of enteric viruses in seawater: relationship to marine water quality in the Florida Keys. Mar Pollut Bull 2004; 48: 698–704.

Yates MV, Gerba CP, Kelly LM. Virus persistence in groundwater. Appl Environ Microbiol 1985; 49: 778–781.

Human Viruses in Water
Albert Bosch (Editor)
© 2007 Elsevier B.V. All rights reserved
DOI 10.1016/S0168-7069(07)17006-3

Chapter 6

Virus Removal During Drinking Water Treatment

Susan Springthorpe, Syed A. Sattar
Centre for Research on Environmental Microbiology (CREM), University of Ottawa, 451 Smyth Road, Ottawa, ON, Canada K1 H 8M5

Introduction

The importance of water as a vehicle for transmission of human viral pathogens is now well recognized (Bosch, 1998; Szewzyk et al., 2000) with the main sources of contamination being from human wastes with various degrees of treatment. Understanding virus removal from water intended for drinking has been made more important because increased urbanization and burgeoning human populations result in increased wastes levels, more reuse of virus-contaminated wastewater and increasing land disposal of sewage sludges (Sattar et al., 1999; Fane et al., 2002; White et al., 2003). When such wastes enter fluvial waters, the impact of contained viral pathogens can be felt far from the original contamination source. Another reason to understand virus reduction during water treatment comes from virus contamination during or subsequent to catastrophic events, including the potential risks to potable water from deliberate contamination with viral bioagents.

Whether human pathogenic viruses can be found in drinking water, and the risk they may pose to human health, depends on the quality of the water source, the multiple-barrier approach employed to treat it, and the nature of the pathogens themselves. Each of these factors has significant variability; influenced by the presence, persistence and aggregation of the viruses in different water environments, and the types and efficiencies of the natural or engineered treatment processes applied.

The purpose of this chapter is to give a brief but critical review of the current state of knowledge on reduction of the human pathogenic virus content in drinking water through treatment processes. Such reduction is comprises two main components—the reduction in virus numbers by processes that remove particulates, and virus inactivations by physical or chemical treatments designed to damage microorganisms and eliminate their infectivity. It is not the intention here to discuss the range of possible human viruses present in water, nor how they can be recovered and detected since these topics have been reviewed in other chapters in this volume and elsewhere (Wyn-Jones and Sellwood, 2001; Fong and Lipp, 2005).

Potential for reduction of viruses during water treatment

Viruses are extremely small and have a very high surface area to volume ratio. Traditionally, virus size is compared with bacterial size only in terms of particle diameters, with the former being 20–300 nm whereas the latter are usually from 0.5 μm to several μm. When considering virus removal, it is important to consider virus size in comparison to bacteria in three dimensions. For a typical small virus with a virion diameter of 30 nm, the volume is approximately $1 \times 10^4 \, nm^3$. For a typical bacterium, 1 μm in diameter and 2 μm in length, the volume is approximately $7 \times 10^9 \, nm^3$. This approximately five to six order of magnitude size differences could lead to potentially very different particle behaviour between viruses and bacteria. However, when this small size is coupled to an overall negative particle charge, it is highly likely, as can be demonstrated, that many virions will become associated with bacterial cell conglomerates and other suspended particulates and will behave as these larger particles do. In fact, many viruses enter source waters already associated with particulates (Wellings et al., 1976). Thus, during conventional water treatment processes the efficiency of coagulation, sedimentation and filtration will also be significant for virus removal. However, some small but undefined fraction of virions may also behave as very small particles and potentially evade the removal process. In considering the disinfection step of water treatment, the small size and large surface area of the virions may make them vulnerable to rapid decontamination, but also assist in their escape from disinfection if they are able to 'hide' within organic matrices or be trapped in biofilms. In summary, the ability of viruses to be particle associated is likely to assist their removal but inhibit their disinfection, whereas the reverse is expected to be true for free virions.

Whereas detection of infectious human viruses in wastewater or sewage-impacted receiving waters is comparatively easy, their detection in drinking water is considerably more difficult, and the magnitude of this difficulty is at least in approximate proportion to the degree of reduction that can be achieved during the water treatment process. If source water can be shown to contain a certain number of infectious virus particles per unit volume (measured as plaque forming units (PFU), or infectious foci, in cell culture), and we then assume a 99% reduction in virus numbers, one would need 100-fold larger sample to detect the same number of viruses in treated water. Frequently, samples from surface water sources for detection of viruses are of 10–100 l; this would suggest that if a 99% reduction were achieved, one would need to sample 1000–10,000 l of finished potable water to have the same potential to detect viruses, assuming that the capture efficiency was equivalent with these larger volumes. Thus, sampling and analysis of potable waters for infectious viruses potentially involves the need to concentrate 1000 l or more and is performed only rarely for experimental studies.

Filtering such large volumes raises many questions about sampling practicality, about what else may be co-concentrated with the viruses, and whether these co-concentrated substances might interfere with virus isolation and or detection. Use of molecular virus detection methods alone fails to confirm virus infectivity, a necessity if risk is to be assessed, and so virus isolation and detection in cell cultures, with or without the assistance of molecular methods, remains the 'gold standard' for virus isolation from water. For virus assay in cell cultures, a secondary concentration step is usually necessary. It is also important to remember that not all waterborne human viruses can be detected in cell culture, and that non-cultivable norovirus, or difficult to culture rotaviruses, may be a higher risk to public health than some cultivable viruses (Lodder and de Roda Husman, 2005). Several additional factors need to be mentioned here: these are not discussed further but may aid the reader in examining the available literature in a critical manner, or in pursuing studies in this field.

- If human pathogenic viruses are found in potable water, it must be ascertained that the sampling apparatus had not been previously used for sampling source waters without being sterilized between samples. Disinfection alone may be insufficient for assurance that drinking water was the source of the viruses found.
- Failure to find viruses in single samples does not mean their absence from the potable water sampled. Sufficient samples need to be taken to have a greater assurance that the results are valid.
- Studies that add viruses to natural waters to assess treatment reductions, perhaps through a pilot treatment system, must be considered to provide only crude estimates because the added virus will not be 'naturally' embedded in the usual particulates. Nevertheless, such 'spiking' studies might provide useful information on behaviour of particular viruses, and pathogenic viruses can even be potentially simulated without hazard (Redman et al., 1997; Caballero et al., 2004). One could speculate that if spiking studies inadequately represent virion embedding in particulates, they might underestimate the contribution of flocculation and sedimentation steps of water treatment to virus removal. For a similar reason, they may overestimate the ability of various disinfection steps/ treatments to inactivate contaminating viruses.
- Ideally, relevant studies should use viruses indigenous to the source water but inevitably, the numbers of human viruses in source waters are usually too low to be able to measure significant reductions. Whether or not indigenous bacteriophages are suitable models for human viruses in drinking water treatment is a matter for discussion; nevertheless, their use might represent one of the best options for true assessment of virus removal potential.

• Virus assays are dependant on interaction between a virus and its host cell. It is therefore important that any material introduced into a host cell culture does not interfere with the ability of the host cell to detect the virus or the ability of the virus to be infectious. In disinfection studies, it is recognized that disinfectant must be neutralized before assay of surviving microorganisms. However, it is not always customary to determine that the neutralized residual does not interfere with growth of the test organism in a growth promotion test with a very low inoculum (positive control). This is particularly important for viruses where a secondary host system is involved in detecting survivors. Similar considerations should be involved in assessing other steps in water treatment, such as coagulation, where chemicals may be carried into the host cultures with the eluted viruses.

With these considerations at the forefront, we will examine what is known from experimental studies of virus removal and disinfection during treatment regimes for provision of potable water.

Virus reduction by removal during water treatment

Virus removal relies on attenuation of virus numbers by flocculation and sedimentation with other organic materials as well as during passage through filters of granular media or membranes. Although virus removal is focused on physical removal processes, concurrent chemical effects and biological decay are also confounded in the overall process measurements at any particular location. Holding of water supplies prior to treatment to permit biological decay to occur could be considered a removal mechanism, but virus persistence, as a function of time and environmental conditions, constitutes a field of study on its own and is not discussed here. There are also a number of point-of-use water treatment devices that claim activity against viruses, either removal or inactivation; these also are considered beyond the scope of this review. Thus, we are focusing on the passage of the virus-contaminated water through conventional treatment plants or processes only.

For groundwater, percolation through soils (sand, clay, loam), granular subsurface matrices and bedrock formations removes unspecified numbers of viruses from diverse land-based sources of virus contamination. Discussion of virus attenuation under these conditions is considered complex and beyond the scope of this review, though it is likely that virus challenge to natural surface and subsurface matrices may have higher virus concentrations in specific localized areas than are found in most surface water sources due to the dilution achieved in the latter. Furthermore, there are potential differences in virus survival in the subsurface as well as differences between bacterial and viral transport mechanisms.

The focus on conventionally engineered water treatment processes can be justified due to the large populations who rely on conventional treatment for their drinking water. Flocculants may be based on aluminum or ferric compounds and may be simple or polymeric. Granulated media may be sand beds (e.g. slow sand filters, bank filtration) or the mixed bed filters used in many water treatment plants. Additional granular activated carbon columns are also present in some locations. Filtration processes are usually preceded by coagulation, flocculation and sedimentation of the particulates in drinking water so that the filters do not block too rapidly. Use of membrane filters as a final removal step for drinking water treatment is becoming more common, especially in smaller plants. Whether viruses are removed by coagulation and sedimentation, filtration through filter beds or by membranes, they end up in a residue that must be disposed of to surface or water sources. Such residues can still be a risk if viruses contained in them survive and are transported to another treatment facility or bathing area.

Relatively few studies have examined the reduction in viral pathogens through a conventional treatment process. As mentioned above, these studies may be of two types: seeding studies with known viruses and/or phages, or removal of indigenous viruses and detection of survivors by a selection of standardized assays. The authors have not combined data from different studies in tabular form due to the wide discrepancies in treatment steps and conditions and are describing studies separately.

Experimental addition of viruses (or seeding studies)

Gerba et al. (2003) used feline calicivirus (FCV) as a surrogate for human noroviruses, and for comparison, three bacteriophages were seeded into a pilot plant. The pilot plant received, in addition to the microbial seeds, alum and polymer. Coagulation occurred in three mixed and one static tank before the floc was allowed to settle in the sedimentation basin. The hydraulic retention time in the mixed tanks was 27 min each and in the static tank and sedimentation basin together was 99 min. The water was then filtered through a mixed bed consisting of granular activated carbon, sand and gravel. The mean range of inactivation measured for the phages was 0.78–$1.97 \log_{10}$ and 1.85–$3.21 \log_{10}$ after sedimentation and filtration, respectively. For the FCV, the corresponding figures were $2.49 \log_{10}$ and $3.05 \log_{10}$. These figures can only be compared at their face value if it is assumed that the recoveries are equivalent for all the viruses and that may not necessarily be the case.

In similar seeding experiments at another pilot plant (Hendricks et al., 2005) the phages used were reduced by approximately $3 \log_{10}$ (MS-2) and $5 \log_{10}$ (Phi-X 174). The viruses, attenuated poliovirus and echovirus 12, were each reduced by just more than $2 \log_{10}$. In multiple experiments on the phages, Hendricks et al. (2005) observed no difference between mono- or dual-filtration media but a marked effect of alum concentration, with greater removals at higher alum concentration. This effect was most marked for Phi-X 174; there was little or no phage removal in the absence of alum. A comparison of phage removal between the experimental pilot plant used and a full-scale plant and pilot plant from a different location showed very different phage removal at the alternate locations (Hendricks et al, 2005); this difference was ascribed to the lower alum concentration. Thus, there are differences in the ability of conventional coagulation, sedimentation and filtration processes to remove viruses that can depend on the efficiency of each of those steps as well as the nature of the virus. Since the hydraulic conditions differ somewhat between plants and between full scale and pilot plants, the real virus reductions in full-scale plants, which are difficult to measure by seeding experiments, are uncertain.

Comparison of coagulants (ferric, alum and polymer) with two human viruses (poliovirus and echovirus) and two phages (MS-2 and PRD-1) showed relatively similar results among different coagulants; the greatest reduction was for polio-virus, which therefore appeared to show the most variation (Bell et al., 2000). MS-2 reduction was consistently higher than PRD-1; based on this data the latter appeared to be more suitable phage to represent the behaviour of the human viruses. Removal of viruses among the plants tested showed an average removal as depicted in Table 1. The relatively large discrepancy between polio- and echoviruses may be real, or it could be an artifact of carry over of coagulant chemical and or other coagulated substances with the virus and the effects of these on interaction between the virus and the host bacteria. Such material could either promote or inhibit infections in the host cells.

In a laboratory study using flocculation and high-rate sedimentation with hep-atitis A virus (HAV) and poliovirus, as well as F+ bacteriophage, Nasser et al. (1995) determined the optimum flocculent dose for virus removal. They showed that humic acid interfered with virus removal whereas addition of a small quantity of (1 mg/l) of a cationic polyelectrolyte improved the performance of alum flocculation for HAV and bacteriophage removal, poliovirus removal was un-affected. Although Nasser et al. (1995) quote removal values from their column experiments; these have not been included as they may not be applicable for removal in treatment plants.

Many smaller plants do not have access to full conventional treatment includ-ing flocculation, sedimentation and filtration, so virus removal can remain a chal-lenge. Yahya et al. (1993) studied virus removal in a pilot plant with relevance to application in small plants using surface water. They used slow sand filtration

Table 1

Removal of viruses by optimized coagulation

Virus	Average $\log_{10}$ virus removal by coagulant		
	Ferric (12 sites)	Alum (11 sites)	Polymer (9 sites)
Poliovirus	2.86	3.39	2.6
Echovirus	1.83	1.53	1.77
MS-2	2.85	3.43	2.89
PRD-1	1.99	1.15	1.39

Data from Bell et al. (2000).

followed by nanofiltration and studied the removal of bacteriophages MS-2 and PRD-1. The slow sand filter removed about 2 log_{10} of each of the phages and there was a 4–6 log_{10} reduction in phages in water from the nanofilters; PRD-1 was removed to a greater extent than MS-2 by both the sand filter and the nanofilter. No human viruses were included, but at least for the nanofilter, where removal is primarily size-dependant, human viruses of comparable size can be expected to be removed in a similar magnitude to MS-2 (HAV, enteroviruses, caliciviruses) or PRD-1 (rotaviruses, reoviruses, adenoviruses).

An alternative to following spiked microorganisms through a full-treatment process can be to challenge individual sections of the treatment process (Huertas et al., 2003). In their pilot-scale investigation of different treatment stages prior to a full plant upgrade, Huertas et al. (2003) used known titres of MS-2 coliphage and found virus removal: 0.86 log_{10} (85.1%) from coagulation-flocculation-sedimentation whereas more than 3 log_{10} were removed by microfiltration or ozonation.

Indigenous virus studies

One of the early studies to examine indigenous virus removal during treatment was performed by Keswick et al. (1984) on a highly polluted water source. After an essentially complete treatment process (clarification, sand filtration and chlorination), virus was still found in 83% of the finished water by culture; details of the virus recovery and removal are given in Table 2. What is obvious from this data is that the water did not meet standards for turbidity and this is reflected in the virus results obtained. Moreover, the apparent increase in rotaviruses detected following clarification and chlorination, and in enteroviruses following clarification, may

Table 2

Viruses and % removal from a contaminated source by water treatment

Season and final treatment given	Turbidity[a] (NTU)	Rotavirus[a] (infectious foci 100/l)	Enterovirus[a] (% positive for enterovirus CPE)	Phages[a] (PFU)
Dry				
None	6.4	610	55	11
Clarified	5.6(12.5)	154(74.3)	35(36.8)	4(63.6)
Finished	1.2(81.3)	40(93.5)	41(25.0)	0(100)
Rainy				
None	26	1745	7	47
Clarified	10(61.5)	3417(0)	29(0)	16(65.7)
Filtered	6.8(73.8)	342(80.4)	15(0)	3(93.6)
Finished	9.6(63.1)	990(43.3)	7(0)	4(91.4)

Data from Keswick et al. (1984).
[a]Mean value and, in parentheses, percent of mean value in raw water.

have reflected a breaking of virus clumps rather than a real increase. It is still likely that some viruses were removed during these processes; however, seasonal differences in both contamination levels and in removal efficacy were seen.

An almost contemporary study in Canada (Payment et al., 1985) used seven treatment plants in the Montreal region. Three plants used a treatment regime consisting of prechlorination, sedimentation, filtration, ozonation and chlorine, two more were similar but lacked the ozonation, one used only filtration and chlorination and one used chlorine alone. The first group showed about 97% of virus removal after chlorination and about 97.5% after sedimentation with further minor reduction after filtration and a final significant drop in virus content after ozonation; some virus was detectable in finished water. The second group that lacked ozonation was reduced to similar levels after sedimentation and had slightly more reduction after filtration and in the finished water; these differences are unlikely to be significant. The last two plants had lower levels of virus contamination to contend with, but virus was found in the finished water of the one that had filtration and chlorination. This shows that if prechlorination is used it can effect significant virus reductions but these reductions are no greater than can be achieved by coagulation and sedimentation alone when this is the first step in the process (see below). In recent years, it has become more common to omit the prechlorination step to avoid the formation of excessive amounts of disinfection byproducts in the treated water.

In a study to determine the most suitable indicator for viruses and parasites in drinking water, Payment and Franco (1993) used indigenous viruses and bacteriophages to determine comparative inactivation levels during treatment at three treatment plants. Their findings are summarized in Table 3 and include contributions from both removal and disinfection steps.

The method used by Payment and Franco (1993) to screen for enteric viruses is an efficient one, but it must be recognized that the use of human immune serum globulin is not exclusive to viruses and might also detect intracellular bacterial pathogens that form infectious foci. Nevertheless, this study is one of very few to use indigenous viruses to measure reduction, and confirms that a log reduction of 1–2 $\log_{10}$ for virus removal after sedimentation is a realistic level. Thus, it can be concluded that flocculation and sedimentation processes perform the majority of virus removal in a full conventional treatment process. The further increments reported here from filtration and chlorination give an overall virus reduction of $>4->7$ $\log_{10}$; incremental reductions from these additional steps cannot be assumed to give the same magnitude of reduction in the absence of the prior processes. Thus, water sources that undergo treatment only through filtration and disinfection or by disinfection alone cannot be assumed to achieve the same level of virus reduction as has been measured for the full conventional treatment process.

A somewhat similar study by Jofre et al. (1995) also showed high overall levels of virus reduction and the potential for *Bacteriodes fragilis* phages to be used as indicators for reduction of human viruses in drinking water. The inclusion of a prechlorination step, as well as other differences among the treatment processes

Table 3

Attenuation of indigenous virus and phages during drinking water treatment

	Plant no. 1	Plant no. 2	Plant no. 3
Geometric mean of infectious foci or PFU · 100/l of source water			
Human enteric viruses (host: MA-104 cells)	2.8×10^2	9.3×10^2	2.6×10^2
Somatic phage (host: *E. coli* CN-13)	1.3×10^5	5.5×10^4	8.4×10^4
Male specific phage (host: Salmonella WG 49)	4.1×10^5	2.4×10^3	6.3×10^4
Somatic phage (host: Salmonella WG 45)	6.3×10^3	2.4×10^2	0.3
Log_{10} reduction after settling			
Human enteric viruses (host: MA-104 cells)	3.2	1.3	3.0
Somatic phage (host: *E. coli* CN-13)	2.1	1.0	2.3
Male specific phage (host: Salmonella WG 49)	2.7	2.0	4.0
Somatic phage (host: Salmonella WG 45)	5.3	2.0	6.6
Log_{10} reduction after filtration			
Human enteric viruses (host: MA-104 cells)	>5.0	5.1	2.5
Somatic phage (host: *E. coli* CN-13)	3.2	3.6	2.1
Male specific phage (host: Salmonella WG 49)	ND	3.8	3.4
Somatic phage (host: Salmonella WG 45)	ND	2.8	5.1
Log_{10} reduction in finished water			
Human enteric viruses (host: MA-104 cells)	>5	>5	>4
Somatic phage (host: *E. coli* CN-13)	>7	>6	>6
Male specific phage (host: Salmonella WG 49)	>7	>6	>6
Somatic phage (host: Salmonella WG 45)	>7	>5	>6

Note: Adapted from Payment and Franco (1993).

were apparent; it is interesting to note that at a plant with prechlorination as well as coagulation and sedimentation but no sand filtration prior to passage through a GAC column and the final disinfection step, a lower overall reduction level for phage was reported. However, the initial contamination levels for this plant were also reduced. It is also important to consider the numerical implications of reported log reductions. The numerical reductions achieved decrease in a logarithmic fashion with each log_{10} reduction in viruses reported. The most important log reductions are the first 1–2 log_{10}, and thus the numerical difference between a 2 log_{10} reduction and a 5 or 6 log_{10} reduction are comparatively small.

In studying virus removal by different steps in a water treatment plant on the River Oise from raw water consistently contaminated with human viruses, Agbalika et al. (1985) obtained reductions after intermittent pre-ozonation and storage of 89%, after coagulation, flocculation and settling of 77% and after sand filtration of 55%. No viruses were recovered from the water after the final treatment stages of post-ozonation and chlorine dioxide treatment.

Natural virioplankton ranges between $<1 \times 10^4$ to $>1 \times 10^8$ (Wommack and Colwell, 2000). Rinto-Kanto et al. (2004) used a staining method to examine visually the removal of virus-like particulates in comparison to those of bacterial size during water treatment. The system used was lake water that first went through bank filtration prior to coagulation, settling and disinfection; the virus-like particles in the lake water differed in numbers at different times of the year. In this case, the bank filtration removed the majority of the virus-like particles with some further removal after coagulation and settling. No further reductions were seen upon disinfection; this is not surprising since no particulates were removed from the system and it is not possible to differentiate between infectious and inactivated viruses by this method.

Stetler et al. (1984) examined removal of indigenous human viruses over a period of a year. The numbers of viruses in the source water was clearly at its highest during the cold months of the year, which may correspond to reduced biological decay mechanisms in the water at that time. These higher numbers during winter were reflected in higher recoveries after sedimentation and filtration steps; virus was only recovered post-filtration from September to March; overall reductions during this period averaged about 1 $\log_{10}$. A variety of treatments was examined by van Olphen et al. (1984). They found riverbank and dune filtration to be effective in removing indigenous human viruses but, as Stetler et al. (1984), were still able to detect some viruses when coagulation and sedimentation was followed by rapid sand filtration; this also was particularly marked in the winter. It is noteworthy that good floc formation during the coagulation step is particularly difficult to achieve in very cold water, and this may have contributed to the data obtained by these two groups.

While the presence of biofilms as a means of trapping and removing indigenous virus during drinking water treatment cannot be addressed with quantitative removal data, it is clearly worth mentioning because of the increasing use of biological filtration, where bacterial biofilms develop on the granular filter media. Such biofilms also form throughout treatment infrastructure and are likely to contain agents of virus decay by biological means. It is not known whether the biofilm represents only an opportunity for biological inactivation of human viruses or if increased risks through trap and release are possible; Skraber et al. (2005) have addressed the risk side of this issue in a recent review.

Pretreatment removal

Dune sand and riverbank filtration are often used as a pretreatment removal steps for poor quality water sources. Dune filtration has been modelled using MS-2 and PRD-1 phages, and found to be an efficient removal system (Schijven et al., 1999). The majority of the removal (3 $\log_{10}$) was achieved in the first 2.4 m with a further reduction of 5 $\log_{10}$ over the subsequent 27 m. After passage of the virus pulse, detachment was very slow. A further laboratory study (Schijven et al., 2003) compared MS-2 removal with that of poliovirus type 1 and coxsackie virus B4; when these results were extrapolated to the field scale it was found that the coxsackie virus was expected to be removed similarly to the MS-2 phage but the poliovirus would be expected to be removed more efficiently based on charge. No actual field trials were done though the efficiency of bank or dune filtration has been confirmed by van Olphen et al. (1984) for human viruses, and others for additional pathogens. Both removal and inactivation contribute significantly to passage of viruses through soils. These issues have been the subject of a recent extensive review (Schijven and Hassanizadeh, 2000). Data was presented to show that the efficiency of virus removal by soil passage declined with distance from the contamination site. Potential reasons for this include variation in the population of test phage, as well as discontinuities and preferential flow in the local soils. The main issues that may need to be addressed for bank filtration are the continued efficiency with the duration of use, and, particularly for soils, preventive maintenance to prevent development of preferential flow channels. Neither of these issues is considered here.

Artificial groundwater recharge is an increasingly common practise as water tables in many areas fall, and it can provide a source of raw water for treatment as a potable source. Also increasingly common is the use of infiltration of surface waters through sand- or riverbanks prior to treatment. In Finland, surface water sources are being replaced by artificial groundwater works (Niemi et al., 2004). These authors studied the removal of bacteriophages through a 5 m column of sand to mimic the percolating phase of infiltration and a riverbank esker of coarse sandy gravel 2 m deep and 18 m long that represented the saturated zone. Water was pumped at 40 l/h to the sand column. The river water feed was spiked with MS-2 for one week at an average concentration of 4.3×10^9 PFU/ml. Samples were taken from four sites based on transport of a conservative sodium chloride tracer. The median MS-2 count was 2.4×10^5 CFU/ml for percolated water was a 96.7% reduction. With the additional reduction in the saturated zone, an overall reduction of 6–7 $\log_{10}$ was achieved.

Slow sand filtration has been a recognized means of removing viruses for many years. McConnell et al. (1984) studied the removal of reoviruses by slow sand filtration as a function of filter maturity and sand filter matrix; no significant differences were found. Infectious virus was not detected in filter effluents, and the removed viruses, based on a radiolabel, were distributed throughout the filter beds but with more in the first few centimetres of the filter; they were no longer infectious to cell cultures.

It would seem a pity not to mention norovirus during discussion of removal of viruses because of its importance as a water-borne pathogen. There have been two laboratory-based studies relevant to norovirus removal by soils. These are discussed by Huffman et al. (2003) in the context of caliciviruses as emerging pathogens in humans. Norovirus removal ranged from low to high depending on the soil type and experimental conditions with a high adsorption to clay loam containing ferric oxides; adsorption to quartz sand was much higher at pH 5, more than 3 $\log_{10}$ removal, than at pH 7, 17% removal.

Removal as the final treatment step

Membrane filters are being increasingly used for water treatment, especially in smaller treatment plants. There are several technologies: microfiltration, with a molecular cut-off size range suited to removal of bacteria and protozoa but not viruses; may nevertheless remove viruses attached to other particulates, and three types of filters with progressively smaller molecular cut-off sizes, that overlap somewhat and which are suited to virus removal—ultrafiltration, nanofiltration and reverse osmosis (RO) membranes. If membrane integrity is maintained, these last three membrane types are likely to have a high efficiency for virus removal based on the principle of size exclusion, as has been demonstrated in numerous studies (e.g. Yahya et al., 1993; Huertas et al., 2003). However, there may be some variation in membrane quality, and Adham et al. (1998) found, in bench studies of composite RO membranes, a wide variation in exclusion of MS-2 phage (from $<2 \log_{10}$ to $>7 \log_{10}$ in virus removal) depending on the selection of membrane. The effect of pore structure and membrane module configuration of virus retention has also been studied (Urase et al., 1996).

Virus inactivation during water treatment

Inactivation processes render viruses non-infectious to the host cell(s) even if they remain present in the treated water. The main inactivation processes used in water treatment include disinfection with chlorine, chloramine or chlorine dioxide, which may be preceded with ozonation or UV irradiation. Special situations may also involve inactivation by heat and or sunlight, though these methods are not usual for water treatment in industrialized nations. Virus inactivation can also be achieved by predation or biological degradation by other microorganisms using extracellular enzymes. However, biological inactivation is not readily quantifiable, nor reproducible, and cannot be considered routinely as an additional barrier that reduces viral contamination. The results of biological degradation are generally confounded as a part of the overall inactivation achieved.

Waterborne viruses are mostly non-enveloped in nature, and it is generally recognized that disinfection of non-enveloped viruses is more difficult than disinfection of vegetative bacteria. Estimates of viral inactivation during the disinfection carried out in water treatment can take different forms. Realistic evaluation of inactivation steps in a treatment process occurs as a part of the treatment train where prior removal steps have reduced the overall viral load such that the ability of the disinfection step to inactivate the target viruses is not being severely challenged. Alternatively, inactivation is measured separately by loading water with challenge organisms to determine not what has been achieved during a treatment process but what the potential for disinfection is by the treatment process. Laboratory studies that demonstrate the ability of the disinfection step to inactivate the virus under ideal conditions hold little relevance for the treatment process itself.

Inactivation of viruses by UV irradiation has received a lot of recent attention. Adenoviruses have been demonstrated to be among the human viral pathogens that are most difficult to inactivate (Nwachuku et al., 2005; Baxter et al., 2007) but a number of other viral pathogens remain to be tested to determine which would be the most suitable challenge for evaluating UV-based water treatment technologies. Inactivation of viruses in 'clean' waters by UV may be significantly different than virus inactivation in water containing particulates due to the protection of virions within the particulates that prevent successful inactivation by UV. This is currently an area of active research and may be of most significance in cold waters with poor floc formation and fluctuating water quality.

Ozone has generally been a successful method for inactivating human viruses (Emerson et al., 1982; Shin and Sobsey, 2003) but ozonation is a relatively rare process due to its high cost. Ozonation also creates additional difficulties due to changes in water chemistry; organic carbon is made more readily available to bacteria entrained in the water, and this tends to promote their downstream regrowth.

Viruses can be disinfected by the chlorine residual usually used for drinking water treatment. However, chlorine reacts readily with organic matter and becomes neutralized. Thus, disinfection can be hampered by the inclusion of virions in clumps or associated with particulate materials (Hejkal et al., 1979). Much less information is available about the ability of viruses to be disinfected with chloramine and chlorine dioxide. Dee and Fogelman (1992) examined the inactivation of indigenous phages in Denver water by monochloramine. They found a wide range of CT values (17 to nearly 5000); phage that escaped in the effluent had a higher CT value than the influent phage. These values are essentially comparable to values obtained for enteric human viruses (Berman and Hoff, 1984), but with a slightly wider range. Monochloramine was also shown to be much less effective as a disinfectant against enteric adenovirus than was chlorine (Baxter et al., 2007). The disinfection of enteric adenovirus and FCV (as a surrogate for noroviruses) by chlorine dioxide has shown that FCV is more resistant to chlorine dioxide than enteric adenovirus for the conditions studied (Thurston-Enriquez et al., 2005).

Estimates for viral inactivation from spiking of samples to be treated with chlorine shows reductions ranging from $\leqslant 1.4 \log_{10}$ in 10 min for phages added to groundwater (Duran et al., 2003), $4 \log_{10}$ for enterovirus and $1.7-3.2 \log_{10}$ for phages, both in 10 min, added to mineral water (Duran et al., 2003). This illustrates how phage are often more resistant to inactivation than human enteroviruses, at least those grown in the laboratory for spiking studies.

Berg et al. (1989) studied the kinetics of poliovirus inactivation in purified as well as drinking water. They found a more rapid inactivation in the drinking water sample studies and suggested that there may be substances present in drinking water that potentiate virus inactivation by chlorine. It is possible that the observations of Berg et al. (1989) could have resulted from changes in susceptibility of the host cells in response to substances in the drinking water, thereby overestimating virus inactivation. Such phenomena have not been investigated in response to virus disinfection studies in drinking water. On the other hand, it has been suggested (Bates et al., 1977) that naturally occurring viruses may be more resistant than those grown in the laboratory that are used for kinetic or spiking studies in drinking water. More work to understand these issues would be desirable but does not appear to have been followed up in recent years.

Risk to humans from viruses in drinking water

Even a single virus particle is sufficient, in principle, to induce disease in humans but this may occur only in a minority of cases. However, means to assess risks to humans from viruses in drinking water form a part of an overall microbial risk assessment (Gale, 2001). Eisenberg et al. (2006) estimated the range of illness for the majority of the population to be from 0.2–18.7 cases per 100,000 population if the total virus reduction was $4\log_{10}$ and 0–0.2 for a total virus reduction of 6 $\log_{10}$. Data used to arrive at these estimates included a source water concentration of 0.93 ± 3.0 infectious virions/l, a water consumption rate of 1.2 ± 1.2 l/day, reduction of 1.99 ± 0.52 $\log_{10}$ by removal (sedimentation and filtration) and 4.0 ± 2.93 $\log_{10}$ due to disinfection as well as data from rotavirus challenge studies (Ward et al., 1986). Such risk estimates must however be taken with some caution since the data used is abstracted from a wide range of non-coherent studies and many assumptions are made. In particular, risk assessment does not usually take into account the uneven temporal and spatial distribution of viral pathogens, nor the degree of viral clumping.

Concluding remarks

The regulatory requirements for virus reduction are usually for 99.99% reduction from source water levels. Although it is clearly desirable to perform site-specific virus reduction measurements using local operating technologies and parameters, this is rarely done. If it were necessary to assume an enteric virus reduction level attributable to an existing process from published studies, it would be imprudent to assume that more than $1–2 \log_{10}$ reduction could be achieved by virus removal. It is clear from a variety of data that turbidity fluctuations in source water can affect influent bacterial concentrations into water treatment plants (Gauthier et al., 2003); it is unclear whether the same would hold true for viruses, or perhaps increased turbidity could contribute to greater association of viruses with particulates and increased removal during treatment. Although correlations of bacteriophages with human viruses are not always seen, and the behaviour of the two groups may be slightly different in some drinking water treatment steps, phages still form better indicators for assessing viral removal and inactivation in drinking water than do bacteria (de Roda-Husman et al., 2005). They are also the only real option and alternative to spiking experiments, especially as the latter cannot be performed with infectious viruses in full-scale treatment plants.

In this review, more attention has been given to the removal of viruses than to their disinfection. This is deliberate because these removal processes are normally at the beginning of the treatment train and effect the largest reduction in virus numbers. Disinfection is left to 'mop up' the remainder of infectious virions that may have escaped the removal process. Although CT values are available for disinfection of many viruses with water disinfectants, and can be developed in the laboratory, it cannot be assumed that identical inactivation kinetics will be achieved when viruses are inactivated during water treatment.

In spite of the apparent high efficiency of virus removal and inactivation through the water treatment process, infectious and or non-cultivable viruses have been detected in tap water (Payment and Armon, 1989; Gerba and Rose, 1990; Lee and Kim, 2002; Haramoto et al., 2004). Whether these viruses represent those that have escaped the treatment, or perhaps, potential infiltration into distribution pipes during pressure transients or pipe breaks, is unknown. LeChevallier et al. (2006) suggest the value of monitoring phage as a potential indicator of distribution system integrity.

References

Adham SS, Trussell RS, Gagliardo PF, Trussell RR. Rejection of MS-2 virus by RO membranes. J Am Water Works Assoc 1998; 90: 130–135.

Agbalika F, Hartemann P, Joret JC, Hassen A, Bourbigot MM. Study of indigenous virus removal at different stages in a drinking water plant treating river water. Water Sci Technol 1985; 17: 211–218.

Bates RC, Shaffer PTB, Sutherland SM. Development of poliovirus having increased resistance to chlorine inactivation. Appl Environ Microbiol 1977; 34: 849–853.

Baxter CS, Hoffmann R, Templeton MR, Brown M, Andrews RC. Inactivation of adenovirus types 2, 5 and 41 in drinking water by UV light, free chlorine and monochloramine. J Environ Eng 2007; 133: 95–103.

Bell K, LeChevallier M, Abbaszadegan M, Amy G, Shahnawaz S, Benjamin M, Ibrahim E. Enhanced and Optimized Coagulation for Particulate and Microbial Removal. Denver, CO: American Water Works Association Research Foundation and the American Water Works Association; 2000.

Berg G, Sanjaghsaz H, Wangwongwatana S. Potentiation of the virucidal effectiveness of free chlorine by substances in drinking water. Appl Environ Microbiol 1989; 55: 390–393.

Berman D, Hoff JC. Inactivation of simian rotavirus SA11 by chlorine, chlorine dioxide and monochloramine. Appl Environ Microbiol 1984; 48: 317–323.

Bosch A. Human enteric viruses in the water environment: a minireview. Int Microbiol 1998; 1: 191–196.

Caballero S, Abad FX, Loisy F, Le Guyader FS, Cohen J, Pinto RM, Bosch A. Rotavirus-like particles as surrogates in environmental persistence and inactivation studies. Appl Environ Microbiol 2004; 70: 3904–3909.

Dee SW, Fogelman JC. Rates of inactivation of waterborne coliphages by monochloramine. Appl Environ Microbiol 1992; 58: 3136–3141.

De Roda-Husman AM, Lodder WJ, Penders EJM, Krom AP, Bakker GL, Hoogenboezem W. Viruses in the Rhine and Source Waters for Drinking Water Production. The Netherlands: Association of River Waterworks; 2005 34.

Duran AE, Muniesam M, Moce-Llivina L, Campos C, Jofre J, Lucerna F. Usefulness of different groups of bacteriophages as model micro-organisms for evaluating chlorination. J Appl Microbiol 2003; 95: 29–37.

Eisenberg JNS, Hubbard A, Wade TJ, Sylvester MD, LeChevallier MW, Levy DA, Colford Jr. JM. Inferences drawn from a risk assessment compared directly with a randomized trial of a home drinking water intervention. Environ Health Perspect 2006; 114: 1199–1204.

Emerson MA, Sproul OJ, Buck CE. Ozone inactivation of cell-associated viruses. Appl Environ Microbiol 1982; 43: 603–608.

Fane SA, Ashbolt NJ, White SB. The implications of system scale for cost and pathogen risk. Water Sci Technol 2002; 46: 281–288.

Fong T-T, Lipp EK. Enteric viruses of humans and animals in aquatic environments: health risks, detection, and potential water quality assessment tools. Microbiol Mol Biol Rev 2005; 69: 357–371.

Gale P. Developments in microbial risk assessment for drinking water. J Appl Microbiol 2001; 91: 191–205.

Gauthier V, Barbeau B, Tremblay G, Millette R, Bernier A-M. Impact of raw water turbidity fluctuations on drinking water quality in a distribution system. J Environ Eng Sci 2003; 2: 281–291.

Gerba CP, Riley KR, Nwachuku N, Ryu H, Abbaszadegan M. Removal of Encephalitozoon intestinalis, calicivirus and coliphages by conventional drinking water treatment. J Environ Sci Health 2003; A38: 1259–1268.

Gerba CP, Rose JB. Viruses in source and drinking water. In: Drinking water microbiology: progress and recent developments (McFeters GA, editor). New York, NY: Springer-Verlag; 1990 Chapter 18.

Haramoto E, Katayama H, Ohgaki S. Detection of norovirus in tap water in Japan by means of a new method for concentrating enteric viruses in large volumes of freshwater. Appl Environ Microbiol 2004; 70: 2154–2160.

Hejkal TW, Wellings FM, LaRock PA, Lewis AL. Survival of poliovirus within organic solids during chlorination. Appl Environ Microbiol 1979; 38: 114–118.

Hendricks DW, Clunie WF, Sturbaum GD, Klein DA, Champlin TL, Kugrens P, Hirsch J, McCourt B, Nordby GR, Sobsey MD, Hunt DJ, Allen MJ. Filtration removals of microorganisms and particles. J Environ Eng 2005; 131: 1621–1632.

Huertas A, Barbeau B, Desjardins C, Galarza A, Figueroa MA, Toranzos GA. Evaluation of *Bacillus subtilis* and coliphage MS2 as indicators of advanced water treatment processes. Water Sci Technol 2003; 47: 255–259.

Huffman DE, Nelson KL, Rose JB. Calicivirus – an emerging contaminant in water: state of the art. Environ Eng Sci 2003; 20: 503–515.

Jofre J, Olle E, Ribas F, Vidal A, Lucena F. Potential usefulness of bacteriophages that infect *Bacteriodes fragilis* as model organisms for monitoring virus removal in drinking water treatment plants. Appl Environ Microbiol 1995; 61: 3227–3231.

Keswick BH, Gerba CP, DuPont HL, Rose JB. Detection of enteric viruses in treated drinking water. Appl Environ Microbiol 1984; 47: 1290–1294.

LeChevallier MW, Karim MR, Weihe J, Rosen JS, Sobrinho J. Coliphage as a potential indicator of distribution system integrity. J Am Water Works Assoc 2006; 98: 87–96.

Lee S-H, Kim S-J. Detection of infectious enteroviruses and adenoviruses in tap water in urban areas of Korea. Water Res 2002; 36: 248–256.

Lodder WJ, de Roda Husman AM. Presence of noroviruses and other enteric viruses in sewage and surface waters in The Netherlands. Appl Environ Microbiol 2005; 71: 1453–1461.

McConnell LK, Sims RC, Barnett BB. Reovirus removal and inactivation by slow sand filtration. Appl Environ Microbiol 1984; 48: 818–825.

Nasser A, Weinberg D, Dinoor N, Fattal B, Adin A. Removal of hepatitis a virus (HAV), poliovirus and MS2 coliphage by coagulation and high rate filtration. Water Sci Technol 1995; 31: 63–68.

Niemi RM, Kytovaara A, Paakkonen J, Lahti K. Removal of F-specific RNA bacteriophages in artificial recharge of groundwater–a field study. Water Sci Technol 2004; 50: 155–158.

Nwachuku N, Gerba CP, Oswald A, Mashadi FD. Comparative inactivation of adenovirus serotypes by UV light disinfection. Appl Environ Microbiol 2005; 71: 5633–5636.

Payment P, Armon R. Virus removal by drinking water processes. CRC Crit Rev Environ Control 1989; 19: 15–31.

Payment P, Franco E. *Clostridium perfringens* and somatic coliphages as indicators of the efficiency of drinking water treatment for viruses and protozoan cysts. Appl Environ Microbiol 1993; 59: 2418–2424.

Payment P, Trudel M, Plante R. Elimination of viruses and indicator bacteria at each step of treatment during preparation of drinking water at seven water treatment plants. Appl Environ Microbiol 1985; 49: 1418–1428.

Redman JA, Grant SB, Olson TM. Filtration of recombinant Norwalk virus particles and bacteriophage MS2 in quartz sand: importance of electrostatic interactions. Environ Sci Technol 1997; 37: 3378–3383.

Rinto-Kanto JM, Lehtola MJ, Vartianen T, Martikainen PJ. Rapid enumeration of virus-like particles in drinking water samples using SYBR green I-staining. Water Res 2004; 38: 2614–2618.

Sattar SA, Tetro J, Springthorpe VS. Impact of changing societal trends on the spread of infections in American and Canadian homes. Am J Infect Control 1999; 27: S4–S21.

Schijven JF, de Bruin HAM, Hassanizadeh SM, de Roda Husman AM. Bacteriophages and clostridium spores as indicator organisms for removal of pathogens by passage through saturated dune sand. Water Res 2003; 37: 2186–2194.

Schijven JF, Hassanizadeh SM. Removal of viruses by soil passage: overview of modeling, processes and parameters. CRC Crit Rev Environ Sci Technol 2000; 30: 49–127.

Schijven JF, Hoogenboezem W, Hassanizadeh M, Peters JH. Modeling removal of bacteriophages MS2 and PRD1 by dune recharge at Castricum, Netherlands. Water Resour Res 1999; 35: 1101–1111.

Shin G-A, Sobsey MD. Reduction of Norwalk virus, poliovirus 1, and bacteriophage MS2 by ozone disinfection of water. Appl Environ Microbiol 2003; 69: 3975–3978.

Skraber S, Schijven J, Gantzer C, de Roda Husman AM. Pathogenic viruses in drinking-water biofilms: a public health risk? Biofilms 2005; 2: 105–117.

Stetler RE, Ward RL, Waltrip SC. Enteric virus and indicator bacteria levels in a water treatment system modified to reduce trihalomethane production. Appl Environ Microbiol 1984; 47: 319–324.

Szewzyk U, Szewzyk R, Manz W, Schleifer K-H. Microbiological safety of drinking water. Ann Rev Microbiol 2000; 54: 81–127.

Thurston-Enriquez JA, Haas CN, Jacangelo J, Gerba CP. Inactivation of enteric adenovirus and feline calicivirus by chlorine dioxide. Appl Environ Microbiol 2005; 71: 3100–3105.

Urase T, Yamamoto K, Ohgaki S. Effect of pore structure of membranes and module configuration on virus retention. J Memb Sci 1996; 115: 21–29.

van Olphen M, Kapsenberg JG, van de Baan E, Kroon WA. Removal of enteric viruses from surface water at eight waterworks in The Netherlands. Appl Environ Microbiol 1984; 47: 927–932.

Ward RL, Bernstein DI, Young EC, Sherwood JR, Knowlton DR, Schiff GM. Human rotavirus studies in volunteers: determination of infectious dose and serological response to infection. J Infect Dis 1986; 154: 871–880.

Wellings FM, Lewis AL, Mountain CW. Demonstration of solids-associated virus in wastewater and sludge. Appl Environ Microbiol 1976; 31: 354–358.

White DC, Gouffon JS, Peacock AD, Geyer R, Biernacki A, Davis GA, Pryor M, Tabacco MB, Sublette KL. Forensic analysis by comprehensive rapid detection of pathogens and contamination concentrated in biofilms in drinking water systems for water resource protection and management. Environ Forensic 2003; 4: 63–74.

Wommack KE, Colwell RR. Virioplankton: viruses in aquatic ecosystems. Microbiol Mol Biol Rev 2000; 64: 69–114.

Wyn-Jones AP, Sellwood J. Enteric viruses in the aquatic environment. J Appl Microbiol 2001; 91: 945–962.

Yahya MT, Cluff CB, Gerba CP. Virus removal by slow sand filtration and nanofiltration. Water Sci Technol 1993; 27: 445–448.

Human Viruses in Water
Albert Bosch (Editor)
DOI 10.1016/S0168-7069(07)17007-5

Chapter 7

Global Supply of Virus-Safe Drinking Water

Ana Maria de Roda Husman[a], Jamie Bartram[b]

[a]*National Institute of Public Health (RIVM), Centre for Infectious Disease Control (CIb), WHO Collaborating Centre for Risk Assessment of Pathogens in Food and Water, Antonie van Leeuwenhoeklaan 9, 3720 BA Bilthoven, The Netherlands*
[b]*World Health Organization, Geneva, Switzerland*

In this chapter the recommendations and guidelines of the World Health Organization (WHO) concerning water, sanitation and health will be evaluated in the light of disease caused by human pathogenic viruses. The focus will be on drinking water safety.

WHO guidelines for drinking water quality

From end product monitoring to prevention

In 1983–1984 and 1993–1997, respectively, the WHO published the first and second editions of the Guidelines for Drinking Water Quality (GDWQ). The development of these Guidelines in the 1980s was a significant departure from previous 'international standards', emphasising their advisory nature to national governments and the importance of their adaptation to take account of national and local sociocultural, environmental and economic circumstances. This philosophy was maintained in the second edition, which provided guidance on many more individual chemicals. In 1995, it was decided to revise the guidelines to account for advancing scientific knowledge with respect to drinking water quality. This especially related to microbial hazards and infectious disease risks. Since the early 1990s, a process of 'rolling revision' of the GDWQ has led to more frequent updating including additional publications regarding the chemical and microbiological aspects of drinking water quality and toxic cyanobacteria in water and addenda to the

Guidelines themselves. All these publications and the Guidelines are freely available on the Internet (http://www.who.int/water_sanitation_health/dwq).

The third edition of the WHO GDWQ was launched in September 2004 at the 'World Water Congress' in Marrakech, Morocco. According to these guidelines, access to safe drinking water is a component of effective policy for health protection. Requirements to ensure drinking water safety include both minimum procedures and specific guideline values as described in the WHO GDWQ, which also describes how to use these requirements. The multiple barrier approach, including source protection, appropriate levels of treatment and protection of water safety during distribution, is a basic principle for the reduction of health consequences by consumption of drinking water.

The guidelines outline a preventive management 'framework for safe drinking water' (Fig. 1). The framework includes 'health-based targets' to assist national authorities who are normally responsible to set the targets for the protection of public health from risks by exposure to drinking water. Water suppliers are responsible to meet these targets by the most appropriate means under local circumstances by specific control measures in the drinking water supply and also by defining management actions in case of regular situations or incidents. Assessing the adequacy of systems, defining and monitoring control measures and establishing management plans are the three components of the so-called 'water safety plans' (WSPs). Achievement of health-based targets may be verified by independent surveillance to assess the safety of the drinking water through additional verification or audit-based approaches. This framework for safe drinking water can be adapted according to environmental, social, economic and cultural circumstances of drinking water provision on the national, regional or local level.

The fourth edition of the WHO GDWQ is planned for 2009 that would include viewpoints on topics such as safe drinking water for vulnerable groups.

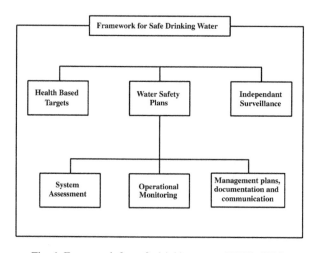

Fig. 1 Framework for safe drinking water (WHO, 2004).

Addenda for this edition have been published in 2006 and 2007 (http://www.who.int/water_sanitation_health/dwq).

Water-related viruses

The WHO GDWQ concerns prevention of waterborne diseases by exposure to microbial, chemical and radiological hazards in the drinking water. Microbial hazards are still considered to be of major concern for providing safe drinking water in both developed and developing countries. Infectious diseases caused by pathogenic bacteria, viruses and parasites (e.g. protozoa and helminths) are the most common and widespread health risk associated with drinking water.

The true waterborne diseases are part of a wider range of water-related disease. Water-related disease may be water-based, water-washed, waterborne or be transmitted by water-related insect vectors (Bradley, 1977). Insect vectors breeding in water or insect vectors only biting near water cause water-related infectious disease. Some viruses are water-related such as dengue virus and the flavivirus causing yellow fever. These viruses are transmitted via the bite of specific mosquitoes. Water-based diseases are typically caused by parasitic worms not virus infections in which waterborne hosts support an essential part of the life cycle of the infecting agent. Waterborne diseases are caused by the ingestion of water contaminated by human and/or animal excreta and secreta containing pathogenic bacteria, viruses or parasites, which are passively carried in the water supplies. Most water-related viruses are waterborne such as poliomyelitis virus, rotavirus, norovirus (NoV) and hepatitis A and E viruses (HAV and HEV, respectively). Water-washed diseases are the result of lack of water. This leads to poor personal hygiene and allows skin or eye infections to develop and spread easily through personal contact or contact with contaminated water. An example of a water-washed virus is adenovirus though some adenoviruses are typically waterborne. Most waterborne diseases can also be water-washed because inadequate personal hygiene facilitates faecal–oral disease transmission.

Besides the well-known water-transmitted viruses (see Chapters 2–4), infection and disease may be associated with (re-)emerging viruses such as the water-related chikungunya virus and emerging viruses such as severe acute respiratory syndrome (SARS) coronavirus and avian influenza virus (see Text box 1). Also, mutation and recombination of existing virus strains may lead to (re-)emergence such as described for vaccine-derived poliovirus (see Chapter 4). As a group, water-transmitted viruses can cause a wide variety of infections and symptoms involving different routes of transmission, routes and sites of infection and routes of excretion. Most water-transmitted virus infections occur asymptomatically or cause diarrhoea and self-limiting gastroenteritis in humans. Some waterborne viruses may cause respiratory infections, conjunctivitis, hepatitis and diseases that have high mortality rates; such as aseptic meningitis, encephalitis and paralysis. In addition, some of these viruses such as specific human enteroviruses have been linked to chronic diseases such as myocarditis and type 1 diabetes mellitus (Witso et al., 2006). As yet other long-term health effects, such as cancer, are not

---

Text box 1
The WHO publication *Emerging Issues in Water and Infectious Disease* provides knowledge and guidance to information sources on the evolution of infectious disease, the emerging waterborne pathogens, new environments and technologies that may influence emergence. It describes scientific advances in water microbiology and changes in human behaviour and vulnerability to broaden awareness and enhance preparedness about the changing world and its associated hazards (http://www.who.int/water_sanitation_health/emerging/emerging.pdf).

---

readily associated epidemiologically with water-transmitted viral disease (Ashbolt, 2004).

Faecal–oral transmission is the most well-known route of infection with waterborne viruses though other excreta (e.g. urine) and secreta (e.g. sweat) may also play a role in virus transmission via water which is less well studied. Transmission through aerosols and droplets derived from contaminated water may also lead to virus infection. A recent example of such environmental transmission of predominantly person-to-person transmitted flu-like viruses was the spread of SARS coronavirus in the large, private apartment complex Amoy Gardens in Hong Kong in 2003 (McKinney et al., 2006).

Since water-transmitted viruses do not grow in water, their numbers are determined by the extent of excretion/secretion into the environment followed by subsequent disintegration (see Chapter 5). The number of water-transmitted viruses in the water is determined by shedding of viruses by symptomatic patients as well as asymptomatically infected humans. Moreover, initially zoonotic viruses may be excreted by both the animal and the human host (see Text box 2).

Subsequent virus disintegration in the environment depends on the specific sensitivity of the virus to inactivation, e.g. by radiation, temperature and its tendency to attach or aggregate with particles (reviewed by John and Rose, 2005). Viruses that are efficiently transmitted via blood or respiratory droplets generally have a limited resistance in the water, such as human immunodeficiency virus (HIV), whereas other, mainly faecal–oral transmission of viruses such as poliovirus, are much more resilient (Moore, 1993). This range of viral resistance in the water is defined by the virus properties such as size, shape, charge, etc. The possible public health threat of (re-)emerging viruses should be assessed based on these particle characteristics. For example, SARS lead to the investigation of previously described respiratory agents such as *Legionella*. The newly identified SARS coronavirus was initially characterised as a respiratory virus without estimating its potential for transmission via water (Ksiazek et al., 2003). Subsequent research showed the common high level of excretion in faeces from infected individuals (Ding et al., 2004). Experimental studies showed stability of infectious virus for

---

Text box 2

The WHO book on *Waterborne Zoonoses: Identification, Causes and Control* provides a critical and balanced assessment of current knowledge about waterborne zoonoses and identifies strategies and research needs for anticipating and controlling future emerging water-related diseases (http://www.who.int/water_sanitation_health/ diseases/zoonoses.pdf).

---

several days in different liquid environments (Duan et al., 2003). This outlines the need for a proper assessment of the potential role of drinking water in SARS coronavirus transmission and in general in the transmission of emerging viruses. Another public health concern arises from the high infectivity as determined for most viruses emphasising the significance of the presence of each infectious virion in water for human use. The probability of infection from exposure to one or a few virus particles in water has been described for several waterborne viruses (Haas et al., 1993; Lindesmith et al., 2003).

Health-based targets

Faecally derived pathogens are currently the principal concerns in setting health-based targets for microbial safety of drinking water. Health-based targets are set by national authorities to protect and improve drinking-water quality such as with respect to the quantity and type of viruses present and, consequently, human health from viral disease. Health-based targets to reduce the waterborne disease burden may take different forms and are best derived using quantitative risk assessment, taking into account local conditions and hazards including epidemiological evidence of waterborne disease. International guidance on good practice is available to assist this (WHO, 2004). Typically performance and health-outcome targets are suitable as types of health-based targets for microbial hazards such as human pathogenic viruses in drinking water (Table 1).

Health outcome target

The health outcome target is the target type for which health effects and drinking-water management are most closely linked, but is not often applicable in practice and other types are more frequently used. Health-outcome targets are most useful in case of a measurable burden of water-transmitted disease, e.g. that associated with the presence of human pathogens in small water supplies both in developing and developed countries. It requires effortless and reliable monitoring of changes in exposure. Several aspects may hamper establishing the attributable risk of waterborne viral disease. First of all, it may be difficult to distinguish viral disease

Table 1

Health-based targets as applicable for microbial, e.g. viral, hazards

Type of target	Nature of target	Typical applications	Assessment
Health outcome			
• Epidemiology based	• Reduction in detected disease incidence or prevalence	• Microbial hazards with high measurable disease burden largely water associated	• Quantitative microbial risk assessment
• Risk assessment based	• Tolerable level of risk from contaminants in drinking water, absolute or as a fraction of the total burden by all exposures	• Microbial hazards in situations where disease burden is low or cannot be measured directly	• Public health surveillance and analytical epidemiology
Performance			
	• Generic performance target for removal of groups of microbes	• Microbial contaminants	• Compliance assessment through system assessment and operational monitoring
	• Customised performance targets for removal of groups of microbes	• Microbial contaminants	• Individually reviewed by public health authority; assessment would then proceed as above

from disease with similar symptoms but caused by other pathogens such as bacteria and parasites. Gastroenteritis may manifest itself with very general complaints such as abdominal cramps, fever, nausea, vomiting, diarrhoea, etc. On the other hand, projectile vomiting is characteristic for specific viruses, i.e. NoV, and bloody diarrhoea is quite uncommon with viral gastroenteritis. The same holds true for hepatitis that may be associated with either one of two specific waterborne viruses (HAV and HEV) or one other virus out of five different virus families but also with alcohol abuse or, though uncommonly, microsporidia infection. Second, it may not be possible to discern water-transmitted disease from disease contracted via other routes such as person to person, utensils or food. And in case of food, the transmission route may involve water such as preparation of lemonade or ice cubes and

---

Text box 3
  The WHO Guidelines for the safe use of wastewater, excreta and grey water describe health-based targets and good management practices to safeguard the health benefits such as better nutrition and food security for households and at the same time control the possible negative health impacts from exposure to hazardous substances such as viruses (WHO, 2006a; http://www.who.int/water_sanitation_health/wastewater/gsuww/en).

---

irrigation or washing of crops (see Text box 3) with virally contaminated water. The virus infections and symptoms resulting from either described route of transmission will be the same. Moreover, in the presence of other obvious high-risk exposures the drinking water may not be considered as the source of infection and disease, for instance in case of bottled water versus fruits or vegetables consumed raw. Nevertheless, if locally food-borne disease encompasses a higher burden as compared with waterborne disease, then it should be given correspondingly greater attention taking also into account costs and impacts of available interventions.

A third aspect hindering the estimation of waterborne viral disease burden is the incubation period; it may take several weeks for some viral diseases such as hepatitis to manifest themselves. The time of peak contamination in the source, failure in treatment or contamination in the distribution may be long gone. Also the drinking water may comply with regulations involving bacterial counts but these may be insufficient with respect to reduction of human pathogenic viruses. And fourth, the primary source may be drinking water but in case of high-level secondary transmission from person to person this may go unnoticed. However, in recent years methodology and awareness have improved greatly resulting in more and more publications on drinking-water disease outbreaks (Poullis et al., 2005). Remarkably few outbreaks were described associated with exposure to enteroviruses in drinking or recreational waters. Recently, large outbreaks with hundreds of cases of meningitis were reported for both types of exposure (Hauri et al., 2005; Amvrosieva et al., 2006). Also, the causative agent of the first reported waterborne outbreak of hepatitis E in Delhi, India in 1955 and 1956 was not identified until the 1990s with the advances in molecular and immunodiagnostics (Worm et al., 2002).

Health outcome-based targets may also be based on the results of quantitative microbial risk assessment (QMRA; see Chapter 8). In these cases, health outcomes are estimated based on information concerning exposure and dose–response relationships. However, with regard to QMRA also there are limitations in the available data and models (see Section *Risk management*). In addition, uncertainties such as short-term fluctuations in water quality may have a major impact on overall health risks—including those associated with background rates of disease and outbreaks—and are a particular focus of concern in expanding application of QMRA. Nevertheless, QMRA may be very useful in directly assessing data gaps if not in actual assessment of risks for consumption of drinking water. The results of a QMRA may be employed as a basis to define water quality targets or may provide the basis for development of performance targets (see Section *Performance target*).

Performance target

In contrast with health outcome targets, performance targets are frequently applied to the control of microbial hazards in piped supplies of any size though there is less accrued experience in their application to source protection or distribution than there is to treatment processes and systems (http://www.who.int/water_sanitation_health/dwq/en/safepipedwater.pdf). Performance targets are also readily applied to wastewater re-use systems (WHO, 2006a). Performance targets aid the selection and use of control measures to achieve acceptable source water quality and treatment efficiency. These control measures must ensure the prevention of pathogens breaching the barriers of source protection, treatment and distribution systems or preventing growth within the distribution system.

Targets for the removal of pathogens preferably are derived from location-specific data on source water quality. The source water may be groundwater or surface water or a mixture of both. Groundwater may be initially free of pathogenic viruses because of age. If water seeps through soil before reaching the groundwater, it may be effectively purified with respect to viruses depending on soil and virus type (Zhuang and Jin, 2003). However, shallow and/or unconfined groundwater may be contaminated through leaking of viruses originating from host ex-/secretions near the source or through flaws in the extraction process (see Text box 4).

In case of surface water, the number and type of human pathogenic viruses depend on the proximity of the viral reservoirs and dilution and inactivation of the viruses between point or diffuse source and the intake location for drinking (water production). Rainwater may also be used as drinking water. This type of source water may be more likely to be microbially contaminated by animal faeces if the rainwater is collected on rooftops with birds and rodents in the vicinity. Hantavirus infection is widely spread among rodents such as different types of mice and rats. On the other hand, poor hygienic and sanitation conditions associated with the design of the reservoir or tap may lead to viruses originating from human faeces. To date no study has been conducted for the presence of human pathogenic viruses in rainwater. Alternatively, seawater may be used as a source for drinking water by

---

Text box 4
  The WHO book on *Protecting Groundwater for Health: Managing the Quality of Drinking Water Sources* is a tool for intersectoral development of strategies to protect groundwater for health providing different points of entry (http://www.who.int/water_sanitation_health/gdwqrevision/groundwater/en/).

use of desalination. However, both bacteriophages and human pathogenic viruses were found to be very prevalent in seawaters (Shuval, 2003). Viruses seem to be readily inactivated in natural seawaters (Wetz et al., 2004). So far nothing is known about the reduction of human pathogenic viruses by desalination but membrane systems such as reverse osmosis and electrodialysis, which are the newest approach to desalination, are known to be very efficient in virus removal. Though desalination remains one of the most expensive ways to produce drinking water, these new advances and the increasing water scarcity may push the use of seawater (Reuther, 2000; http://www.who.int/water_sanitation_health/gdwqrevision/desalination/en/).

The quality of source water with respect to human pathogenic viruses thus highly depends on the circulation of particular viruses in the human population and with respect to zoonotic viruses in the animal population. Specific clinical data or source water quality data may not be available for human pathogenic viruses or for the location. In both developing and developed regions, viral disease is often classified as non-bacterial, probably viral, but never confirmed. In absence of specific data, national or regional data and/or data on viruses such as bacterial viruses (bacteriophages) may serve as input for setting performance targets. In this respect, free data sharing among researchers and risk managers is a key to successful reduction of the environmental spread of pathogenic viruses and reduced burden of water-transmitted viral disease.

Depending on the specific source water quality, the required performance target as in reduction of viruses by treatment can be set. These requirements concern control measures that must ensure both sufficient and robust virus reduction. The treatment efficiency that is needed to produce drinking water of acceptable quality with respect to viruses can be estimated by means of a system assessment (see Section *Risk management*). Alternatively in case of a high disease burden the efficiency of point-of-use treatment may be assessed with a surveillance programme. A multiple barrier approach including several treatment steps will facilitate a constant drinking water quality. However, besides knowledge of source water quality this requires trained operators and straightforward risk management actions (see Section *System assessment*). Moreover, the viral water quality often varies rapidly and over a wide range (Westrell et al., 2006). Treatment processes should be evaluated with respect to handling of short-term fluctuations as well as their potential to be the source of short-term fluctuations.

The role of the distribution system in the fate of human pathogenic viruses in drinking water so far has not been the focus of many studies (Skraber et al., 2005). The question is whether human pathogenic viruses benefit from their transport in the distribution system, for instance by the presence of biomass or biofilms preventing their inactivation. Or that growth in the distribution system may be considered as an additional treatment step aiding virus reduction in drinking water before consumption. Performance requirements are also important in certification of devices for drinking water treatment and for pipe installation that prevents ingress. Leaks or human error leading to unintended incorrect connections may be an important attributor to virus-associated disease outbreaks (Jardine et al., 2003).

The choice of an index virus to be able to target the group of human pathogenic viruses is very useful not only to indicate source water quality but also to estimate the reduction of viruses in the treatment of the source water and virus dynamics in the distribution system. Including different viruses reflecting the diverse characteristics with respect to their replication rates, natural reduction and reduction by treatment will improve the eventual drinking water quality by optimising the drinking water supply from source to tap. For viruses transmitted by the faecal–oral route, drinking water is only one vehicle of transmission. Contamination of food, hands, utensils and clothing can also play a role, particularly when domestic sanitation and hygiene are poor. Since water is the most important transmission route for some viruses whereas others are mainly transmitted via food or person to person, improvements in the quality and availability of water, in excreta disposal and in general hygiene are all important in reducing faecal–oral transmission of viruses. In combination with target microbes representing other pathogen groups, i.e. bacteria, parasites and helminths, performance targets can be developed that encompass both control challenges and health significance for a broad range of pathogens. In addition to source water quality, treatment efficiency and distribution, disease burden and dose–response relationships for specific pathogens serve as the basic parameters for the derivation of performance targets. Performance targets may be derived in relation to exposure to specific pathogens; however, account should be taken of the impact of short-term peaks in virus concentration on overall exposure and therefore the risk of drinking water disease.

Reference level of risk

To be able to establish the possible public health priority of reduction of water-transmitted viral disease as compared with other burdens of disease, a reference level of risk for the comparison of diseases is useful. A common denominator for disease burden enables a consistent approach for dealing with each hazard. Water-transmitted disease may be compared with disease contracted via other pathways. Similarly, viral diseases may be compared with disease associated with other pathogen groups or with each other. The hazard identification may be hampered, though, by the equivalence of disease symptoms of viral and non-viral diseases as explained above for gastroenteritis and hepatitis that may have alternative causes. A common metric taking into account differing probabilities, severities and duration of effects is the disability adjusted life year (DALY) widely used by WHO in the GDWQ. The range of differing severities, including acute, delayed and chronic effects and both morbidity and mortality associated with the specific viral pathogen have to be known. In addition, infection with a water-transmitted virus may have more severe disease outcome in vulnerable individuals as compared with the general population. This, for instance, holds true for HEV infection with a relatively high mortality rate among pregnant women (Rab et al., 1997).

In two ways viruses transmitted via water may be specifically significant as compared with other routes of transmission or other pathogens. One, major changes in viruses leading to a possibly more virulent, infectious or pathogenic virus type may result from recombination; this in contrast with other pathogens. In particular recombination may occur if the host is exposed to multiple virus variants from the same species. These double infections are more likely to occur as a result of consumption of drinking water or water-contaminated foods but not of person-to-person transmission. Two, consumption of virus-contaminated drinking water from a reservoir or served by a piped distribution reaches a relatively large part of the population rapidly leading to many exposures of humans and perhaps animals. And more generally, by means of the described approach employing a reference level of risk it may become evident that waterborne viral disease is a far more important public health challenge than frequently appreciated.

Water safety plans

Hazard analysis and critical control point (HACCP) has been a breakthrough in the control of food safety initially applied for the safety of foods for the manned space programme in the 1960s. The principles of HACCP are based on developing an understanding of the system, prioritising risks and ensuring that appropriate control measures are in place to reduce risks to an acceptable level. These same principles have been long applied in drinking water safety management and more recently have been refined and tailored to the context of drinking water (Havelaar, 1994) following the application of HACCP by several water utilities in the world (see example for Australia in http://www.who.int/wsportal/wsp/en/). Such a systematic management approach to water safety by the water suppliers was translated by WHO into a so-called WSP that incorporates system protection and process control. The WSP objectives are to ensure safe drinking water through good water supply practice. This involves preventing contamination of source waters, treating the water to the extent necessary to meet health-based targets and preventing contamination during storage, distribution and handling of drinking water. A WSP comprises system assessment and design, operational monitoring and management plans, including documentation and communication, concerning all aspects of the drinking water supply from catchment to consumer. The WHO GDWQ asks for the inclusion of management plans as an integral part of the WSPs. These plans should describe actions to be taken during normal operation or incident conditions. The system assessment should be documented (including upgrade and improvement). Furthermore, monitoring and communication plans and supporting programmes should be described. In this way the management and control of drinking water supply are safeguarded and continued through systematic and detailed assessment and prioritisation of hazards and the operational monitoring of barriers or control measures. Moreover, in case of emergencies plans of action are in place with consent of the people involved, which can be readily initiated, and these plans have been rehearsed in emergency drills. The control measures as described in the WSP will be effective to protect public health from water-related viral disease. WSP issues, specifically with respect to human pathogenic viruses in drinking water, are discussed below.

System assessment

Risk assessment has been used for decades, initially for decision making in psychotherapy and vaccination strategies, and today for management of pandemics such as influenza and AIDS. In 1993, the application of this tool was first described for exposure to pathogenic viruses in treated drinking water, since it is difficult to estimate risks posed from exposure to low levels using epidemiology (Haas et al., 1993). According to the US National Academy of Sciences, the successive steps of the risk assessment include the hazard identification, exposure assessment, dose–response modelling and risk characterisation to estimate the probability and consequences of infection—here specifically for individuals consuming various levels of viruses in the drinking water. Risk assessment should be considered as an ongoing process to allow advances in research and technology to be fed back into the risk characterisation. As stated in the WHO GDWQ, a system assessment establishes the capacity of a drinking-water supply to deliver water of a quality that meets the health-based targets. This also includes the assessment of design criteria for new systems and upgrading of existing systems.

Hazard identification

The word *virus* translated from Latin means slimy liquid, slime or poison, especially of snakes' venom; any harsh taste or smell. Clearly, a virus as intended in this book is unwanted in drinking water and should be considered a hazard: a biological agent that has the potential to cause harm. A number of issues have to be taken into account to determine those viruses that may be water-related and need to be considered in the specific system assessment.

Human and animal sources. The viruses that may cause disease in humans may be derived from humans or animals. More specifically, numerous human pathogenic viruses are shed in human and animal excreta (i.e. faeces, urine, vomitus), secreta (i.e. saliva, tears, semen, mucus, sweat) and blood of infected individuals. In addition, viruses may target specific host organs such as the liver or the heart and therefore these organs will harbour viruses. Porcine HEVs phylogenetically related to human HEV strains were shown to be shed in different bodily fluids and tissues of pigs such as faeces and liver (De Deus et al., 2007), but it is yet unknown if HEV is shed in human fluids other than serum and faeces. Viruses account for the highest infection risks through accidental exposure of health care workers (Tarantola et al., 2006), but often a broad investigation of viruses in bodily fluids of humans has not been done leaving broad knowledge gaps (Table 2).

Table 2

Incomplete but informative list of virus types present in human excreta and secreta

	Faeces	Urine	Saliva	Tears	Sweat	Reference
Adenovirus	+	+	+	+	?	Ramsay et al. (2002), Hatakeyama et al. (2006), Kaye et al. (2005)
Aichivirus	+	?	?	?	?	Yamashita et al. (2000)
Astrovirus	+	?	?	?	?	Guix et al. (2005)
Enterovirus	+	+	+	?	?	Ramsay et al. (2002), Muir et al. (1993)
Hepatitis A virus	+	?	+	?	?	Mackiewicz et al. (2004)
Hepatitis C virus	+	?	+	?	+	Beld et al. (2000), Ortiz-Movilla et al. (2002)
Hepatitis E virus	+	?	?	?	?	Singh et al. (1998)
Norovirus	+	?	+	?	?	Herrmann et al. (1985)
Parvovirus	?	?	+	?	?	Ramsay et al. (2002)
Picobirnavirus	+	?	?	?	?	Cascio et al. 1996, Banyai et al. (2003)
Polyomavirus	+	+	?	?	?	Hatakeyama et al. (2006), Berger et al. (2006)
Reovirus	+	?	?	?	?	Giordano et al. (2002)
Rotavirus	+	?	?	?	?	Bowdre (1983)
Sapovirus	+	?	?	?	?	Phan et al. (2006)
SARS coronavirus	+	+	+	?	+	Wang et al. (2004), Ding et al. (2004)
TT virus	+	−	+	+	−	Matsubara et al. (2000)

Note: +, studied and confirmed; −, studied but not detected; ?, no studies included in publications shown by search engine NCBI PubMed.

Typically numerous viruses are shed in human faeces (reviewed by Carter, 2005). Most of these viruses replicate in the gastrointestinal tract and are referred to as enteric viruses (see Chapters 2 and 3). Some viruses are renowned for their excretion in urine such as polyomavirus JC but also human papillomavirus (HPV), cytomegalovirus, hepatitis B viruses (HBV) and more recently SARS coronaviruses. The most common viruses to be secreted in saliva are HBV, HPV and cytomegalovirus, but recently HAV, a well-known food- and waterborne virus, was also associated with human saliva (Mackiewicz et al., 2004). In addition, saliva and throat wash of SARS patients were positive for the presence of the specific coronavirus (Wang et al., 2004). Hepatitis C viruses replicate in sweat glands leading to virus release in sweat of patients (Ortiz-Movilla et al., 2002). Torque teno virus (TTV) DNA was detected in tears as well as in saliva, breast milk, semen, blood, faeces and vaginal fluid whereas no evidence was found for the presence of the virus in urine and sweat.

For viruses shed in bodily fluids and organs the probability of being transmitted via water is largely determined by their circulation in the population, their ability to reach water resources and their inactivation rate in water. These issues may be unknown besides the existing knowledge gaps on virus shedding in the absence of specific research efforts or lack of access to the results by means of international, peer-reviewed journals or other accessible publications. Also, epidemiological evidence for viruses to be water-related may be lacking but this may be a research bias as explained in Section *Health outcome target*.

Drinking water outbreaks. If information on specific virus types causing waterborne outbreaks exists (see Chapters 2 and 3) then outbreak data should be collected including the number of patients, their exposure and the number and type of viruses in the contaminated water to characterise the hazard and determine a dose–response relation (the latter instead of or in addition to human volunteer or animal studies). Numerous waterborne outbreaks of mainly NoV-associated gastroenteritis have been described originating from contaminated drinking water but often the descriptions are anecdotical or at best an epidemiological association was determined between exposure of the patients and the common water source. With the improvement of detection methods with respect to sensitivity and specificity the causative agent can be detected more often in the water and typed with similar sequences in the patients and the common water source. Especially screening of patient samples for the virus followed by specific design of the detection method for monitoring the environment enhances the chance of success in confirming the link (Duizer and Koopmans, 2006; Hoebe et al., 2004). In some instances, the cause of the drinking water contamination is resolved and the drinking water was found to be contaminated by sewage through pump failure or blockage of a sewage system. In other cases, inadequate or failing treatment processes lead to insufficient removal of viruses from source waters for drinking water production. Only recently enteric cytopathic human orphan (ECHO) virus type 30 has been identified as an important waterborne agent of meningitis in different parts of the world.

The possibility for enteroviruses to be transmitted via drinking water (Amvrosieva et al., 2006) or recreational water (Hauri et al., 2005) was not recognised previously though numerous enterovirus-associated meningitis outbreaks have been described for decades. These data confirm the value of outbreak investigations.

Selection and prioritisation of viral hazards. To be able to select and prioritise human pathogenic viruses to be included in the system assessment, criteria concerning the extent of the health risk, the waterborne transmission and the quantitative risk assessment need to be set. First of all, location-specific information on the prevalence of the virus infection and disease in the population is essential though it may be substituted by, for instance, knowledge on human pathogenic viruses in sewage or surface waters. Furthermore, viral disease outcomes, ranging from asymptomatic through severe defects to death (Table 3), largely determine the health-based targets, since prevention of a high disease burden will be a more effective intervention as compared with prevention of disease with an overall low burden (see Section *Health-based targets*). The burden of viral disease involves the severity of disease as well as prevalence. For instance, if overall more individuals experience episodes of mild, limited viral gastroenteritis in a specific region then patients with a more severe health outcome, viral hepatitis, should also be considered in setting the health-based targets. Besides being dependent on the pathogenicity of the virus, the severity of viral disease also depends on the susceptibility of the host. The disease burden of rotavirus in high-income countries is much lower than in low-income countries with a much higher susceptible fraction of the population and more severe outcome of the infection (Havelaar and Melse, 2003; http://www.who.int/water_sanitation_health/bathing/en/).

Table 3

Disease outcomes of human pathogenic viruses

Disease outcome	Causative viral agent	Reference
Conjunctivitis	Adenovirus, enterovirus	Tavares et al. (2006)
Gastroenteritis	Adenovirus, aichivirus, astrovirus, enterovirus, norovirus, picobirnavirus, rotavirus, sapovirus	Glass et al. (2001)
Hepatitis	Hepatitis A–G virus, TT virus	Luo et al. (1999)
Leukoencephalopathy	Polyomavirus	Khalili et al. (2006)
Meningitis	Enterovirus, reovirus	Johansson et al. (1996)
Myocarditis	Adenovirus, enterovirus, parvovirus	
Poliomyelitis	Enterovirus	
SARS	SARS coronavirus	

The extent of the health risk associated with a specific virus is also determined by the ability to treat the virus infection and the availability of the treatment. Though the area of antiviral treatment is rapidly advancing, there is no cure for many viruses and in many countries there are no means to administer drugs to people in need. Vaccination strategies to prevent viral disease have been very successful. Smallpox was eradicated in 1978 and the eradication of poliomyelitis is in sight. (Re-)emergence of vaccine-preventable viral diseases should be monitored such as disease caused by vaccine-derived polioviruses, which can be efficiently transmitted via water (see Chapter 4). The possibility of a virus to spread epidemically and/or endemically is important in waterborne transmission because of the level of viruses circulating in the asymptomatic and symptomatic population and the total burden of viruses shed into the environment, and if relevant, also those viruses derived from animals. In case of HEV it was found that the consumption of liver and blood of domestic pigs, wild boar and/or deer leads to infections with HEV variants causing hepatitis E in the exposed individuals (Li et al., 2005). The human HEV strains were phylogenetically related to the animal variants (Lu et al., 2006). Moreover, HEV transmission is primarily water- and food-borne but may also be transmitted via blood transfusion, contact with sewage or animals or vertical transmission (Singh et al., 2003; Vaidya et al., 2003; Li et al., 2005). These routes of HEV transmission appear to be more efficient as compared with person-to-person transmission (Somani et al., 2003). Other possible zoonotic viruses such as some porcine NoV and sapoviruses (SaV) are genetically or antigenically related to human strains (Wang et al., 2006). Recombinants within NoV and SaV occur for human and pig strains, and the high prevalence and sub-clinical infection rate of these viruses in pigs raise questions of whether pigs may be reservoirs for human strains or for the emergence of new human and porcine recombinants. The occurrence of zoonotic virus strains stresses the need to take animal reservoirs into consideration regarding source protection and when assessing the health risk for drinking water.

Environmental surveillance.   In addition to disease surveillance in the population to whom the (produced) drinking water is delivered, environmental surveillance can aid the identification of the viral hazards by typing and quantification. Data resulting from studies on human pathogenic viruses in sewage and surface waters are accumulating. Raw sewage waters were shown to contain human enteric viruses such as enteroviruses, reoviruses, rotaviruses, NoV and HEV (Clemente-Casares, 2003; Villena et al., 2003; Lodder and de Roda Husman, 2005). In most studies presence/absence data were shown but the virus concentrations were not determined. Though some viruses such as specific entero- and adenoviruses can be cultured, there is no susceptible cell line for many other human pathogenic viruses. In this case, the virus concentration may be estimated from molecular data collected by (semi-)quantitative and/or real-time polymerase chain reaction (PCR). Typically 10^6 viral genomes can be detected in raw sewage as compared with 10^3 infectious viruses (Schvoerer et al., 2001; Laverick et al., 2004; Lodder and de Roda

Husman, 2005; van den Berg et al., 2005). In these studies the applied sewage-treatment processes approximately resulted in a limited 1–2 $\log_{10}$-units reduction as assessed by culture or molecular techniques. In general, primary and secondary sewage-treatment processes do not efficiently reduce the virus concentration in contrast with tertiary processes (Fleischer et al., 2000; Schvoerer et al., 2001; Gehr et al., 2003; van den Berg et al., 2005; Myrmel et al., 2006). Bacteriophages may be useful in assessing reduction of human viruses by wastewater treatment processes (Lucena et al., 2004). Wastewater treatment is warranted to reduce viral load in wastewaters discharging onto receiving waters (Godfree and Farrell, 2005). However, since human viruses enter wastewater at high numbers wastewater will still contain viruses even after treatment. Therefore, in addition to wastewater overflows treated effluents will have to be taken into account as a significant source of viral pollution for wastewater used in agriculture and for surface waters used for aquaculture and fishery, recreational purposes or as a source for drinking water production. Moreover, both raw and treated sewage may contaminate groundwater through leakage (Abbaszadegan et al., 1999; Borchardt, 2003; Borchardt et al., 2004).

Common sources of faecal pollution, directly or indirectly contaminating surface waters with human pathogenic viruses, include raw and treated sewage, wash-off of animal manure, faeces from wildlife such as waterfowl or deer, grazing animals and vermin in and around reservoirs, backflow from unprotected connections and sewer cross connections. In river water samples enteroviruses, reoviruses, rotaviruses, HAV, astroviruses, TTV and NoV were detected in different countries around the world but these results are biased since most researches were done in Europe (Matsuura et al., 1993; Gilgen et al., 1997; Schvoerer et al., 2000; Hot et al., 2003; Denis-Mize et al., 2004; Hörman et al., 2004; Haramoto et al., 2005, Villar et al., 2006). Viruses originating from discharge of treated-sewage water onto the receiving surface waters used for drinking water production (or recreational purposes) pose a health risk (Schernewski and Julich, 2001). Most papers on viruses in surface waters document presence/absence data only but to be able to estimate the infectious risk by consumption of drinking water quantitative data are necessary. In some studies, the virus concentrations were determined in river water varying from 0.001 to 10 infectious entero- and reoviruses per litre to 5 to 5000 genomes for NoV and rotaviruses per litre of river water (Haramoto et al., 2005; Lodder and de Roda Husman, 2005). Location-specific data on concentration and types of viruses in source waters and inactivation rates under the specific conditions in the specified waters are preferred for the hazard identification. In the absence of such data, virus data from other locations that may be considered comparable may also be employed. For instance, if treated and untreated urban sewage are discharged then it could be assumed that virus types such as enteroviruses and rotaviruses may be present and virus concentrations could be contracted from available studies. If there is sewage discharge from slaughterhouses, faecal pollution from wildlife such as wild boar or deer or wash-off of animal manure HEV and SaV could be present and may be carried to receiving waters. In case of rodent vermin hantaviruses could be expected.

Exposure assessment

In some regions in the world source waters that are used directly as drinking water receive human and animal excreta and secreta originating from bathers, animals, boats or indirectly from, for instance, wastewater discharges. In this case no reduction of viruses will take place before consumption. In other regions, the source waters are treated before use as drinking water. To be able to assess the ability of the system of concern to provide safe drinking water the viral hazards in the drinking water should be studied quantitatively.

Pathogenic viruses in drinking water. A few studies have shown the presence of human pathogenic viruses in untreated or treated drinking water used in different regions around the world (Leclerc et al., 2000). As described surface waters may be contaminated with viruses originating from a diversity of sources. Similar human and animal sources may contaminate groundwater that is insufficiently protected from external influences. This may be the case with semi- or unconfined groundwater from sandy saturated zones or from limestone or marl or with pre-treated, artificially infiltrated surface water or with bank filtered water. Groundwater was shown to be positive for viruses in 8 or 16% of the samples analysed (Borchardt, 2003; Fout et al., 2003). Reovirus, rotavirus, enterovirus, HAV and NoV were detected by molecular methods (Fout et al., 2003). None of the samples was positive in case cell culture was applied to detect infectious enterovirus (Fout et al., 2003). Another study revealed virus positivity in well waters by molecular (50%) and cell culture methods (5%; Borchardt et al., 2004). The presence of NoV genomes could not be confirmed in mineral waters whether spring water or finished product (Lamothe et al., 2003). Other studies confirmed virus-contaminated groundwater to be the source of outbreaks of acute gastroenteritis among hundreds of people (Parshionikar, 2003; Kim et al., 2005). Interestingly in one of the studies the contamination appeared to be transient since none of the sequential well water samples were virus positive (Borchardt, 2003). This may be due to short-term fluctuations in pathogen load caused by epidemics in the human or animal population, seasonal variations and/or incidents. The latter may encompass upstream (raw or treated) wastewater discharges, building of upstream sewage treatment plant or discharge, digging, drilling or maintenance near production site, injecting manure near unconfined source or discharge of polder waters near intake. Also heavy rainfall events, high river-/stream discharge and flow, high groundwater level, flooding of catchment/production site, thawing of (faecally contaminated) ice on reservoirs, frost leading to high numbers of birds on reservoirs and high numbers of birds (or other game) may be associated with peaks in faecal contamination of groundwater and also surface water.

Evidence has been published on the presence of human pathogenic viruses in treated water since drinking water treatment processes may have a limited capacity for virus reduction. Infectious rotavirus and enterovirus were detected in 83% of treated drinking water samples associated with rainfall but not with the presence of bacterial indicators (Keswick et al., 1984). In the African region, using cell culture, Ali et al. (2004) reported the presence of cultivable enteroviruses in 7 out of 30 finished drinking water samples at concentrations ranging from 5 to 33 plaque-forming units per litre. In Asian and African countries, integrated cell culture PCR enabled the detection of enteroviruses (Vivier et al., 2004), adenoviruses (Van Heerden et al., 2003) or both (Lee and Kim, 2002; Lee and Jeong, 2004) in drinking water. With the use of only molecular methods the presence of several pathogenic viruses such as rotavirus (Divizia et al., 2004; van Zyl et al., 2004), astrovirus (Gofti-Laroche et al., 2003) and NoV (Kukkula et al., 1999; Haramoto et al., 2004) were detected in drinking water in European, African and Asian countries. In another study in Thailand, no HAV was detected in tap waters, but in Brazil, 20% of river and tap waters were HAV positive (Kittigul, 2006; Villar et al., 2006). In Canada, outbreaks were identified associated with the presence of HAV and NoV in drinking water (Schuster et al., 2005). In case of direct evidence for the presence of pathogenic viruses in treated or untreated drinking water by either molecular or cell culture or other techniques, the quantity and host specificity should really be determined to be able to perform QMRA.

Pathogenic viruses in source water and virus reduction by treatment. The absence of pathogenic viruses in finished drinking waters does not exclude their presence but may merely indicate a lack of sensitivity of the detection techniques and/or the representativeness of the sampling. As shown with the recent advances in virus concentration and detection techniques more and more often human viruses are detected in tap water. However, at low dose viruses are still significant with respect to infectious and disease risk by exposure to drinking water because of their infectivity and pathogenicity and because of the large exposed population. An alternative approach in case of low virus concentrations in the finished waters is their detection in source waters combined with the assessment of the efficiency of drinking water treatment processes with respect to virus reduction. This approach ideally involves location-specific, quantitative information on the virus concentration in the source waters and virus reduction by treatment to estimate the virus concentration in the drinking water. In addition, information on the virus analysis, such as virus recovery rate of the method, infectivity and host specificity of the virus, are needed. Moreover, the local habits of drinking water consumption of the consumers should be known.

The recovery of the applied concentration–detection procedure should be determined to be able to accurately estimate the actual virus concentration. Loss of virus particles during the analysis will lead to underestimation of the infectious risk. Virus recovery rates depend on the employed concentration and detection methods, the volume of water to be concentrated, the turbidity of the water and the characteristics of the target virus (Senouci et al., 1996; Hill et al., 2005; Olszewski et al., 2005; Polaczyk et al., 2006). In order to determine the recovery rate of the sample analysis the procedure that is followed includes analysis of the viruses in the natural water sample and analysis of the same sample seeded with a specific dose of a known virus. This virus commonly involves the use of cultivable viruses such as specific enterovirus types or bacteriophages. These viruses are assumed to behave similarly to each other but differences in recovery between human and bacterial viruses have been reported (Hill et al., 2005). Moreover, these model viruses often have undergone several passages through the susceptible cell line and are therefore considered cell line adapted strains, which will behave differently as compared with naturally occurring human pathogenic viruses in the environment. However, in the absence of alternative approaches experiments with model viruses need to be done preferably coinciding with each sample analysis since virus recovery rates have been found to be highly variable even when the virus and water type are kept constant (Denis-Mize et al., 2004). In addition, in seed experiments often high virus levels are used to easily recover the model virus. But studies have shown that at the natural low dose virus recovery is lower and more variable (Fuhrman et al., 2005). Though many studies were undertaken in the past years, further optimisation of methods with respect to virus recovery is still warranted.

Since virus genomes may be present in the source water long after the virus particle has lost its ability to infect a host cell knowledge on the infectivity of the target virus is necessary to limit overestimation of the virus concentration and therefore the infectious risk. On the one hand, it may be argued that the detection of genomes of human pathogenic viruses in the drinking water is a necessary condition for the presence of infectious viruses. The employment of culture and molecular techniques in the detection of human pathogenic viruses in drinking water both encompass advantages and disadvantages. Susceptible cell lines are known for only some waterborne viruses and not for NoV (Duizer et al., 2004) and such cell lines may not be suitable for implementation as routine and robust diagnostic tools. Also, cell culture assays are laborious and expensive. On the other hand, plaque assays give quantitative results assuming that one plaque originates from one initial infectious virus particle. The latter may be another point of discussion (Teunis et al., 2005).

Often virus strains and types belonging to one virus family have a long history of developing their own host specificity (Fields et al., 2002). These genetically related viruses may be detected by the use of the same molecular or culture technique but not each virus strain will be capable to infect the human host. In case of the detection of human and animal, but non-zoonotic virus strains the infectious risk would be overestimated. Examples of such virus families are enteroviruses and adenoviruses (Fong and Lipp, 2005).

A few studies have been done on human pathogenic viruses at the intake point of drinking water suppliers. During an extensive study period of 9 years, 9% of source waters for drinking water production were positive for the presence of reovirus at relatively low titres but enterovirus or adenovirus were not detected (Sedmak et al.,

2005). Rotaviruses were detected in source waters by use of molecular techniques (Van Zyl et al., 2004). The use of indicators for the presence of human pathogenic viruses in source waters has been reported with highly variable success with respect to correlation between index and indicator organisms (Bosch, 1998). The QMRA ideally involves location-specific, quantitative information on the virus concentration in the source waters that is representative for the time period including potentially wide quality fluctuations within it. Alternatively, literature data may be used but as shown such data are scarce.

Also location-specific, quantitative information is needed on virus reduction by treatment. Treatment may include a multiple barrier approach and/or point-of-use treatment in the home. Indicators such as somatic phages or F-specific phages with similar characteristics (size, surface charge, shape etc.) may be used in experiments to determine the efficiency of virus reduction (Table 4). These viruses that prey on specific bacterial hosts occur at higher numbers, which facilitates their detection after treatment (Payment, 1991). The ease and relative cost of phage detection is another advantage over human pathogenic viruses. Comparative analyses of the index and indicator virus reduction by the specific treatment step should show first that these viruses behave similarly as was performed by Payment and Franco (1993).

Field data on pathogen removal or inactivation derived from the applied treatment plant is the most significant input for the QMRA but may be difficult because of low-level contamination or lack of methodology. Some field studies have been described. Analysis of viruses in drinking water produced by coagulation, sedimentation, sand filtration and chlorination showed 83% of the samples positive for infectious rotavirus and/or enterovirus (Keswick et al., 1984). The overall reduction

Table 4

Characteristics of human pathogenic viruses and their possible indicators (Ferguson et al., 2003)

Virus type	Size (nm)	pI	Phage type	Size (nm)	pI
Poliovirus 1	29	7.2	PM2	60	7.3
Poliovirus 2	29	4.5–6.5	PM2	60	7.3
Coxsackie virus A/B	29	6.6-8.2	ΦX174, PM2	26–32, 60	6.6, 7.3
ECHO virus	29	5.3–6.4	ΦX174, Qβ	26–32, 24	6.6, 5.3
Hepatitis A virus	27	2.8	MS2	26	3.9
Hepatitis E virus	30	–	–	–	–
Norovirus	25	4.9♯	Qβ, PRD1	24, 63	5.3, 4.2
Sapovirus	34	–	–	–	–
Reovirus	75	3.9	MS2	26	3.9
Rotavirus	70	–	–	–	–
Astrovirus	29	–	–	–	–
Adenovirus	60–80	–	–	–	–

Note: ECHO, enteric cytopathogenic human orphan; Reo, respiratory enteric orphan; –, unknown; nm nanometer; pI, isoelectric point; ♯, determined for Norwalk-like virus particles.

of enteroviruses for similar processes in a 1-year study at 7 treatment plants was 99.97% with different coxsackie virus types B, polioviruses and echovirus identified in the finished drinking waters at an average concentration of 0.0006 infectious viruses per litre (Payment et al., 1985). Though in another study including 6 waterworks 11 out of 55 samples of partially treated waters were positive for different types of enteroviruses, no viruses were detected in 100 samples of finished drinking water (van Olphen et al., 1984). The evaluation of enteric virus and bacteriophages removal at three treatment plants did not show any positive results for viral presence in finished waters (Payment and Franco, 1993). Conventional treatment was insufficient for the reduction of enteroviruses in case of highly polluted intake water (Ali et al., 2004). Pilot plants on site are usually very similar to the operational plant and less troublesome for the retrieval of data because of the smaller scale. The removal of feline calicivirus and different phages was evaluated during conventional drinking water treatment ranging from 1.85 to 3.21 $\log_{10}$ units (Gerba et al., 2003). Laboratory experiments, for instance, with the specific source water and sand from an applied rapid sand filter will also generate useful data. Comparison of the removal of viruses and phages for a point-of-use treatment unit showed (more than) 5 $\log_{10}$-units reduction for somatic phages and MS2 coliphages and poliovirus, HAV, adenoviruses, rotaviruses and astrovirus (Grabow et al., 1999). If facilities for collection of field, pilot plant or laboratory data are missing, literature data on similar systems as determined by specific control parameters may be employed but though not complete, the overview above shows again that limited data are available on human pathogenic viruses.

Fate and behaviour of viruses during transport and storage. Human pathogenic viruses may be present in tap water depending on the concentration in the source water and in case of treatment on the efficiency of the treatment processes with respect to virus reduction. The concentration of pathogenic viruses in the drinking water can be used to assess the health risk associated with the consumption of tap water, as has been done for different viruses (see Section *Risk characterisation*). However, the previous studies have not taken into account the dynamics of human pathogenic viruses meaning the variations in the prevalence and infectivity of viruses during transport and storage. First of all, though human and animal viruses do not grow during transport and storage, sanitation and hygiene problems may cause (re-)introduction of pathogens in the drinking water that may survive for the duration of storage as was shown for bacteriophages. In case of a piped distribution system, the virus concentration in tap water is determined by the viral burden of the source water on one hand and reduction by applied treatment processes (if this is the case) and during transport on the other hand. Reduction of viruses in drinking water may be caused if temperatures are high, i.e., most viruses are readily inactivated at over 20°C (John and Rose, 2005). Exposure of the viruses to, for instance, sunlight as well as other environmental factors will decrease the numbers of viruses rapidly (Lytle and Sagrapanti, 2005). Again (re-)introduction of human pathogenic viruses may occur during maintenance and repair practices

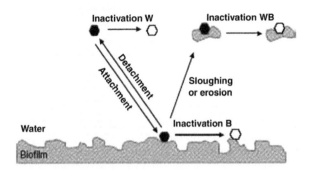

Fig. 2 Parameters that control virus behaviour in drinking water distribution systems. Viral inactivation was considered as follows. 'Inactivation W' corresponds to the inactivation of virus in the water phase; 'Inactivation B' corresponds to the inactivation of virus in biofilm; 'Inactivation BW' corresponds to the inactivation of virus entrapped or attached to biofilm and released into the water phase by sloughing or erosion. Solid hexagon, free or particle-associated virus that is able to infect target cells; open hexagon, free or particle-associated virus that is not able to multiply anymore. From Skraber et al. (2005).

(Havelaar, 1994). Second, biofilms may play an important role with respect to virus concentrations in the finished waters (Fig. 2; Skraber et al., 2005). Since it has been shown that viruses can attach to biofilms, there is a possibility that drinking water biofilms accumulate pathogenic viruses present in the water entering the distribution system acting as an additional removal process. On exposure to single viruses or sloughs containing viruses released from biofilms the consumer could be infected. Data on fate and behaviour of human pathogenic viruses during transport and storage are extremely limited but this information is of importance with respect to intervention measures.

Consumption of drinking water. Here consumption of drinking water is discussed as route of transmission for human viruses disregarding exposure of the skin, eyes, ears and other body parts. Besides estimation of the concentration of virus in drinking water the assessment of the health risk posed by consumption of drinking water requires estimation of the volume of drinking water ingested and the response of the host to the ingested dose of viruses. Since waterborne viruses are readily inactivated at high temperatures the risk relates to the use of unboiled drinking water, e.g. excluding the consumption of tea but including the water used for washing salad and for making ice cubes.

Global data on the consumption of drinking water are limited especially for the developing regions. In studies carried out in Canada, The Netherlands, the United Kingdom and the USA, the average daily per capita consumption was usually found to be less than 2 l, but there was considerable variation between individuals in different countries and within the same country. The daily median intake of cold tap water in The Netherlands was 0.052 l, in Germany 0.15 l, in the United Kingdom 0.475 l, in Sweden 0.8 l and in Australia 0.5–1 l (Mons et al., 2005;

Westrell et al., 2006). Another study in The Netherlands showed a daily median intake of cold tap water of 0.153 l, threefold higher than the study mentioned above (Teunis et al., 1996). As water intake will vary with climate, physical activity and culture, the above studies, which were conducted in temperate zones, can give only a limited view of consumption patterns throughout the world. At temperatures above 25°C, e.g. there is a sharp rise in fluid intake, largely to meet the demands of an increased sweat rate (Howard and Bartram, 2005). Other factors such as changes in cultural behaviour because of public awareness through information to the public or changes in the drinking-water demand or supply may also play a role in the habits of drinking water consumption. In the WHO GDWQ with respect to microbial hazards, per capita daily consumption of 1 l of unboiled water was assumed.

Dose–response modelling

The number of virus particles necessary to initiate an infection or even disease in the susceptible host after ingestion may be as low as a few particles or even a single particle (Ward et al., 1986; Lindesmith et al., 2003). The host may not experience symptoms though often acute symptoms will occur. These symptoms may be mild, severe and/or life-threatening (see Chapters 2, 3 and 4). A virus infection may become chronic and/or health effects may be delayed, so-called sequelae. Each of these health outcomes may result from infection with one and the same virus variant. For instance, infection with coxsackie virus variant B4 may pass unnoticed, cause gastroenteritis, cause myocarditis or diabetes or lead to infant death. The dose–response relation describes the quantitative relation between the intensity of exposure, i.e. the dose (here pathogenic viruses in drinking water), and the frequency of the occurrence of the adverse health effect within the exposed population of hosts, i.e. the response (here number of cases with specific symptoms in the human population). Available dose–response data have been obtained mainly from clinical studies including healthy adult volunteers but (premature) infants were included in some other studies (Teunis et al., 1996). The dose–response relations have been determined for oral ingestion of some specific variants of waterborne rotavirus and enteroviruses during volunteer studies (Table 5).

As shown in Table 5 the probability of infection after consumption of one infectious virus may range from 0.000714 for poliovirus type 1 LSc2ab to 0.388 for poliovirus type 1 sm. The number of viruses that cause infection in 50% of the human volunteers may vary from 1.411 poliovirus type 1 sm to 69,300 poliovirus type 1 LSc2ab. These findings demonstrate the variability in infectivity not just between virus families but also between virus types and even between virus variants affecting the QMRA. On the other hand, it may reflect differences in exposed populations (from neonates to adults) and ways of exposure (ingestion to bathing). In case of a specific human pathogenic virus variant detected in a known fraction of the drinking water than specific dose–response relation should be used. However, often exposure is determined for a broad class of pathogens, like enterovirus concentrations as determined by use of a specific susceptible cell line (most commonly

Table 5

Infectivity of specific waterborne virus variants

Organisms	Symptom scored	$P^*_{inf}(1.0)$	ID_{50}
Rotavirus[a]	Excretion	2.65×10^{-1}	6.11
Echovirus 12[b]	Excretion	1.76×10^{-3}	1.05×10^3
Poliovirus			
1 sm	Excretion	3.88×10^{-1}	1.411
1 LSc2ab	Excretion	7.14×10^{-4}	6.93×10^4
1	Excretion	9.10×10^{-3}	76.2
3 Fox (infants)	Excretion	1.90×10^{-1}	5.513
3 Fox (premature)	Excretion	2.66×10^{-1}	5.05

Source: From Teunis et al. (1996).
Note: P^*_{inf}, conditional probability of infection; ID_{50}, number of microorganisms required to cause infection in 50% of experimentally infected animals, a measure of infectivity.
[a]Administered in pH-buffered solution.
[b]Rejected at the 95% level, within 99% confidence range for the deviance from maximum possible likelihood.

Buffalo green monkey, BGM). For such a large family of viruses a dose–response relation can be defined, but this should include an additional level of variation: differences in infectivity (and/or pathogenicity) between related virus strains. A hierarchic dose–response model could be applied, accounting for variability both within and between pathogenic virus strains (Teunis, personal communication).

Besides viral characteristics, such as high infectivity and recombination, the host characteristics largely determine the adverse health effects. Genetic predisposition may render an individual susceptible or insusceptible to a specific virus infection as described for HIV (Paxton et al., 1996) and NoV (Lindesmith et al., 2003). Other host susceptibility factors include nutrition, age, pregnancy and mmune status. Some of these factors are preventable such as nutrition whereas genetic background cannot be altered.

Risk characterisation

The concentration of pathogenic viruses in the drinking water (either directly determined or derived from location-specific data on human pathogenic viruses in source water and virus reduction by treatment) in combination with the dose–response relation can be used to assess the health risk associated with the consumption of drinking water. Several QMRA studies were performed to estimate the risk for exposure to drinking water contaminated with adenovirus, rotavirus and coxsackie virus (Regli et al., 1991; Haas et al., 1993; Gerba et al., 1996; Crabtree et al., 1997; Mena et al., 2003; Van Heerden et al., 2005). These studies involve static QMRA models ignoring the dynamics of a viral disease process in the exposed population with respect to the different transmission routes for the same pathogen, immunity, secondary transmission and genetic predisposition. These factors are included in a dynamic QMRA model (Eisenberg et al., 1998) but these have not been applied to date for estimation of the health risk associated with the consumption of virally contaminated drinking water. Anyway, the degree of confidence in the final risk estimate will largely depend on the assumptions, the uncertainties and variability of the data on which the QMRA is built and these need to be determined. Moreover, a level of tolerable risk needs to be set by the risk managers. For human pathogenic viruses to date the most common tolerable risk level was established at less than one infection in a population of 10,000 persons per year but this is to be discussed and decided by all parties involved in drinking water safety. More recently the WHO GDWQ use a reference level of risk of 10^{-6} DALY per person per year.

Risk management

Given the assessment of the risk of viral disease from consumption of drinking water, the risk managers have to select and implement appropriate control measures. These may involve cost-effectiveness reasoning. After implementation of a selected intervention, the effectiveness needs to be evaluated by surveillance of viral disease. In addition, environmental surveillance analysing the numbers of viral pathogens in the drinking water or alternatively in the source waters may be helpful in evaluating source protection interventions. Interventions may vary from additional research in case the risk estimate is unacceptably uncertain. In case of an unacceptable health risk the intervention may concern an additional treatment step in the process specifically designed to reduce human pathogenic viruses from the source water. Alternatively, the intervention may involve enhanced source water protection by barring virus-positive animal hosts or reducing contamination in distribution.

Besides a system assessment, the WSPs of each drinking water supply should include operational monitoring. Control measures need to be identified that control identified risks and ensure health-based targets are met. For each control measure identified, an appropriate means of operational monitoring should be defined that will ensure that any deviation from required performance is detected in time. Operational monitoring normally focuses on simple, cheap and rapid tests and as such microbial testing is rarely applicable.

For purposes of verification as opposed to operational monitoring, E. coli bacteria are measured. In case of the presence of this faecal indicator, the drinking water could be contaminated with human pathogenic viruses. The appropriate management action could, for instance, be a boiling water advisory. It would be inappropriate to perform retesting of the same sample or resampling because the basis for E. coli testing as a verification measure is the good performance of the test as laid down in standard operating procedures and its use as indicator. Retesting and resampling would take even more time and this should be spent on finding the fault or source of contamination to be able to protect public health. Again end product monitoring is useful in this way but system assessment and operational monitoring can elucidate the actual ability of the drinking water system to provide safe drinking water also with respect to human pathogenic viruses.

Risk communication

Communication of the viral risk estimate and the proposed risk management measures contributes to the improvement of the drinking water safety with respect to human pathogenic viruses. Dissemination involves not only risk managers (policy makers and inspectors) and risk assessors but consumers and suppliers of drinking water alike. For this purpose tools such as brochures and websites should be attuned and transparent. In The Netherlands, a webtool for the assessment of risk by non-experts was launched simultaneously with the Inspectorate guideline for analysis of microbial safety of drinking water (De Roda Husman and Medema, 2006) to be used by Dutch policy makers and inspectors and will be released for use by other interested parties.

Concluding remarks and recommendations

The global supply of safe drinking water should be considered with an open mind to the possible viral threats. The potential transmission of (re-)emerging human pathogenic viruses should be assessed based on research. For instance, foot-and-mouth disease viruses were considered to be solely respiratory pathogens and therefore their potential as possible waterborne pathogens was disregarded. Not that these viruses display a major threat to the drinking water supply or human health (Schijven et al., 2005) but animal health is endangered and at least the risk should be assessed instead of hypothesised. Recently this has received a great deal of attention for the emerging disease SARS: the associated coronavirus was evaluated with respect to possible shedding and transmission routes including water. The same is true for avian influenza viruses which occasionally infect humans from their animal reservoirs mostly by close contact; the potential risk of waterborne transmission was estimated (WHO, 2006b). Theoretically, viruses could be considered as biocolloids with specific properties such as size, shape, structure, charge, composition, genome, etc. These viral characteristics determine their behaviour in the environment, resistance to natural inactivation and treatment and disinfection processes. For each (re-)emerging virus these properties may be known or could be assessed predicting the effectiveness of possible intervention measures for prevention of waterborne disease. In this respect, free data sharing among researchers and risk managers in times of emergencies and disasters as encouraged by WHO is one of the keys to successful reduction of the environmental spread of pathogenic viruses and reduced burden of water-transmitted viral disease.

References

Abbaszadegan M, Stewart P, LeChevallier M. A strategy for detection of viruses in groundwater by PCR. Appl Environ Microbiol 1999; 65: 444–449.

Ali MA, Al-Herrawy AZ, El-Hawaary SE. Detection of enteric viruses, Giardia and Cryptosporidium in two different types of drinking water treatment facilities. Water Res 2004; 38: 3931–3939.

Amvrosieva TV, Paklonskaya NV, Biazruchka AA, Kazinetz ON, Bohush ZF, Fisenko EG. Enteroviral infection outbreak in the Republic of Belarus: principal characteristics and phylogenetic analysis of etiological agents. Cent Eur J Publ Health 2006; 14(2): 67–73.

Ashbolt NJ. Microbial contamination of drinking water and disease outcomes in developing regions. Toxicology 2004; 198(1–3): 229–238.

Banyai K, Jakab F, Reuter G, Bene J, Uj M, Melegh B, Szucs G. Sequence heterogeneity among human picobirnaviruses detected in gastroenteritis outbreak. Arch Virol 2003; 148: 2281–2291.

Beld M, Sentjens R, Rebers S, Weel J, Wertheim-van Dillen P, Sol C, Boom R. Detection and quantitation of hepatitis C virus RNA in feces of chronically infected individuals. J Clin Microbiol 2000; 38(9): 3442–3444.

Berger JR, Miller CS, Mootoor Y, Avdiushko SA, Kryscio RJ, Zhu H. JC virus detection in bodily fluids: clues to transmission. Clin Infect Dis 2006; 43: e9–e12.

Borchardt MA. Incidence of enteric viruses in groundwater from household wells in Wisconsin. Appl Environ Microbiol 2003; 69(2): 1172–1180.

Borchardt MA, Haas NL, Hunt RJ. Vulnerability of drinking-water wells in la Crosse, Wisconsin, to enteric-virus contamination from surface water contributions. Appl Environ Microbiol 2004; 70: 5937–5946.

Bosch A. Human enteric viruses in the water environment: a minireview. Int Microbiol 1998; 1: 191–196.

Bowdre JH. Viral gastroenteritis and laboratory detection of rotavirus. Am J Med Technol 1983; 49(9): 665–668.

Bradley D. Health aspects of water supplies in tropical countries. In: Water, Wastes and Health in Hot Climates (Feachem R, McGarry M, Mara D, editors). Chichester: Wiley; 1977; pp. 3–17.

Carter MJ. Enterically infecting viruses: pathogenicity, transmission and significance for food and waterborne infection. J Appl Microbiol 2005; 98(6): 1354–1380.

Cascio A, Bosco M, Vizzi E, Giammanco A, Ferraro D, Arista S. Identification of picobirnavirus from faeces of Italian children suffering from acute diarrhea. Eur J Epidemiol 1996; 12(5): 545–547.

Clemente-Casares P. Hepatitis E virus epidemiology in industrialized countries. Emerg Infect Dis 2003; 9: 448–454.

Crabtree KD, Gerba CP, Rose JB, Haas CN. Waterborne adenovirus: a risk assessment. Water Sci Technol 1997; 35: 1–6.

De Deus N, Seminati C, Pina S, Mateu E, Martin M, Segales J. Detection of hepatitis E virus in liver, mesenteric lymph node, serum, bile, and faeces of naturally infected pigs affected by different pathological conditions. Vet Microbiol 2007; 119(2–4): 105–114.

Denis-Mize K, Fout GS, Dahling DR, Francy DS. Detection of human enteric viruses in stream water with RT-PCR and cell culture. J Water Health 2004; 2(1): 37–47.

De Roda Husman AM, Medema GJ. Dutch Environmental Inspectorate guideline 'Assessment of the microbial safety of drinking water'. Artikelcode 2006; 5318.

Ding Y, He L, Zhang Q, Huang Z, Che X, Hou J, Wang H, Shen H, Qiu L, Li Z, Geng J, Cai J, Han H, Li X, Kang W, Weng D, Liang P, Jiang S. Organ distribution of severe acute respiratory syndrome (SARS) associated coronavirus (SARS-CoV) in SARS patients: implications for pathogenesis and virus transmission pathways. J Pathol 2004; 203(2): 622–630.

Divizia M, Gabrieli R, Donia D, Macaluso A, Bosch A, Guix S, Sanchez G, Villena C, Pinto RM, Palombi L, Buonuomo E, Cenko F, Leno L, Bebeci D, Bino S. Waterborne gastroenteritis outbreak in Albania. Water Sci Technol 2004; 50: 57–61.

Duan SM, Zhao XS, Wen RF, Huang JJ, Pi GH, Zhang SX, Han J, Bi SL, Ruan L, Dong XP, SARS Research Team. Stability of SARS coronavirus in human specimens and environment and its sensitivity to heating and UV irradiation. Biomed Environ Sci 2003; 16(3): 246–255.

Duizer E, Schwab KJ, Neill FH, Atmar RL, Koopmans MP, Estes MK. Laboratory efforts to cultivate noroviruses. J Gen Virol 2004; 85: 79–87.

Duizer E, Koopmans M. Tracking foodborne viruses: Lessons from noroviruses. In: Emerging foodborne pathogens (Motarjemi A, editor). Boca Raton, USA: CRC Press; 2006; pp. 77–110.

Eisenberg JN, Seto EY, Colford Jr JM, Olivieri A, Spear RC. An analysis of the Milwaukee Cryptosporidiosis outbreak based on a dynamic model of the infection process. Epidemiology 1998; 9(3): 255–263.

Ferguson C, de Roda Husman AM, Altavilla N, Deere D, Ashbolt N. Fate and transport of surface water pathogens in watersheds. Crit Rev Environ Sci Technol 2003; 33(3): 299–361.

Fields BN, Knipe DM, Howley PM, editors. Fields Virology. 4th ed. Philadelphia, PA: Lippincott Williams and Wilkins; 2002.

Fleischer J, Schlafmann K, Otchwemah R, Botzenhart K. Elimination of enteroviruses, other enteric viruses, F-specific coliphages, somatic coliphages and E. coli in four sewage treatment plants of Southern Germany. J Water Supply Res Technol 2000; 49: 127–137.

Fong TT, Lipp EK. Enteric viruses of humans and animals in aquatic environments: health risks, detection, and potential water quality assessment tools. Microbiol Mol Biol Rev 2005; 69(2): 357–371.

Fout GS, Martinson BC, Moyer MW, Dahling DR. A multiplex reverse-transcription-PCR method for detection of human enteric viruses in groundwater. Appl Environ Microbiol 2003; 69(6): 3158–3164.

Fuhrman JA, Liang X, Noble RT. Rapid detection of enteroviruses in small volumes of natural waters by real-time quantitative reverse transcriptase PCR. Appl Environ Microbiol 2005; 71(8): 4523–4530.

Gehr R, Wagner M, Veerasubramanian P, Payment P. Disinfection efficiency of peracetic acid, UV and ozone after enhanced primary treatment of municipal wastewater. Water Res 2003; 37: 4573–4586.

Gerba CP, Rose JB, Haas CN, Crabtree KD. Waterborne rotavirus: a risk assessment. Water Res 1996; 30: 2929–2940.

Gerba CP, Riley KR, Nwachuku N, Ryu H, Abbaszadegan M. Removal of *Encephalitozoon* intestinalis, calicivirus, and coliphages by conventional drinking water treatment. J Environ Sci Health A Tox Hazard Subst Environ Eng 2003; 38(7): 1259–1268.

Gilgen M, Germann D, Luthy J, Hubner P. Three-step isolation method for sensitive detection of enterovirus, rotavirus, hepatitis A virus, and small round structured viruses in water samples. Int J Food Microbiol 1997; 37: 189–199.

Giordano MO, Martinez LC, Isa MB, Ferreyra LJ, Canna F, Pavan JV, Paez M, Notario R, Nates SV. Twenty year study of the occurrence of reovirus infection in hospitalized children with acute gastroenteritis in Argentina. Pediatr Infect Dis J 2002; 21(9): 880–882.

Glass RI, Bresee J, Jiang B, Gentsch J, Ando T, Fankhauser R, Noel J, Parashar U, Rosen B, Monroe SS. Gastroenteritis viruses: an overview. Novartis Foundation Symposium; 2001; 238; pp. 5–19. Discussion; pp. 19–25.

Godfree A, Farrell J. Processes for managing pathogens. J Environ Qual 2005; 34(1): 105–113.

Gofti-Laroche L, Gratacap-Cavallier B, Demanse D, Genoulaz O, Seigneurin JM, Zmirou D. Are waterborne astrovirus implicated in acute digestive morbidity (E.M.I.R.A. study)? J Clin Virol 2003; 27: 74–82.

Grabow WO, Clay CG, Dhaliwal W, Vrey MA, Muller EE. Elimination of viruses, phages, bacteria and cryptosporidium by a new generation Aquaguard point-of-use water treatment unit. Zentralbl Hyg Umweltmed 1999; 202(5): 399–410.

Guix S, Bosch A, Pinto RM. Human astrovirus diagnosis and typing: current and future prospects. Lett Appl Microbiol 2005; 41(2): 103–105.

Haas CN, Rose JB, Gerba C, Regli S. Risk assessment of virus in drinking water. Risk Anal 1993; 13(5): 545–552.

Haramoto E, Katayama H, Ohgaki S. Detection of noroviruses in tap water in Japan by means of a new method for concentrating enteric viruses in large volumes of freshwater. Appl Environ Microbiol 2004; 70(4): 2154–2160.

Haramoto E, Katayama H, Oguma K, Ohgaki S. Application of cation-coated filter method to detection of noroviruses, enteroviruses, adenoviruses and torque tenoviruses in the Tamagawa river in Japan. Appl Environ Microbiol 2005; 71(5): 2403–2411.

Hatakeyama N, Suzuki N, Yamamoto M, Kuroiwa Y, Hori T, Mizue N, Tsutsumi H. Detection of BK virus and adenovirus in the urine from children after allogeneic stem cell transplantation. Pediatr Infect Dis J 2006; 25(1): 84–85.

Hauri AM, Schimmelpfennig M, Walter-Domes M, Letz A, Diedrich S, Lopez-Pila J, Schreier E. An outbreak of viral meningitis associated with a public swimming pond. Epidemiol Infect 2005; 133(2): 291–298.

Havelaar AH. Application of HACCP to drinking-water supply. Food Control 1994; 5: 145–152.

Havelaar AH, Melse JM. Quantifying public health risks in the WHO Guidelines for Drinking-Water Quality: a burden of disease approach. Report 734301022/2003; Bilthoven, the Netherlands: RIVM; 2003.

Herrmann JE, Nowak NA, Blacklow NR. Detection of Norwalk virus in stools by enzyme immunoassay. J Med Virol 1985; 17(2): 127–133.

Hill VR, Polaczyk AL, Hahn D, Narayanan J, Cromeans TL, Roberts JM, Amburgey JE. Development of a rapid method for simultaneous recovery of diverse microbes in drinking water by ultrafiltration with sodium polyphosphate and surfactants. Appl Environ Microbiol 2005; 71(11): 6878–6884.

Hoebe CJ, Vennema H, de Roda Husman AM, van Duynhoven YT. Norovirus outbreak among primary schoolchildren who had played in a recreational water fountain. J Infect Dis 2004; 189(4): 699–705.

Hörman A, Rimhanen-Finne R, Maunula L, von Bonsdorff CH, Torvela N, Heikinheimo A, Hanninen ML. Campylobacter spp., Giardia spp., Cryptosporidium spp., noroviruses, and indicator organisms in surface water in southwestern Finland, 2000–2001. Appl Environ Microbiol 2004; 70: 87–95.

Hot D, Legeay O, Jacques J, Gantzer C, Caudrelier Y, Guyard K, Lange M, Andreoletti L. Detection of somatic phages, infectious enteroviruses and enterovirus genomes as indicators of human enteric viral pollution in surface water. Water Res 2003; 37: 4703–4710.

Howard G, Bartram J. Effective water supply surveillance in urban areas of developing countries. J Water Health 2005; 3(1): 31–43.

Johansson PJ, Sveger T, Ahlfors K, Ekstrand J, Svensson L. Reovirus type 1 associated with meningitis. Scand J Infect Dis 1996; 28(2): 117–120.

Jardine C, Hrudey S, Shortreed J, Craig L, Krewski D, Furgal C, McColl S. Risk management frameworks for human health and environmental risks. J Toxicol Environ Health B Crit Rev 2003; 6(6): 569–720.

John DE, Rose JB. Review of factors affecting microbial survival in groundwater. Environ Sci Technol 2005; 39(19): 7345–7356.

Kaye SB, Lloyd M, Williams H, Yuen C, Scott JA, O'Donnell N, Batterbury M, Hiscott P, Hart CA. Evidence for persistence of adenovirus in the tear film a decade following conjunctivitis. J Med Virol 2005; 77(2): 227–231.

Keswick BH, Gerba CP, DuPont HL, Rose JB. Detection of enteric viruses in treated drinking water. Appl Environ Microbiol 1984; 47(6): 1290–1294.

Khalili K, Gordon J, White MK. The polyomavirus, JCV and its involvement in human disease. Adv Exp Med Biol 2006; 577: 274–287.

Kim SH, Cheon DS, Kim JH, Lee DH, Jheong WH, Heo YJ, Chung HM, Jee Y, Lee JS. Outbreaks of gastroenteritis that occurred during school excursions in Korea were associated with several waterborne strains of norovirus. J Clin Microbiol 2005; 43(9): 4836–4839.

Kittigul L. Detection and characterization of hepatitis A virus in water samples in Thailand. J Appl Microbiol 2006; 100(6): 1318–1323.

Ksiazek TG, Erdman D, Goldsmith CS, Zaki SR, Peret T, Emery S, Tong S, Urbani C, Comer JA, Lim W, Rollin PE, Dowell SF, Ling AE, Humphrey CD, Shieh WJ, Guarner J, Paddock CD, Rota P, Fields B, DeRisi J, Yang JY, Cox N, Hughes JM, LeDuc JW, Bellini WJ, Anderson LJ, SARS Working Group. A novel coronavirus associated with Severe acute respiratory syndrome. New Engl J Med 2003; 348(20): 1953–1966.

Kukkula M, Maunula L, Silvennoinen E, von Bonsdorff CH. Outbreak of viral gastroenteritis due to drinking water contaminated by Norwalk-like viruses. J Infect Dis 1999; 180: 1771–1776.

Lamothe GT, Putallaz T, Joosten H, Marugg JD. Reverse transcription-PCR analysis of bottled and natural mineral waters for the presence of noroviruses. Appl Environ Microbiol 2003; 69(11): 6541–6549.

Laverick MA, Wyn-Jones AP, Carter MJ. Quantitative RT-PCR for the enumeration of noroviruses (Norwalk-like viruses) in water and sewage. Lett Appl Microbiol 2004; 39: 127–136.

Leclerc H, Edberg S, Pierzo V, Delattre JM. Bacteriophages as indicators of enteric viruses and public health risk in groundwaters. J Appl Microbiol 2000; 88(1): 5–21.

Lee HK, Jeong YS. Comparison of total culturable virus assay and multiplex integrated cell culture-PCR for reliability of waterborne virus detection. Appl Environ Microbiol 2004; 70: 3626–3632.

Lee SH, Kim SJ. Detection of infectious enteroviruses and adenoviruses in tap water in urban areas in Korea. Water Res 2002; 36: 248–256.

Li TC, Chijiwa K, Sera N, Ishibashi T, Etoh Y, Shinohara Y, Kurata Y, Ishida M, Sakamoto S, Takeda N, Miyamura T. Hepatitis E virus transmission from wild boar meat. Emerg Infect Dis 2005; 11(12): 1958–1960.

Lindesmith L, Moe C, Marionneau S, Ruvoen N, Jiang X, Lindblad L, Stewart P, LePendu J, Baric R. Human susceptibility and resistance to Norwalk virus infection. Nat Med 2003; 9: 548–553.

Lodder WJ, de Roda Husman AM. Presence of noroviruses and other enteric viruses in sewage and surface waters in the Netherlands. Appl Environ Microbiol 2005; 71(3): 1453–1461.

Lu L, Li C, Hagedorn CH. Phylogenetic analysis of global hepatitis E virus sequences: genetic diversity, subtypes and zoonosis. Rev Med Virol 2006; 16(1): 5–36.

Lucena F, Duran AE, Moron A, Calderon E, Campos C, Gantzer C, Skraber S, Jofre J. Reduction of bacterial indicators and bacteriophages infecting faecal bacteria in primary and secondary wastewater treatments. JAppl Microbiol 2004; 97: 1069–1076.

Luo KX, Zhang L, Wang SS, Nie J, Yang SC, Liu DX, Liang WF, He HT, Lu Q. An outbreak of enterically transmitted non-A, non-E viral hepatitis. J Viral Hepat 1999; 6(1): 59–64.

Lytle CD, Sagrapanti JL. Predicted inactivation of viruses of relevance to biodefense by solar radiation. J Virol 2005; 79(22): 14244–14252.

Mackiewicz V, Dussaix E, Le Petitcorps MF, Roque-Afonso AM. Detection of hepatitis A virus RNA in saliva. J Clin Microbiol 2004; 42(9): 4329–4331.

Matsubara H, Michitaka K, Horiike N, Yano M, Akbar SM, Torisu M, Onji M. Existence of TT virus DNA in extracellular body fluids from normal healthy Japanese subjects. Intervirology 2000; 43(1): 16–19.

Matsuura K, Ishikura M, Nakayama T, Hasegawa S, Morita O, Katori K, Uetake H. Ecological studies on reovirus pollution of rivers in Toyama Prefecture. II. Molecular epidemiological study of reoviruses isolated from river water. Microbiol Immunol 1993; 37: 305–310.

McKinney KR, Gong YY, Lewis TG. Environmental transmission of SARS at Amoy gardens. J Environ Health 2006; 68(9): 26–30.

Mena KD, Gerba CP, Haas CN, Rose JB. Risk assessment of waterborne coxsackie virus. J Am Water Works Assoc 2003; 95: 122–131.

Mons M, Blokker M, van der Wielen J, Medema G, Sinclair M, Hulshof K, Dangendorff F, Hunter P. Estimation of the consumption of cold tap water for microbiological risk assessment. EU project MICRORISK (contract EVK1-CT-2002-00123); 2005. http://217.77.141.80/clueadeau/microrisk/uploads/finalreportwp5.pdf, retrieved in November, 2006.

Moore BE. Survival of Human Immunodeficiency Virus (HIV), HIV-Infected lymphocytes, and poliovirus in water. Appl Environ Microbiol 1993; 59(5): 1437–1443.

Muir P, Nicholson F, Jhetam M, Neogi S, Banatvala JE. Rapid diagnosis of enterovirus infection by magnetic bead extraction and polymerase chain reaction detection of enterovirus RNA in clinical specimens. J Clin Microbiol 1993; 31(1): 31–38.

Myrmel M, Berg EM, Grinde B, Rimstad E. Enteric viruses in inlet and outlet samples from sewage treatment plants. J Water Health 2006; 4(2): 197–209.

Olszewski J, Winona L, Oshima KH. Comparison of 2 ultrafiltration systems for the concentration of seeded viruses from environmental waters. Can J Microbiol 2005; 51(4): 295–303.

Ortiz-Movilla N, Lazaro P, Rodriguez-Inigo E, Bartolome J, Longo I, Lecona M, Pardo M, Carreno V. Hepatitis C virus replicates in sweat glands and is released into sweat in patients with chronic hepatitis C. J Med Virol 2002; 68(4): 529–536.

Parshionikar SU. Waterborne outbreak of gastroenteritis associated with a norovirus. Appl Environ Microbiol 2003; 69(9): 5263–5268.

Paxton WA, Martin SR, Tse D, O'Brien TR, Skurnick J, VanDevanter NL, Padian N, Braun JF, Kotler DP, Wolinsky SM, Koup RA. Relative resistance to HIV-1 infection of CD4 lymphocytes from persons who remain uninfected despite multiple high-risk sexual exposure. Nat Med 1996; 2(4): 412–417.

Payment P. Fate of human enteric viruses, coliphages, and Clostridium perfringens during drinking-water treatment. Can J Microbiol 1991; 37(2): 154–157.

Payment P, Franco E. Clostridium perfringens and somatic coliphages as indicators of the efficiency of drinking water treatment for viruses and protozoan cysts. Appl Environ Microbiol 1993; 59(8): 2418–2424.

Payment P, Trudel M, Plante R. Elimination of viruses and indicator bacteria at each step of treatment during preparation of drinking water at seven water treatment plants. Appl Environ Microbiol 1985; 49(6): 1418–1428.

Phan TG, Trinh QD, Yagyu F, Sugita K, Okitsu S, Muller WE, Ushijima H. Outbreak of sapovirus infection among infants and children with acute gastroenteritis in Osaka City, Japan during 2004–2005. J Med Virol 2006; 78(6): 839–846.

Polaczyk AL, Roberts JM, Hill VR. Evaluation of 1MDS electropositive microfilters for simultaneous recovery of multiple microbe classes from tap water. J Microbiol Meth 2006 doi:10.1016/j.mimet.2006.08.007.

Poullis DA, Attwell RW, Powell SC. The characterization of waterborne-disease outbreaks. Rev Environ Health 2005; 20(2): 141–149.

Rab MA, Bile MK, Mubarik MM, Asghar H, Sami Z, Siddiqi S, Dil AS, Barzgar MA, Chaudhry MA, Burney MI. Water-borne hepatitis E virus epidemic in Islamabad, Pakistan: a common source outbreak traced to the malfunction of a modern water treatment plant. Am J Trop Med Hyg 1997; 57(2): 151–157.

Ramsay M, Reacher M, O'Flynn C, Buttery R, Hadden F, Cohen B, Knowles W, Wreghitt T, Brown D. Causes of morbiliform rash in a highly immunised English population. Arch Dis Child 2002; 87: 202–206.

Regli S, Rose JB, Haas CN, Gerba CP. Modelling the risk from Giardia and viruses in drinking water. J Am Water Works Assoc 1991; 83: 76–84.

Reuther CG. Saline solutions: the quest for fresh water. Environ Health Perspect 2000; 108(2): A78–A80.

Schernewski G, Julich WD. Risk assessment of virus infections in the Oder estuary (southern Baltic) on the basis of spatial transport and virus decay simulations. Int J Hyg Environ Health 2001; 203: 317–325.

Schijven JF, Rijs GB, de Roda Husman AM. Quantitative risk assessment of FMD virus transmission via water. Risk Anal 2005; 25(1): 13–21.

Schuster CJ, Ellis AG, Robertson WJ, Charron DF, Aramini JJ, Marshall BJ, Medeiros DT. Infectious disease outbreaks related to drinking water in Canada, 1974–2001. Can J Pub Health 2005; 96(4): 254–258.

Schvoerer E, Bonnet F, Dubois V, Cazaux G, Serceau R, Fleury HJ, Lafon ME. PCR detection of human enteric viruses in bathing areas, waste waters and human stools in Southwestern France. Res Microbiol 2000; 151: 693–701.

Schvoerer E, Ventura M, Dubos O, Cazaux G, Serceau R, Gournier N, Dubois V, Caminade P, Fleury HJ, Lafon ME. Qualitative and quantitative molecular detection of enteroviruses in water from bathing areas and from a sewage treatment plant. Res Microbiol 2001; 152: 179–186.

Sedmak G, Bina D, Macdonald J, Couillard L. Nine-year study of the occurrence of culturable viruses in source water for two drinking water treatment plants and the influent and effluent of a Wastewater Treatment Plant in Milwaukee, Wisconsin (August 1994 through July 2003). Appl Environ Microbiol 2005; 71(2): 1042–1050.

Senouci S, Maul A, Schwartzbrod L. Comparison study on three protocols used to concentrate poliovirus type 1 from drinking water. Zentralbl Hyg Umweltmed 1996; 198(4): 307–317.

Shuval H. Estimating the global burden of thalassogenic diseases: human infectious diseases caused by wastewater pollution of the marine environment. J Water Health 2003; 1(2): 53–64.

Singh S, Mohanty A, Joshi YK, Deka D, Mohanty S, Panda SK. Mother-to-child transmission of hepatitis E virus infection. Indian J Pediatr 2003; 70: 37–39

Singh V, Singh V, Raje M, Nain CK, Singh K. Routes of transmission in the hepatitis E epidemic of Saharanpur. Trop Gastroenterol 1998; 19(3): 107–109.

Skraber S, Schijven JF, Gantzer C, de Roda Husman AM. Pathogenic viruses in drinking water biofilms: a public health risk? Biofilms 2005; 2: 1–13.

Somani SK, Aggarwal R, Naik SR, Srivastava S, Naik S. A serological study of intrafamilial spread from patients with sporadic hepatitis E virus infection. J Viral Hepat 2003; 10(6): 446–449.

Tarantola A, Abiteboul D, Rachline A. Infection risks following accidental exposure to blood or body fluids in health care workers: a review of pathogens transmitted in published cases. Am J Infect Control 2006; 34(6): 367–375.

Tavares FN, Costa EV, Oliveira SS, Nicolai CC, Baran M, da Silva EE. Acute hemorrhagic conjunctivitis and coxsackie virus A24v, Rio de Janeiro, Brazil, 2004. 1. Emerg Infect Dis 2006; 12(3): 495–497.

Teunis PFM, Heijden OG van der, Giessen JWB van der, Havelaar AH. The dose-response relation in human volunteers for gastro-intestinal pathogens. RIVM Report 284550002 1996. Freely available from www.rivm.nl.

Teunis PFM, Lodder WJ, Heisterkamp SH, de Roda Husman AM. Mixed plaques: statistical evidence how plaque assays may underestimate virus concentrations. Water Res 2005; 39: 4240–4250.

Vaidya SR, Tilekar BN, Walimbe AM, Arankalle VA. Increased risk of hepatitis E in sewage workers from India. J Occup Environ Med 2003; 45(11): 1167–1170.

Van den Berg H, Lodder W, van der Poel W, Vennema H, de Roda Husman AM. Genetic diversity of noroviruses in raw and treated sewage water. Res Microbiol 2005; 156: 532–540.

Van Heerden J, Ehlers MM, Van Zyl WB, Grabow WO. Incidence of adenoviruses in raw and treated water. Water Res 2003; 37: 3704–3708.

Van Heerden J, Ehlers MM, Vivier JC, Grabow WO. Risk assessment of adenoviruses detected in treated drinking water and recreational water. J Appl Microbiol 2005; 99(4): 926–933.

Van Olphen M, Kapsenberg JG, van de Baan E, Kroon WA. Removal of enteric viruses from surface water at eight waterworks in the Netherlands. Appl Environ Microbiol 1984; 47: 927–932.

Van Zyl WB, Williams PJ, Grabow WO, Taylor MB. Application of a molecular method for the detection of group A rotaviruses in raw and treated water. Water Sci Technol 2004; 50(1): 223–228.

Villar LM, de Paula VS, Diniz-Mendes L, Lampe E, Gaspar AM. Evaluation of methods used to concentrate and detect hepatitis A virus in water samples. J Virol Meth 2006; 137(2): 169–176.

Villena C, El-Senousy WM, Abad FX, Pinto RM, Bosch A. Group A rotavirus in sewage samples from Barcelona and Cairo: emergence of unusual genotypes. Appl Environ Microbiol 2003; 69: 3919–3923.

Vivier JC, Ehlers MM, Grabow WO. Detection of enteroviruses in treated drinking water. Water Res 2004; 38: 2699–2705.

Wang QH, Souza M, Funk JA, Zhang W, Saif LJ. Prevalence of noroviruses and sapoviruses in swine of various ages determined by reverse transcription-PCR and microwell hybridization assays. J Clin Microbiol 2006; 44(6): 2057–2062.

Wang WK, Chen SY, Liu IJ, Chen YC, Chen HL, Yang CF, Chen PJ, Yeh SH, Kao CL, Huang LM, Hsueh PR, Wang JT, Sheng WH, Fang CT, Hung CC, Hsieh SM, Su CP, Chiang WC, Yang JY, Lin JH, Hsieh SC, Hu HP, Chiang YP, Wang JT, Yang PC, Chang SC, SARS Research Group of the National Taiwan University/National Taiwan University Hospital. Detection of SARS-associated coronavirus in throat wash and saliva in early diagnosis. Emerg Infect Dis 2004; 10(7): 1213–1219.

Ward RL, Bernstein DI, Young EC, Sherwood JR, Knowlton DR, Schiff GM. Human rotavirus studies in volunteers: determination of infectious dose and serological response to infection. J Infect Dis 1986; 154(5): 871–880.

Westrell T, Andersson Y, Stenström TA. Drinking water consumption patterns in Sweden. J Water Health 2006; 04: 511–522.

Wetz JJ, Lipp EK, Griffin DW, Lukasik J, Wait D, Sobsey MD, Scott TM, Rose JB. Presence, infectivity, and stability of enteric viruses in seawater: relationship to marine water quality in the Florida Keys. Mar Pollut Bull 2004; 48(7–8): 698–704.

WHO (World Health Organization). Guidelines for drinking water quality. Geneva: World Health Organization; 2004. http://www.who.int/water_sanitation_health/dwq

WHO (World Health Organization). Guidelines for the safe use of wastewater, excreta and greywater. Geneva: World Health Organization; 2006a. ISBN, 92 4 154686 7.

WHO (World Health Organization). Review of latest available evidence on risks to human health through potential transmission of avian influenza (H5N1) through water and sewage. 2006b. Freely available from http://www.who.int/water_sanitation_health/emerging/h5n1background.pdf.

Witso E, Palacios G, Cinek O, Stene LC, Grinde B, Janowitz D, Lipkin WI, Ronningen KS. Natural circulation of human enteroviruses: high prevalence of human enterovirus A infections. J Clin Microbiol 2006; 44(11): 4095–4100.

Worm HC, van der Poel WH, Brandstatter G. Hepatitis E: an overview. Microb Infect 2002; 4(6): 657–666.

Yamashita T, Sugiyama M, Tsuzuki H, Sakae K, Suzuki Y, Miyazaki Y. Application of a reverse-transcription-PCR for identification and differentiation of Aichi virus, a new member of the picornavirus family associated with gastroenteritis in humans. J Clin Microbiol 2000; 38(8): 2955–2961.

Zhuang J, Jin Y. Virus retention and transport through Al-oxide coated sand columns: Effects of ionic strength and composition. J Contam Hydrol 2003; 60(3–4): 193–209.

Human Viruses in Water
Albert Bosch (Editor)
© 2007 Elsevier B.V. All rights reserved
DOI 10.1016/S0168-7069(07)17008-7

Chapter 8

Waterborne Viruses: Assessing the Risks

Kristina D. Mena
*University of Texas Health Science Center at Houston, School of Public Health,
1100 N. Stanton Street, Suite 110, El Paso, TX 79902, USA*

Introduction

The diversity and frequency of human health consequences associated with exposures to hazards in the environment is traditionally documented during outbreaks of disease resulting from human exposure to an extensive amount of a hazard. During such situations, susceptible populations and resulting diseases may be identified as well as particular places or media associated with transmission. In lieu of an outbreak investigation, a risk assessment approach may be undertaken to address a hazard or a group of hazards to identify individuals most vulnerable, the range of resulting human health outcomes, and likely places and routes of exposure. Human viruses have been well documented as waterborne pathogens and waterborne transmission of disease continues to occur in the United States despite the application of water treatment technologies. The human health impact associated with exposure to these viruses can range from asymptomatic infections to gastroenteritis and other acute health outcomes, to even more serious health consequences, such as myocarditis, hepatitis, and aseptic meningitis. Economic costs can range widely depending on length and severity of the resulting condition (Crabtree, 1996).

Risk assessment is the first component of the risk analysis process, followed by risk management and risk communication (NRC, 1984). Results of risk assessments may then be used to guide policy makers in allocating resources for mitigation strategies or additional research, as well as direct the scientific information released to the public regarding a hazard. A four-step risk assessment framework was initially developed by the National Research Council (NRC) to address environmental exposures to chemicals (NRC, 1983), which includes hazard identification, dose-response assessment, exposure assessment, and risk characterization. Assessments may be qualitative or quantitative depending upon the information available, with the latter the more ideal approach. This framework has since been applied worldwide to address the human health impact from exposure to microorganisms in the environment, including waterborne viruses (Gerba and Haas, 1988; Regli et al., 1991; Gerba and Rose, 1993; Haas et al., 1993; Gerba et al., 1996; Crabtree et al., 1997; Mena, 2002; Haas et al., 1999; Mena et al., 2003; van Heerden et al., 2005a and 2005b). In addition, the framework has been modified to address issues specific to microorganisms such as survivability of a virus in the environment and virus susceptibility to treatment and—more recently—the impact of immunity and multiple exposures. The following sections will discuss each phase of the risk assessment approach in addressing the human health impact from exposure to waterborne viruses as well as data needs and risk management implications.

Quantitative microbial risk assessment (QMRA)

The quantitative microbial risk assessment (QMRA) process has been applied,.in some form, to address the human health impacts of waterborne hazards (including viruses) for nearly 20 years. Since its initial inception (NRC, 1983) and application, the process has been reviewed by a variety of experts to address its utility and appropriateness in evaluating microbial hazards and to suggest modifications (ILSI, 1996; ILSI, 2000). The ILSI (1996) document presents the risk assessment process as a three-component framework—problem formulation, analysis (including characterization of exposure and human health effects), and risk characterization—simulating the traditional approach used for ecological risk assessments. A subsequent review of this framework's usefulness and applicability for addressing both waterborne and food-borne microbial hazards took place in 1999 by a team of experts and suggestions for modification of the framework were documented (ILSI, 2000). Nevertheless, whether following the original four-step NRC paradigm or a subsequent modification, pertinent issues when addressing microbial hazards—including viruses—emerge and the difference in the various risk assessment processes is in the packaging and therefore potential understandability by the interpreter. With any of the structured frameworks, an emphasis is on the dynamic (iterative) nature of the process to allow for flexibility in applying risk assessment to diverse settings. For the purpose of this chapter, the following subsections will resemble the NRC framework (1983) yet will capture the issues important for risk assessments of waterborne viruses conducted using any venue.

Waterborne viruses as human health threats: identifying the hazard

Viruses as causative agents of waterborne disease are well documented worldwide. It is believed, however, that the true disease burden placed on a society by

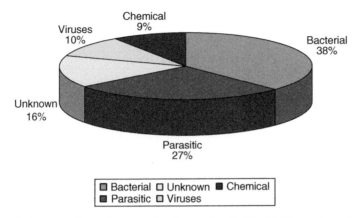

Fig. 1 Etiological agents of waterborne outbreaks associated with drinking water and recreational water in the United States, 2001–2002 (CDC, 2004).

waterborne viruses is underestimated due to the underreporting of their occurrence and the challenges with identifying specific etiological agents during epidemiological investigations of disease. Fig. 1 shows the breakdown of causative agents of waterborne outbreaks in the United States for 2001–2002 for drinking water and recreational water combined with viruses reported to be the responsible agent for approximately 10% of the cases (Centers for Disease Control and Prevention, 2004). In nearly 16% of the documented waterborne outbreaks, the causative agent was never identified. It is speculated that many of these may be due to viruses (particularly norovirus) based on the human health outcomes observed, therefore emphasizing the underestimation of the public health impact of viruses as waterborne agents. In addition, infectious disease within communities due to any waterborne pathogen is often not documented—both in the United States and worldwide—further contributing to this underestimation. Recent attempts have been undertaken to theoretically estimate the incidence of endemic waterborne gastroenteritis that may be caused by viruses or other waterborne pathogens (Colford et al., 2006; Messner et al., 2006).

A component of conducting the hazard identification in a QMRA is identifying the range of potential human health outcomes associated with exposure. Table 1 lists some waterborne viruses and associated human health effects. Infection does not necessarily mean that clinical illness will result. The occurrence of clinical illness may range from less than 1% (for poliovirus) (Evans, 1982) to greater than 95% (for hepatitis A virus in adults and coxsackievirus type B) (Mena, 2002). When clinical symptoms do occur, the human health impacts range from acute gastroenteritis (associated with noroviruses, for example) to more chronic outcomes like obesity (adenoviruses) and insulin-dependent diabetes mellitus (coxsackievirus), to potentially more severe consequences such as hepatitis (hepatitis A and E viruses) and kidney disease (BK and JC viruses).

Table 1

Waterborne viruses and associated human health effects

Virus	Human health effects
Adenovirus	Gastroenteritis, respiratory infections, eye infections, intussusception, asthma, obesity, meningoencephalitis, hemorrhagic cystitis
Astrovirus	Gastroenteritis
BK virus	Hemorrhagic cystitis, kidney disease
Enterovirus	
Coxsackievirus	Gastroenteritis, respiratory infections, aseptic meningitis, myocarditis, insulin-dependent diabetes mellitus, encephalitis
Echovirus	Gastroenteritis, respiratory infections, aseptic meningitis, myocarditis
Poliovirus	Aseptic meningitis, paralysis
Hepatitis A virus	Infectious hepatitis
Hepatitis E virus	Infectious hepatitis
JC virus	Kidney disease
Norovirus	Gastroenteritis
Reovirus	Respiratory infections, gastroenteritis
Rotavirus	Gastroenteritis

The occurrence of clinical symptoms and the severity of human health outcomes depend on a variety of factors including the type of virus, host characteristics (such as age, immune and nutritional status), and the presence of other illnesses. Hepatitis A virus, for example, is more severe in older individuals with clinical illness apparent in only about 5% of infected children (Lednar et al., 1985). Individuals with compromised immune systems, such as the very young and old, transplant recipients, HIV-infected and cancer patients, and pregnant women, are more susceptible to adverse outcomes when exposed to particular waterborne viruses, such as adenovirus, BK virus, and JC virus. Depending upon host factors, the type of virus, and resulting clinical outcome, death may occur from exposure to waterborne viruses. Reported mortality ratios for waterborne viruses may be overestimated since they are actually case-fatality ratios reflecting hospitalized cases, but they reportedly range from 0.0001% (noroviruses) to as high as 0.94% (coxsackievirus) (Mena, 2002). Adenovirus and hepatitis E virus have been associated with relatively high mortality in transplant patients (60%, Zahradnik et al., 1980 and Shields et al., 1985) and pregnant women (greater than 8%, Boccia et al., 2006), respectively.

Examining exposure

Several factors are addressed when assessing exposure. The amount (concentration) of virus in the exposure (such as drinking water, recreational water, aerosolized droplets), the volume of the exposure, and the frequency and duration of the exposure (such as a one-time event or a repeated exposure for a certain number of days) are all considered. There are many ways an individual may encounter waterborne viruses including via drinking water, recreational water (lakes and swimming pools), through hygiene practices such as hand-washing and showering, and through contact with water reuse applications. The venue of exposure (such as within the home, hospitals, schools, recreational areas for children, cruise ships) is also characterized for its impact in the overall risk characterization, for it may direct subsequent risk management strategies as well as provide insight into *who* may be exposed (as addressed in the previous section).

Waterborne viruses may be transmitted through a variety of mechanisms including through direct ingestion from drinking water or recreational water, through inhalation of aerosolized droplets, through person-to-person contact, or through transmission routes involving several media, such as food contaminated by pathogen-laden water or fomites. The media involved and transmission route may impact the virus' survivability and infectivity. For example, the amount of microorganism necessary to initiate infection via inhalation is generally less than the number necessary when transmission is through ingestion. In addition, the extent of person-to-person transmission has been documented during viral waterborne outbreaks with secondary attack rates reportedly as high as 90% for poliovirus, approximately 75% for hepatitis A virus and coxsackievirus, and 30% for norovirus (Gerba and Rose, 1993).

One of the primary objectives of the exposure assessment step is to determine the amount of pathogen (e.g., virus) in the exposure. The viruses listed in Table 1 have been isolated from water supplies, including finished (drinking) water and surface water (see Chapter 5); however, in some instances, the specific virus is not identified, such as when enteroviruses as a group are isolated. Having quantitative data as input in the QMRA is essential, but often waterborne virus occurrence studies utilize molecular methodologies for detection and report virus occurrence as presence or absence.

Another issue regarding virus occurrence data and QMRA is if and how to incorporate non-detects reported in virus monitoring data sets. Traditionally, QMRA has been conducted utilizing the data that are positive for the microorganism of interest. However, with the application of more rapid, cost-effective detection methodologies, more of a monitoring approach has been undertaken to address pathogen occurrence in waters, therefore resulting in data sets with substantial "zeroes" or, at the very least, "non-detects" due to limits of detection of the laboratory method. How to appropriately account for these non-detects (or zeroes) in QMRA has yet to be determined. The survivability of the virus to environmental stressors and its susceptibility to inactivation by water treatment are also pertinent components of the exposure assessment (see Chapters 5 and 6). In addition, the recovery efficiency during laboratory detection should be incorporated. Furthermore, depending upon the availability and source of the data (e.g., viral data from surface water used for drinking water), it may be appropriate, when addressing drinking water exposures and risks, to consider the log reduction of viruses presumably achieved due to water treatment.

Once the virus concentration in a medium has been determined, the amount of medium in the exposure is defined. For drinking water exposure, the United States Environment Protection Agency (USEPA) has traditionally used a 2-l/person/day exposure level for the general population (Roseberry and Burmaster, 1992). However, a more recent report of water intake within the general population as well as subpopulations (e.g., children, pregnant women, etc.) states that a more appropriate estimate of (total) water consumption is 1.2 l/person/day (USEPA, 2004). An exposure of 130 ml/swim has been suggested as an exposure (ingestion) volume for this recreational activity as well as inhalation rates for aerosolized droplets of 15 m^3/day for children and 20 m^3/day for adults (Covello and Merkhofer, 1993).

The roles of multiple exposures and immunity in QMRA are not completely understood. The four-step NRC risk assessment paradigm focuses on the probability of an individual becoming infected or ill due to a one-time exposure, with an assumption that each exposure is statistically independent of another. In addition, it is assumed that any effects of secondary (person-to-person) transmission and preexisting immunity to a pathogen (such as a virus) are either inconsequential or balanced by the other. Subsequently, to address these issues, a more *dynamic* approach has been proposed that approaches risk assessment at the population level—as opposed to the *static* (individual) approach—and incorporates epidemiological data to categorize individuals into various epidemiological states (e.g., susceptible, exposed, diseased, carrier, post-infected, etc.) (Soller et al., 2004).

Assessing dose-response

Once exposure to a waterborne virus has occurred, the probability of the individual becoming infected, developing illness, or dying depends on several factors associated with the virus itself and the host (as described earlier in this chapter). This component of QMRA describes the *relationship* between the dose of the virus and the human health consequence (be it infection, illness, or death). The probability of clinical illness developing, however, does not appear to be related with the exposure dose via ingestion, at least not with rotavirus—a virus with high infectivity (Ward et al., 1986).

In QMRA, dose-response models developed from human-feeding or human-inhalation studies using volunteers have provided the template for risk assessment models for specific microorganisms, including viruses. Such studies have been conducted for many waterborne viruses including rotavirus (Ward et al., 1986), poliovirus (Koprowski et al., 1956; Katz and Plotkin, 1967; Minor et al., 1981), echovirus (Schiff et al., 1984), coxsackievirus (Suptel, 1963; Couch et al., 1965), adenovirus (Couch et al., 1966), and hepatitis A virus (Ward et al., 1958). It is understood that the usefulness of the information derived from these studies may be limited since the studies involved healthy volunteers (as opposed to a population with a range of susceptibilities) and also incorporated relatively high doses of viruses (rather than lower amounts that may be found in community water supplies). With infection the human health endpoint of interest in many of the studies, the information therefore can contribute to a perhaps conservative dose-response assessment.

Haas (1983) evaluated mathematical models for their appropriateness in representing the microorganism–host interaction, including viruses. The models evaluated by Haas (1983) assume that the viruses are distributed randomly in the exposure and that at least one virus is present in the exposure and is capable of initiating an infection within a particular site within the host. Two mathematical models—the exponential and the beta-Poisson—have been identified to represent the microorganism–host interaction and have been applied to address waterborne viruses in several QMRAs. The exponential model:

$$P_i = 1 - \exp(-rd)$$

where P_i is the probability of infection, r represents the fraction of viruses that survive and are able to infect the host at a specific location within the body, and d is the number of ingested or inhaled viruses (determined through the exposure assessment). The beta-Poisson model assumes a more heterogeneous interaction between the virus and the host, therefore consisting of two dose-response parameters, α and β:

$$P_i = 1 - \left(1 + d/\beta\right)^{-\alpha}$$

where P_i is the probability of infection, α and β are parameters related to the dose-response curve, and d is the number of ingested or inhaled viruses. These two models have been extensively described elsewhere (Haas et al., 1999) and have been evaluated for their goodness-of-fit to viral dose-response data using the method of maximum likelihood (Haas et al., 1993; Haas et al., 1999); and associated dose-response parameters for specific viruses have been defined. Table 2 lists the best-fit

Table 2

Best-fit model and defined parameters for waterborne viruses

Waterborne Virus	Best-Fit Model	Defined Parameters
Adenovirus (type 4)	Exponential	$r = 0.4172$
Coxsackievirus (type B4)	Exponential	$r = 0.007752$
Echovirus (type 12)	Exponential	$r = 0.012771$
Echovirus (type 12)	Beta-Poisson	$\alpha = 0.374,\ \beta = 186.69$
Hepatitis A virus	Exponential	$r = 0.548576$
Poliovirus (type 1)	Exponential	$r = 0.009102$
Poliovirus (type 3)	Beta-Poisson	$\alpha = 0.409,\ \beta = 0.788$
Rotavirus	Beta-Poisson	$\alpha = 0.26,\ \beta = 0.42$

Source: Regli et al., 1991; Haas et al., 1999.

model—either exponential or beta-Poisson—for specific waterborne viruses and associated defined parameters. Both of these models are *non-threshold* models meaning that it is assumed that only one pathogenic virus would be necessary to (potentially) initiate infection. Threshold models have been evaluated but have not shown to appropriately represent the virus–host interaction (Haas et al., 1999).

In addition to infection, the probability of illness and death can also be determined by multiplying the probability of infection (P_i) with appropriate morbidity and case-fatality ratios, with the assumption that the probability of illness is conditional to the probability of infection and the probability of death is conditional to the probability of illness. Calculating these probabilities as described above is based on a single exposure event to a dose (d). To determine probabilities associated with more than one exposure event such as over the course of a year, the following equation may be applied:

$$P_{year} = 1 - (1 - P_i)^{365}$$

Characterizing the risks

By compiling and integrating the information collected and interpreted during the other components of the QMRA process, human health risks associated with exposure to waterborne viruses can then be determined either as point estimates of risk or as risk distributions. For the factors considered throughout the QMRA process—those related to exposure and/or dose-response—a range of values may be incorporated as opposed to a single input. It is of great importance to include all values and assumptions made throughout the entire process so that resulting risk estimates may be interpreted accordingly. In addition, due to the heterogeneity of characteristics among the waterborne virus group, it is ideal to conduct QMRAs on *specific* viruses rather than viral groups (such as a risk assessment of the enterovirus group as a whole).

Uncertainty and variability are inherent in any QMRA and should be addressed during the risk characterization. *Uncertainty* refers to what is not known (i.e., an unknown value for a QMRA input) and *variability* refers to being inconstant or changeable (i.e., a QMRA input better represented by a range of values). A sensitivity analysis can be conducted to determine the extent to which a QMRA input influences the risk output. A previous analysis has shown that factors related to exposure contribute greater variability to the overall risk estimate than other factors, such as those related to dose-response (Haas et al., 1993). The potential to underestimate or overestimate risks exists in QMRA depending on the data used in the assessment further emphasizing the need to explain all inputs and assumptions made throughout the process. One approach to address this possibility of over- or underestimating risks during the assessment is to input a lower and upper value (e.g., using upper and lower 95% confidence limits for a certain parameter like dose-response, for example, or use minimum and maximum microorganism concentrations/volume) to then compute a range of risk estimates. In some situations, a "worst-case scenario" approach may be appropriate to provide conservative risk estimates that are protective of all populations (particularly, the immunocompromised). In general, however, defining inputs and assumptions that reflect realistic situations are the ideal.

Monte Carlo analysis is a simulation tool that can be applied to use probability distributions for each parameter rather than point estimates to capture the different possible combinations of input data and determine the greatest source of variability and uncertainty among these parameters. This stochastic approach involves defining probability distributions for the risk assessment input(s) of interest followed by a simulation where random values are incorporated. Through thousands of iterations (deterministic calculations) a final probability distribution of the risk is subsequently determined (based on the input of "realistic" values). Statistical software programs are commercially available to be used for such analyses.

It is also pertinent to understand the source (quality) of the data to fully appreciate where and to what extent risk calculations may be over- or underestimated. For example, for ethical reasons, dose-response studies are only conducted using healthy volunteers and results may not reflect the immunocompromised populations that may be more severely affected. Therefore, dose-response parameters that are defined and used in risk assessments may not represent the appropriate value for such subpopulations. In addition, water volume intake studies have addressed different segments of the population such as those in specific age groups or pregnant women, for example, so it is important to know which volume intake corresponds with which subgroup of interest.

Table 3 lists some of the exposure-related factors that may be defined or assumed during a QMRA and how each factor affects the risk output. As discussed previously, the parameters that may be included when determining exposure (d) include recovery efficiency during detection, % of infectious viruses present in the exposure, water treatment efficiency (log reduction), and water intake volume (liters). The infectivity parameter (% infectious viruses) and water intake volume are directly proportionate to the calculated risk estimate, i.e., when these parameters decrease, the risk estimates decrease. However, as the recovery efficiency during detection improves (increases) and as water treatment efficiency

Table 3

Defined assumptions for determining exposure (d) in a risk assessment of viruses in water and impact on risk estimate

Assumptions	Change	Risk Estimate
Recovery efficiency of detection method	↑	↓
% of infectious viruses	↓	↓
Log removal of viruses due to water treatment	↑	↓
Water intake (liters)	↓	↓

(log reduction) increases, risk estimates decrease. For QMRAs that have been conducted for waterborne viruses (Gerba et al., 1996; Crabtree et al., 1997; Mena et al., 2003), it was assumed that all (100%) of the viruses in a given exposure are capable of initiating infection, which provides some level of conservatism in the estimated risks. In addition, traditionally QMRAs have used the past USEPA default value of 2 l/person/day for drinking water exposure, which may have been too high (USEPA, 2004) resulting in an overestimation of risk.

Building models: a regulatory tool?

Table 4 provides annual risk of infection estimates for specific waterborne viruses at various (hypothetical) concentrations. The appropriate dose-response model (either the exponential or beta-Poisson) and defined dose-response parameter(s) were used as provided in Table 2, and a volume of 1.2 l/person/day was used as the assumed amount of water ingested (USEPA, 2004). Annual risks of infection ranged from approximately 3:10,000 (3×10^{-4}) for coxsackievirus type B4 and poliovirus type 1 to nearly 3:100 (3×10^{-2}) for rotavirus when considering the lowest concentrations (0.01 viruses per 100 l) assumed in this exercise.

When considering the highest concentrations, all of the annual risk of infection estimates closely approach 1 (9.9×10^{-1}). The USEPA recommends that microbial risks of infection for drinking water should not exceed 1:10,000 (1×10^{-4}) per year (Macler, 1993). Assuming the concentration data presented in Table 4 represents source water quality, one could determine the treatment (log reduction of viruses) required to meet this USEPA recommendation. When considering the lowest concentration of rotavirus in this exercise, for example, a two-log (99%) virus reduction during treatment would need to be achieved to meet the 1:10,000 goal. Due to the variation introduced by the parameters used throughout the QMRA process, caution is warranted when attempting to draw overall conclusions that may lead to impractical or costly implications based on a comparison to one value. However,

Table 4

Annual risks of infection associated with exposure to selected waterborne viruses at various concentrations[a]

	Number of viruses per 100 l				
	0.01	0.1	1	10	100
Adenovirus	1.81×10^{-2}	1.67×10^{-1}	8.39×10^{-1}	9.99×10^{-1}	9.99×10^{-1}
Coxsackievirus (type B4)	3.39×10^{-4}	3.39×10^{-3}	3.34×10^{-2}	2.88×10^{-1}	9.66×10^{-1}
Hepatitis A	2.37×10^{-2}	2.14×10^{-1}	9.10×10^{-1}	9.99×10^{-1}	9.99×10^{-1}
Poliovirus (type 1)	3.98×10^{-4}	3.98×10^{-3}	3.91×10^{-2}	3.28×10^{-1}	9.81×10^{-1}
Rotavirus	2.66×10^{-2}	2.37×10^{-1}	9.31×10^{-1}	9.99×10^{-1}	9.99×10^{-1}

[a]Using model and parameters defined in Table 2 and assuming an exposure of 1.2 l/person/day (USEPA, 2004).

QMRA results can be used as a guide when considering water treatment efficacy and options.

QMRA has many useful applications. Individual components of QMRA may uncover pertinent characteristics of specific waterborne viruses, such as occurrence in certain types of water or geographical locations, or resistance to a specific form of treatment. In addition, the QMRA process can identify data gaps thereby highlighting future research directions. In some situations, it may be appropriate to utilize QMRA to compare risks associated with exposure to different hazards. Perhaps the most beneficial aspect of QMRA is its flexibility in that it can be applied in a variety of settings to a diverse population and can address a wide range of hazards. QMRA can be used to assess the public health impact of waterborne viruses on populations throughout the world and evaluate human health concerns at both the individual and the community level. In a regulatory setting, the ultimate challenge (and potential pitfall) in building models for policy-making purposes ensures that the model is appropriate for the question to be answered.

Conclusion

The QMRA process developed over the past several years continues to evolve to better address not only individual-level risks but also risks associated with communities by incorporating epidemiological data of populations when available. Goals to improve QMRA continue to focus on factors related to exposure assessment, including better and more frequent quantitative detection of specific viruses in various types of water. Emphasis should also be placed on defining QMRA's role in the regulatory arena to appropriately interpret QMRA results in a way that directs effective risk management strategies. Although QMRA provides the pathway for science to influence policy, its application as a regulatory tool has yet to be fully realized.

References

Boccia D, Guthmann J-P, Klovstad H, Hamid N, Tatay M, Ciglenecki I, Nizou J-Y, Nicand E, Guerin PJ. High mortality associated with an outbreak of hepatitis E among displaced persons in Darfur, Sudan. Clinical Infectious Diseases 2006; 42: 1679–1684.

Centers for Disease Control and Prevention. Surveillance for Waterborne-disease Outbreaks Associated with Recreational Water—United States, 2001-2002 and Surveillance for waterborne-disease outbreaks associated with drinking water—United States, 2001-2002. In: Surveillance Summaries, October 22, 2004. *MMWR* 2004;53(No. SS-8).

Colford JM, Roy S, Beach MJ, Hightower A, Shaw SE, Wade TJ. A review of household drinking water intervention trials and an approach to the estimation of endemic waterborne gastroenteritis in the United States. J Water Health 2006; 4(Supp. 2): 71–88.

Couch RB, Cate T, Gerone P, Fleet W, Lang D, Griffith W, Knight V. Production of illness with a small-particle aerosol of coxsackie A21. J Clin Invest 1965; 44(4): 535–542.

Couch RB, Cate TR, Gerone PJ, Fleet WF, Lang DJ, Griffith WR, Knight V. Production of illness with a small-particle aerosol of adenovirus type 4. Bacteriol Rev 1966; 30: 517–528.

Covello ET, Merkhofer MW. Risk Assessment Methods. New York: Plenum Press; 1993.

Crabtree, K.D. (1996). Risk assessment of viruses in water. PhD Dissertation, The University of Arizona.

Crabtree KD, Gerba CP, Rose JB, Haas CN. Waterborne adenovirus: a risk assessment. Water Sci Technol 1997; 35: 1–6.

Evans AS. Epidemiological concepts and methods. In: Viral Infections of Humans (Evans AS, editor). New York: Plenum Press; 1982; pp. 3–13.

Gerba CP, Haas CN. Assessment of risks associated with enteric viruses in contaminated drinking water. In: Chemical and Biological Characterization of Sludges, Sediments, Dredge Spoils, and Drilling Muds, ASTM STP 976 (Lichtenberg JJ, Winter JA, Weber CI, Fradkin L, editors). Philadelphia: American Society for Testing and Materials; 1988; pp. 489–494.

Gerba CP, Rose JB. Estimating viral disease risk from drinking water. In: Comparative Environmental Risk Assessment (Cothern CR, editor). Boca Raton: Lewis Publishers; 1993; pp. 117–135.

Gerba CP, Rose JB, Haas CN, Crabtree KD. Waterborne rotavirus: a risk assessment. Water Res 1996; 30: 2929–2940.

Haas CN. Estimation of risk due to low doses of microorganisms: a comparison of alternative methodologies. Am J Epidemiol 1983; 118: 573–582.

Haas CN, Rose JB, Gerba C, Regli S. Risk assessment of virus in drinking water. Risk Anal 1993; 13(5): 545–552.

Haas CN, Rose JB, Gerba CP. Quantitative Microbial Risk Assessment. New York: John Wiley & Sons, Inc.; 1999.

ILSI. International Life Sciences Institute. A conceptual framework to assess the risks of human disease following exposure to pathogens. Risk Anal 1996; 16(6): 841–848.

ILSI. International Life Sciences Institute. (2000). Revised framework for microbial risk assessment. An ILSI Risk Science Institute Workshop Report. Washington, DC: ILSI Press.

Katz M, Plotkin SA. Minimal infective dose of attenuated poliovirus for man. American JPublic Health Nations Health 1967; 57(10): 1837–1840.

Koprowski H, Norton TW, Jervis GA, Nelson TL, Chadwick DL, Nelsen DJ, Meyer KF. Clinical investigations on attenuated strains of poliomyelitis virus; use as a method of immunization of children with living virus. J Am Med Assoc 1956; 160(11): 954–966.

Lednar WM, Lemon SM, Kirkpatrick JW, Redfield RR, Fields ML, Kelley PW. Frequency of illness associated with epidemic hepatitis A virus infections in adults. Am JEpidemiol 1985; 122: 226–233.

Macler B. Acceptable risk and U.S. microbial drinking water standards. In: Safety of Water Disinfection (Craun GF, editor). Washington, D.C: ILSI Press; 1993; pp. 619–626.

Mena KD. Environmental exposure to viruses: a human health risk assessment methodology. In: The Encyclopedia of Environmental Microbiology (Bitton G, editor). New York: John Wiley & Sons, Inc.; 2002; pp. 2741–2752.

Mena KD, Gerba CP, Haas CN, Rose JB. Risk assessment of waterborne coxsackievirus. J Am Water Works Assoc 2003; 95(7): 122–131*.

Messner M, Shaw S, Regli S, Rotert K, Blank V, Soller J. An approach for developing a national estimate of waterborne disease due to drinking water and a national estimate model application. J Water Health 2006; 4(Supp. 2): 201–240.

Minor TE, Allen CI, Tsiatis AA, Nelson DB, D'Alessio DJ. Human infective dose determination for oral poliovirus type 1 vaccine in infants. J Clin Microbiol 1981; 13: 388.

NRC. National Research Council. Risk Assessment in the Federal Government: Managing the Process. Washington, D.C: National Academy Press; 1983.

NRC. National Research Council. Science and Judgment in Risk Assessment. Washington, D.C: National Academy Press; 1984.

Regli S, Rose JB, Haas CN, Gerba CP. Modeling the risk from *Giardia* and viruses in drinking water. J Am Water Works Assoc 1991; 83: 76–84.

Roseberry AM, Burmaster DE. Lognormal distributions for water intake by children and adults. Risk Anal 1992; 12: 99–104.

Schiff GM, Stefanovic GM, Young EC, Sander DS, Pennekamp JK, Ward RL. J Infect Dis 1984; 150(6): 858–866.

Shields AF, Hackman RC, Fife KH, Corey L, Meyers JD. Adenovirus infections in patients undergoing bone-marrow transplantation. New Engl J Med 1985; 312(9): 529–533.

Soller JA, Olivieri AW, Eisenberg JNS, Sakaji R, Danielson R. (2004). Evaluation of Microbial Risk Assessment Techniques and Applications. Report 00-PUM-3. Water Environment Research Foundation. United Kingdom: IWA Publishing.

Suptel EA. Pathogenesis of experimental coxsackie virus infection. Arch Virol 1963; 7: 61–66.

USEPA. United States Environmental Protection Agency. Estimated Per Capita Water Ingestion and Body Weight in the United States—An Update. Washington, D.C: USEPA Office of Water, Office of Science and Technology; 2004.

van Heerden J, Ehlers MM, Vivier JC, Grabow WOK. Risk assessment of adenovirus detected in treated drinking water and recreational water. J Appl Microbiol 2005a; 99: 926–933.

van Heerden J, Ehlers MM, Grabow WOK. Detection and risk assessment of adenoviruses in swimming pool water. J Appl Microbiol 2005b; 99: 1256–1264.

Ward R, Krugman S, Giles J, Jacobs M, Bodansky O. Infectious hepatitis: studies of its natural history and prevention. New Engl J Med 1958; 258(9): 402–416.

Ward RL, Bernstein DL, Young EC, Sherwood JR, Knowlton DR, Schiff GM. Human rotavirus studies in volunteers: determination of infectious dose and serological response to infection. J Infect Dis 1986; 154(5): 871–877.

Zahradnik JM, Spencer MJ, Porter DD. Adenovirus infection in the immunocompromised patient. Am J Med 1980; 68: 725–732.

Human Viruses in Water
Albert Bosch (Editor)
DOI 10.1016/S0168-7069(07)17009-9

Chapter 9

The Detection of Waterborne Viruses

Peter Wyn-Jones

Institute of Geography and Earth Sciences, University of Wales, Aberystwyth, UK

Viruses in water are usually present in concentrations too low for detection by direct analysis. Virological investigation of water samples is therefore nearly always a multi-stage process involving concentration of viruses present followed by an appropriate detection procedure. The exception is analysis of sewage, where viruses may be present in sufficiently high numbers to be detectable without concentration.

The volume of water analysed and the degree of concentration required will depend on the number of viruses likely to be present and therefore on the origin of the sample. While viruses in sewage may require minimal concentration (or none at all) to render them detectable, those in treated drinking water or groundwater may require several thousand-fold concentration to make detection likely. It is often possible to find viruses in 100 ml of unconcentrated inlet (i.e. raw) sewage, whereas several hundred litres of drinking water may have to be processed. It is common, for instance, to take 10 l samples of water from recreational sites which may be subject to sewage effluent pollution and which will require concentration of about thousand-fold. The final volume of concentrate will be influenced by (a) the minimum volume achievable by the concentration technique and (b) the volume required by the detection procedure(s) and any replicates thereof. In practise, final concentrate volumes of about 5–10 ml are usually produced.

There are several approaches to detection of viruses. Part or all of the concentrate may be inoculated into cell cultures to detect infectious cytopathogenic virus, and if this is done in a quantitative fashion the virus can be enumerated, the count being reported as plaque-forming units (pfu), the tissue culture infectious dose (TCD_{50}), or most probable number (MPN) units. The virus may be isolated and identified from the cell cultures. Viruses that multiply without producing an identifiable cytopathic effect (c.p.e.) in culture may sometimes be detected by immunoperoxidase or immunofluorescence staining. The concentrate may also be analysed by molecular biological procedures (usually polymerase chain reaction (PCR) or real-time-PCR (RT-PCR)). The problem then is that such techniques do not usually detect the infectious virus, and novel approaches have been made recently to meet this challenge.

Concentration methods

It is common for concentration to comprise at least two stages. The first stage will reduce the volume to between 100 and 400 ml, and the second stage will reduce it to 2–10 ml. Supplementary stages may be added to remove cytotoxic or PCR-inhibitory compounds.

Block and Schwartzbrod (1989) defined a number of criteria that an ideal concentration method must fulfil to be of practical use. The method should:

- be technically easy to accomplish in a short time;
- have a high virus recovery rate;
- concentrate a range of viruses;
- provide a small volume of concentrate;
- not be costly;
- be capable of processing large volumes of water; and
- be repeatable (within a laboratory) and be reproducible (between laboratories).

No single method fulfils all these requirements.

The properties of viruses are most often exploited in their concentration, and the general approaches to concentration derived from them are shown in Table 1. Numerous methods based on these approaches have been devised for the concentration of viruses from water and the principal ones are summarised in Table 2. These have been reviewed extensively by Wyn-Jones and Sellwood (1998) in respect of enteroviruses and by Wyn-Jones and Sellwood (2001) for other virus groups. The virology of waterborne disease is discussed in Percival et al. (2004).

Table 1

General approaches to virus concentration

Property	Technique applicable
Ionic charge	Adsorption/elution
Particle size	Ultrafiltration
Density and sedimentation coefficient	Ultracentrifugation

Table 2

Summary of concentration techniques for viruses in water and related materials

Technique	Method	Water quality	Initial volume	Relative virus content	Recovery	Capital cost	Recurrent cost	Secondary concentration required?	Comments
Adsorption/elution	Gauze pads	Sewage or effluent	Large	High	Low to medium	Nil	Very low	No	Not quantitative
	Electronegative membranes	All waters	1–100 l	Low to medium	50–60% with practise	Medium	Medium	Yes	High volumes require dosing pumps
	Electropositive membranes	All waters	1–100 l	Low to medium	50–60% with practise	Medium	High	Yes	No pre-conditioning required
	Electronegative cartridges	Any low turbidity	1–50 l	Low to medium	Variable: higher with clean waters	Low	Low	Yes	Clogs more quickly than membranes
	Electropositive cartridges	All waters	1–100 l	Low to medium	Variable	Medium	High	Yes	Wide range of viruses
	Glass wool	All waters	1–100 l	Low to medium	Variable	Low	Very low	Yes	No pre-conditioning required
	Glass powder	All waters	<100 l	Any	20–60%	Medium	Low	For volume > 100 l	Special apparatus
Entrapment: ultrafiltration	Alginate membranes	Clean only	Low	High	Good	Low	Low	No	Very slow. Clogs rapidly if turbid
	Single membranes	Clean	Low	Any	Variable	Medium	Low	No	Slow
	Tangential (= cross) flow and hollow fibres	Treated effluents or better	High	Low	Variable	High	Medium	Sometime	Pre-filter for turbid waters
	Vortex flow	Treated effluents or better	High	Low	Unknown	High	Medium	Unknown	Undeveloped yet
Hydroextraction	PEG or sucrose	Any	Low	High	Variable (toxicity)	Negligible	Very low	No	High virus loss in wastewaters
Ultracentrifugation		Clean	Low	High	Medium	High	Medium	No	Wide range, but usually impractical
Other techniques	Iron oxide flocculation	All	Low	Any	Variable	Low	Low	No	Toxic to cells
	Biphasic partition	All	<7 l	Any	Variable	Low	Low	No	
	Immunoaffinity and magnetic beads	Unknown	Low	Low	High	High	Low	No	New method

Concentration based on ionic charge: adsorption/elution

The development of virus adsorption/elution methods, suitable for the recovery of viruses from waters, stems from the work of Melnick and his colleagues in Houston, TX (e.g. Wallis and Melnick, 1967a,b,c; Wallis et al. (1970). In general terms, a virus-containing sample is brought into contact with a solid matrix to which the virus will adsorb under specific conditions of pH and ionic strength. Once the virus is adsorbed, the water in which it was originally suspended is discarded. The virus is then released from the matrix by elution into a smaller volume of fluid, though this is usually still too large to be analysed directly. Choice of adsorbing matrix, eluting fluid and processing conditions will be influenced by the nature of the sample and by experience, but elution is commonly done using a solution containing beef extract or skimmed milk, both at high pH, which displaces the virus from the adsorbing matrix into the eluant. Eluants comprising basic amino acids (glycine, lysine) are also used. The USEPA Standard Method (2007) for the concentration of waterborne viruses is based on an adsorption/elution procedure, quoted in the Information Collection Rule (ICR) and at http://www.epa.gov/microbes/about.htm.

Adsorption to electronegative membranes and cartridges

The popularity of membranes, made of cellulose acetate or nitrate, is due to their availability in various pore sizes, configurations and compositions. The virus is bound to the filter by electrostatic attractive forces, and not by size exclusion. It is possible to get good recoveries of the virus accompanied by good flow rates and a minimum of filter clogging even from turbid waters, and many solids-associated virus can be recovered. In its simplest form, a virus-containing sample is passed under positive pressure or vacuum through a cellulose nitrate membrane 142 or 293 mm in diameter and of mean pore diameter 0.45, 1.2 or 5 μm (Fig. 1). For waters containing particulate material a pre-filter is used upstream of the membrane.

Since viruses and the filter materials are both negatively charged at neutral pH the water sample must be conditioned to allow electrostatic binding of virus particles to the filter matrix. The water sample is adjusted to pH 3.5 and Al^{3+} or Mg^{2+} ions may be added, though opinion is divided as to whether metal ions are needed at all when using cellulose nitrate membranes.

Negatively charged filters may also be used in tube form. Balston filters are epoxy resin-bound glass fibre filters with an 8 μm nominal pore diameter. They were originally used for concentration of viruses from tap water (Jakubowski et al., 1974) and have since been employed for concentration of viruses from river water (e.g. Morris and Waite, 1980) and other waters. Their recoveries are as good as membrane filters, they are less expensive and can be obtained in sterile cartridges in disposable form. However, they are prone to clogging, cannot be used with even moderately turbid water and according to Gerba (1987), cannot be used at high flow rates. Because of problems of clogging of membrane or tube filters, the

Fig. 1 Membrane filtration.

processing of seawater samples in this way is limited to a maximum of 20 l before filters have to be changed (Block and Schwartzbrod, 1989).

One way of overcoming the problem of clogging without having to change membranes or tubes frequently is to increase the surface area of filtration by the use of larger cartridge filters, where sheets of negatively charged pleated filter material approximately 25 cm wide are rolled and used in 30 cm cartridge holders. These were evaluated by Farrah et al. (1976) who used fibreglass membrane material in a pleated format. Seeded poliovirus was recovered from 378 l volumes of seawater with 53% efficiency. The authors reported that the filters could be regenerated up to five times by soaking for 5 min in 0.1 M NaOH.

Papaventsis et al. (2005) reported a modification of the use of negatively charged membranes wherein they were able to culture sewage-derived enteroviruses directly from the filter without elution, thus reducing the total time required for analysis.

Generally, recovery rates are as variable with negatively charged filter media as with any other kind. Block and Schwartzbrod (1989), citing Beytout et al. (1975) considered cellulose nitrate membranes relatively efficient insofar as they give 60% recovery of virus; the same authors recorded glass fibre filters giving a poor average yield from wastewater but 70% recovery with river water. Payment and Trudel (1979), using glass fibre filters, reported 38–58% recovery of 10^2–10^6 pfu seeded in 100 ml–1000 l volumes. Few studies have been done on recovery efficiencies from marine waters in a controlled way; however controlled studies have been done to evaluate the recovery efficiency of the method using drinking water.

It is not usually possible to conduct studies where the virus is deliberately added to water systems, however Hovi et al. (2001) in assessing the feasibility of environmental poliovirus surveillance added poliovirus type 1 into the Helsinki sewers and recovered it over a period of 4 days by taking samples at downstream locations and concentrating 100-fold by polymer two-phase separation.

Adsorption to electropositive membranes and cartridges

Positively charged filters adsorb virus from water and other materials without the need for prior conditioning of the sample. Initial work was done in the USA by Sobsey and Jones (1979) and by Hou et al. (1980). They adsorb virus in the pH range 3–6; at pH values above 7 the adsorption falls off rapidly, so the pH still needs to be carefully controlled. These properties make the use of positively charged filters attractive, not only for the convenience of not having to condition the sample but also because it makes possible the concentration of other viruses such as rotavirus and coliphages, which are sensitive to the low pH conditions needed for adsorption to negatively charged media. Keswick et al (1983) reported that type 1 poliovirus and rotavirus SA11 survived at least 5 weeks on electropositive filters at 4°C, which makes them useful for on-site concentration. They are used in the same way as electronegative materials. The virus is eluted from the filter and secondary concentration is carried out as for the electronegative types.

Recoveries from positively charged filters are similar to those from negatively charged ones; Sobsey and Jones (1979) reported 22.5% recovery using a two-stage procedure in the concentration of poliovirus from drinking water. The original positively charged material, Zeta-plus Series S, is made of a cellulose/diatomaceous earth/ion-exchange resin mixture. Sobsey & Glass (1980) compared these Virozorb 1 MDS filters with Filterite (fibreglass) pleated cartridge filters for recovery of poliovirus from 1000 l tap water and obtained recoveries of about 30% with both types. The advantages of these filters lie in the large volumes they can handle without the need for conditioning the sample. Elution from the filter still needs to be carried out at pH 9 or above, which limits their use to viruses stable below that pH, though Bosch et al. (1988) successfully concentrated rotavirus in this way. Organic materials in the sample, especially fulvic acid, were reported to interfere more with virus recovery from Virozorb cartridges than from glass-fibre materials (Sobsey and Hickey, 1985; Guttman-Bass and Catalano-Sherman, 1986). Such filters are used extensively in the USA for concentration of many types of viruses from treated drinking water to sewage effluent (e.g. Sedmak et al., 2005). A different electropositive material (MK) is cheaper but its recoveries were reported to be not as good as 1 MDS in comparative tests (Ma et al., 1994). Improvements to poliovirus and norovirus recovery from tap water samples by coating of electropositive Zetapor filters by passage of $AlCl_3$ prior to filtration was reported by Haramoto et al. (2004).

During the 2002/2003 outbreak of severe acute respiratory syndrome (SARS) attention was focused on possible transmission of the SARS-corona virus (SARS-CoV) in sewage since SARS-CoV RNA had been found in the stools of affected patients. Electropositive filters were used to concentrate the virus from sewage (Wang et al., 2005) and SARS-CoV RNA was recovered from sewage concentrates.

Advances in membrane technology have also resulted in charge-modified nylon membranes being available for concentration of viruses from water. Gilgen et al (1995, 1997) described the use of positively charged nylon membranes coupled with ultrafiltration for the concentration of a variety of enteric viruses prior to detection by RT-PCR. Other nylon membranes are also available which are made in various pore sizes, which would permit passage of the virus (0.45, 1.2 and 3 µm) and have a positive surface charge over the pH range 3–10, which would promote strong binding of negatively charged particles. Although nylon filters have been shown to bind viruses in freshwater samples, adsorption from marine samples is very poor and they would not be used for seawater (Sellwood, personal communication). Their low cost and ease of use suggest that further evaluative research should be done. Triple-layered polyvinylidene fluoride (PVDF) membranes and cartridges have been used in industry for the removal of polio and influenza viruses from pharmaceutical products (AranhaCreado et al., 1997), though whether the viruses can be recovered from the filter is not known.

A recent advance in the use of positively charged filters has been the use of membranes (disc or pleated) consisting of "nano-alumina" fibres approximately 2 nm in diameter bound into a support matrix of cellulose, polyester and glass fibre. Such filters carry a high electropositive charge and are claimed to bind $6\log_{10}$ MS2 phage in the pH range 5–9 with no conditioning of the water and to have a high flow rate. The virus may be eluted from the filter using a high-protein fluid such as beef extract at pH 9 (see below). Originally intended as water purification devices, these filters have been considered for use as filters to meet the USEPA drinking water standard. There are no reported peer-reviewed studies on their performance with animal viruses.

The need to determine the presence of *Cryptosporidium* and *Giardia* as well as viruses in water samples has led some workers to attempt the simultaneous concentration of both types of microorganism (e.g. Watt et al., 2002).

Adsorption to glass wool

Glass wool is an economic alternative to microporous filters. It is used in a column and provided it is evenly packed to an adequate density, adsorption of viruses appears at least as efficient as with other filter types. An advantage of the method is that the virus will adsorb to the filter matrix at or near neutral pH, and without the addition of cations, which makes it suitable for viruses sensitive to acid, however, elution still has to be done at high pH.

The technique was pioneered in France principally by Vilaginès and co-workers (e.g. Vilaginès et al., 1988), who applied it to the concentration of a range of viruses from surface, drinking and waste waters. Glass wool packed into holders (Fig. 2) at a density of $0.5\,\text{g/cm}^3$ is washed through in sequence with HCl, water, NaOH and

Fig. 2 Glass wool filtration.

finally with water again to neutral pH before the sample is passed through the filter. Different sizes of filter can be prepared according to the type of water and flow rate.

In the French studies sample sizes ranged from 100 to 1000 l for drinking waters, 30 l for surface waters and 10 l for wastewaters. The only pre-treatment necessary was dechlorination of drinking waters. Surface water samples were filtered at 50 l/h in a 42 mm diameter filter holder. The virus was eluted from the filter with 0.5% beef extract solution and secondary concentration done by organic flocculation.

Recovery efficiency of approximately 10^2 pfu poliovirus seeded into 400 l drinking water averaged 74% (SD 18.9%). For surface waters the recovery rate was 63% and 57%, respectively. Clogging of the filters was reduced by lowering the flow rate to 50 l/h.

Other viruses were also concentrated during field evaluation of the method; adenoviruses and reoviruses were also recovered, though as expected enteroviruses predominated. Vilaginès et al. (1993) also reported a survey of two rivers over a 44-month period and concluded that the technique was robust enough to be used for routine monitoring of surface waters.

Glass wool has been used in many other laboratories; Hugues et al. (1991) found it more sensitive than the glass powder method; it was used by Wolfaardt et al. (1995) to concentrate small round-structured viruses (SRSVs, now noroviruses) from spiked sewage and polluted water samples prior to detection by RT-PCR, by Ehlers et al. (2005) for concentration of enteroviruses from sewage and treated drinking water, and by van Heerden et al. (2005) to recover human adenoviruses from 2001 treated drinking water samples and 251 river water samples. Adsorption to the filters was done on site and the filters transported to the laboratory for elution.

Adsorption to glass powder

Glass beads constitute a fluidised bed and so have the advantage that the filter matrix cannot become clogged as with glass-fibre systems. Sarrette et al. (1977) first developed this technique, which was extended by Schwartzbrod and Lucena-Gutierrez (1978). The method gives a low eluate volume, which may not need secondary concentration prior to further analysis. A disadvantage is the complexity of the apparatus.

Other adsorbents

A range of viruses can be concentrated from different waters using talc (magnesium silicate) mixed with celite (diatomaceous earth) (e.g. Sattar and Westwood, 1978; Ramia and Sattar, 1979; Sattar and Ramia, 1979).

Baggi and Peduzzi (2000) reported a simple and inexpensive (though relatively insensitive) method for concentration of rotaviruses from surface waters and sewage, which involved addition of 200 µl (*sic*) SiO_2 per litre of conditioned water sample, settling or centrifugation of the silica and elution of virus from the pellet.

Dahling et al. (1985); Lahke and Parhad (1988) and Chaudhiri and Sattar (1986) used powdered coal as an adsorbent with a view to transferring the virus concentration and water purification technology to developing countries.

The same kind of matrix in a more refined state was used as granular activated carbon by Jothikumar et al. (1995) for the first stage concentration of enteroviruses, hepatitis E virus (HEV) and rotaviruses. Using RT-PCR as a detection method, these authors reported 74% recovery of poliovirus 1.

Entrapment

Entrapment, or size exclusion, refers to those techniques in which the virus in a sample is bound to a filter matrix principally by virtue of its size rather than by any charges on the particle, though in practice electrostatic effects can also exert an effect.

Ultrafiltration

Variations in technique involve passing the sample through capillaries (e.g. Rotem et al., 1979), membranes (e.g. Divizia et al., 1989a,b) and hollow fibres (Belfort et al., 1982) with pore sizes that permit passage of water and low molecular mass solutes but exclude viruses and macromolecules, which become concentrated on the membrane or fibre. Most laboratories use membranes or fibre systems with cut-off levels of 30–100 kDa. In systems in which the fluid passes directly through the filter, non-filterable components quickly clog the filter or precipitate at the membrane surface, thus this type of filter is only useful for small volumes (<1000 ml). Some ultrafilters employ tangential flow or vortex flow (VFF), which reduces clogging. Tsai et al. (1993) used VFF for processing inshore water samples in Southern California. Fifteen litres of each sample were concentrated to 100 ml using a 100 kDa cut-off membrane and the samples were further concentrated to 100 µl using Centriprep and Centricon units at $1000 \times g$.

The minimum "dead" volume (e.g. 10–15 ml, Divizia et al., 1989a) is the final volume of concentrate. If this is small enough then it may be analysed or it may have to be further processed by secondary concentration. Hill et al. (2005) showed that it was possible to concentrate viruses and other microorganisms simultaneously using hollow fibre technology, and used sodium polyphosphate to minimise adhesion of organisms to the filter. Rutjes et al. (2005) used a membrane ultrafilter of 10 kDa cut off for secondary concentration of enteroviruses following primary concentration by adsorption/elution; the starting primary concentrate volumes were approximately 650 ml (raw sewage) and 1800 ml (river water).

Some workers have experienced binding of the virus to the membrane rather than just the prevention its passage through it. In these cases the virus was eluted by backwashing with glycine buffer or beef extract and the eluate reconcentrated by organic flocculation. Some authors have even reported differences in binding between related viruses. Divizia et al. (1989b) for example noted that hepatitis A virus (HAV) was recovered with 100% efficiency though poliovirus was recovered very poorly under standard conditions, but this improved if the membranes were pre-treated with different buffers. Further, recovery was best if the virus was eluted with beef extract at neutral (not high) pH.

The advantages of ultrafiltration are principally that the sample requires no pre-conditioning and that a wide range of viruses can therefore be recovered, including those sensitive to the pH changes necessary in most adsorption/elution procedures, and also bacteriophages (e.g. Nupen et al., 1981; Urase et al., 1994). Efficiency of recovery is usually good, though as with all methods it is variable. Surface water samples may take a long time to process if they are turbid; Nupen et al. (1981) were able to filter 50 l volumes but this took about 40–72 h depending on the sample. Systems have high capital cost, though disposable cartridges have recently become available. The technique is sometimes seen as an advance on the adsorption/elution technique (e.g. Grabow et al., 1984; Muscillo et al., 1997).

Ultracentrifugation

Ultracentrifugation is a catch-all method capable of concentrating all viruses in a sample provided sufficient g-force and time are used. Differential ultracentrifugation allows separation of different virus types. A number of studies have been reported, including one in which virus from a polluted well was recovered (Mack et al., 1972), and one where viral numbers in natural waters were as high as $2.5 \times 10^8/$ml, 10^3–10^7 times as high as had been found by plaque assay (Bergh et al., 1989). However the limited volumes that can be processed, even using continuous flow systems, together with the high capital costs and lack of portability of the equipment, limit its usefulness in concentrating viruses directly from natural waters. It does find a use as a secondary concentration method however. Murphy et al. (1983), in an investigation of a gastroenteritis outbreak associated with polluted drinking water, concentrated 5 l samples of borehole water to 50 ml using an ultrafiltration hollow fibre system and followed this by ultracentrifugation to pellet the virus for electron microscopical examination. They were thus able to detect rapidly rotaviruses, adenoviruses and SRSVs (noroviruses), as well as enteroviruses, which were confirmed by cell culture.

In an investigation to detect HEV in sewage, Pina et al. (1998) concentrated viruses and removed suspended solids from 40 ml samples by differential ultracentrifugation; Vaidya et al. (2002) used the same protocol to detect HEV and HAV in sewage samples. Le Cann et al. (2004) concentrated astroviruses from sewage samples by ultracentrifugation and extracted the RNA from the pellets.

Other methods

Many other methods exist, though none satisfies all the requirements given above by Block and Schwartzbrod (1989). These include hydroextraction with hygroscopic solids (Wellings et al., 1976; Ramia and Sattar, 1979), iron oxide flocculation (Rao et al., 1968; Bitton et al., 1976), two-phase separation (Lund and Hedstrom, 1966) and freeze-drying (Bosch et al., 1988; Kittigul et al., 2001).

Affinity columns were used by Schwab et al. (1996) in a broad-based antibody-capture technique for a variety of viruses and Myrmel et al. (2000) described the separation of noroviruses in this way. An important attribute of this method is that it acts as a clean-up stage to remove RT-PCR inhibitors. Cromeans et al. (2004) reported the preparation and use of a soluble Coxsackie virus-adenovirus (sCAR) receptor immobilised to magnetic beads for the concentration of Coxsackie and adenoviruses from water sample concentrates. The receptor, which neutralised Coxsackie virus B3, also reacted with other Coxsackie B types. The group also reported the use of a neutralising monoclonal antibody for immunocapture of the same viruses.

Secondary concentration

Where proteinaceous eluant fluids are used, the most commonly used secondary concentration technique is that of Katzenelson et al. (1976); the pH of the primary eluate is reduced to 3.5–4.5, which causes isoelectric coagulation (flocculation) of the protein. The virus adsorbs to the floc, which is deposited by centrifugation and dissolved in 5–10 ml neutral phosphate buffer. If the concentrate is to be inoculated into cell cultures it is common to filter it through a 0.22 μm pore diameter filter to remove contaminating bacteria.

Secondary concentration can also be accomplished using two-phase separation, usually with polyethylene glycol (PEG)/NaCl, or PEG and dextran T40. Rutjes et al. (2006) compared two-phase separation (PEG and dextran T40) with ultrafiltration for secondary concentration of noroviruses from water following primary concentration by adsorption/elution, and found ultrafiltration to be better, the techniques being assessed by estimation of the recovered norovirus RNA.

If molecular biological analysis is to be done, the volume may be reduced to about 1 ml by dialysis, in spin-columns or microconcentrators with a M_r cut-off of 100,000 KDa.

Gilgen et al. (1997) developed a protocol for analysis of bathing waters and drinking water which used filtration through positively charged membranes followed by ultrafiltration as a secondary concentration step, and Huang et al. (2000) used positively charged membranes followed by beef extract elution and PEG precipitation for the concentration of caliciviruses in water.

Table 2 summarises the methods for virus concentration from different water types.

Detection and enumeration of waterborne viruses

Detection and enumeration are conveniently considered together since for many viruses they are performed simultaneously. Detection may be done by infectivity-based methods where the virus undergoes at least partial multiplication in cell culture, or it may be done by techniques based on properties other than infectivity. Most important in this latter category are the molecular biological techniques, especially the PCR. Enumeration by molecular means may be semi-quantitative, such as end-point dilution assays or, increasingly, by real-time PCR for enumerating genome copies of a target virus, though the relationship between numbers of infectious units and genome copies depends on many variables.

Detection of virus infectivity is traditionally done by inoculating cell cultures with part or all of the concentrate and allowing the virus to multiply in the cells so that they are killed. The c.p.e. of many enteroviruses and some other types is visible to the naked eye. If a range of cell cultures is inoculated under liquid assay it should be possible to detect polio, Coxsackie B, echo viruses, as well as some adenoviruses and reoviruses. HAV may also be detected this way but only after prolonged incubation of cultures, and it is therefore not an approach used in routine waterborne HAV detection.

Cell culture

The line most favoured for enumeration of water-associated enteroviruses is the Buffalo green monkey (BGM) line first described in a water context by Dahling et al. (1974). This was reported to give higher plaque assay titres of poliovirus, Coxsackie viruses B, some echovirus and reoviruses than obtained in rhesus or grivet monkey kidney cells. Morris (1985) examined ten cell lines for their ability to grow enteroviruses isolated from wastewater effluent. Eighty-two percent of isolates were positive in BGM cells, 73% in rhabdomyosarcoma (RD) cells and 64% in chimpanzee liver cells. BGM was also the most sensitive in the number of plaques counted.

Dahling and Wright (1986) carried out an extensive set of experiments to optimise the BGM line in respect of a number of assays for waterborne viruses, and made recommendations in respect of many cell culture and assay parameters, as well as doing a comparative virus-isolation study involving BGM cells and nine other cell lines. This work has become the accepted basis for many standard methods on detection of water-associated viruses.

Other cell lines have been investigated for their ability to support the growth of enteric viruses. Most of these studies have been directed at growing the more fastidious agents like rotaviruses and astroviruses, but Patel et al. (1985) carried out a large survey on the susceptibility of a range of lines to different enteroviruses, including all 31 serotypes of echovirus; they found that two lines, HT-29 and SKCO-1, had a markedly wider sensitivity for enteroviruses than primary monkey kidney (PMK) or RD cell cultures. They require a high seed density and do not grow quickly however, and perhaps this is why they have not found greater favour, along with CaCO2 cells (Fogh et al., 1977), which are of similar origin, in the detection of waterborne enteric viruses generally. This latter line, along with RD, BGM and human epithelial type 2 (HEp-2) cells, were used by Sedmak et al. (2005) in the detection of infectious reoviruses, enteroviruses and adenoviruses in a range of water types.

A549 cells, derived from human lung tissue, support the growth of some adenoviruses derived from water; they have also been used in the integrated cell culture-PCR technique (see below) for rapid detection of infectious adenoviruses (Greening et al., 2002).

There are two approaches to the enumeration of virus infectivity, plaque assay and liquid culture assay.

Plaque assay

The plaque assay is most frequently used for the enumeration of infectious waterborne enteroviruses. All the concentrate should be tested. In both cases plaques develop following incubation and may be counted as they become visible, in the case of enteroviruses usually after about 3 days. One plaque is taken as being the progeny of one infectious unit of the virus; this may be the same as one virus particle, but is unlikely given the aggregation of virions and their association with both organic and inorganic particulate matter.

Monolayer plaque assay

The virus concentrate is inoculated on to preformed monolayers in petri dishes or flasks and the cells are reincubated under an agar overlay until a c.p.e. is seen.

Plaques are counted daily starting at day 2. Since viruses multiply at different rates counting is continued after the first appearance of plaques. Echoviruses, for example, take longer to form plaques, if they do at all. The UK (Standing Committee of Analysts SCA, 1995) method recommends counting plaques for 2–5 days; the USEPA (2007) method suggests counting should continue for 12 days or until no new plaques appear between counts; Block and Schwartzbrod (1989) recommended 6–14 days.

Suspended-cell plaque assay

The suspended-cell assay (Cooper, 1967) increases the sensitivity of the ordinary plaque assay by five to eight times (Dahling and Wright, 1988). Five times as many cells are used, suspended in the agar instead of being in a layer underneath it and thus many more adsorption sites are available to any virus present. No prior establishment of monolayers or fluid changes are required since cells and concentrate are added to the culture vessels at the same time. It can only be used where the virus is liberated into the medium. The USEPA method recommends that the suspended-cell assay should be used where the level of indigenous virus is likely to be less than 5 pfu/ml.

Liquid assays

Cells under liquid media may support the growth of more viruses than cells growing in or under agar. Many enteroviruses, especially some echoviruses, do not form plaques and so will not be detected under agar; some viruses take a long time to produce a c.p.e. and agar cultures may have deteriorated too far to be useful. In these cases cells growing under liquid medium are used. Virus multiplication produces cell degeneration and often a c.p.e. characteristic of the infecting virus, so some idea may be gained of the agent at hand.

Most probable number assay

Lee and Jeong (2004) analysed source, finished, and tap water samples for enteroviruses and adenoviruses in a comparative study of MPN titres, obtained by normal observation of c.p.e. and MPN titres, obtained by integrated cell-culture PCR (ICC-PCR, see page 196). They found that by normal observation of c.p.e. 15% of cultures were positive, all from source water samples, and that titres ranged from 3.3–21 MPN/100 l water. In contrast, MPN by ICC-PCR gave 21% cultures positive and a narrower range of titres for source waters (4.5–10.2 MPN/100 l water). Target viruses were also found in the finished and tap waters (0–0.9 MPN/ 100 l). The range of viruses detected by ICC-PCR will be limited by the primers used, and in this study re-resting of the c.p.e.-positive dishes with reovirus-specific primers revealed 89% of cultures positive. The MPN approach can thus be extended beyond the simple scoring of c.p.e.-positive cultures, but the limits of the detection system need to be kept in mind.

End-point dilution assay (TCD_{50})

Serial dilutions of the concentrate are inoculated into cell cultures and each culture is scored positive or negative after incubation. The titre is calculated (e.g. by the method of Reed & Muench, 1938) as the logarithm of the dilution of the virus producing a c.p.e. in 50% of the cultures. Though the method is simple and economic, its precision is difficult to evaluate. It is the least favoured of the three methods described.

Choice of assay method

Table 3 shows a comparison of assay methods in agar and under liquid media. It will be seen that there is no clear-cut best method. Plaque assays have greater advantages of individualising the pfu and providing entities (plaques), which are countable and directly related to the number of viruses (or aggregates). For many users this is an easier concept to grasp than the more abstract MPN or TCD_{50}. The MPN is more reliable than the others provided the number of cultures inoculated per dilution exceeds 30 (Block and Schwartzbrod, 1989).

Several comparative studies have been done on methods for the detection of enteroviruses in water. Morris and Waite (1980), for example, concluded that

Table 3

Characteristics of cell-culture assay methods

Attribute	Liquid	Agar
Range of viruses detected	Wide range possible	Non-plaquing viruses not detected
Blind passage	Blind passage possible to increase titres to detectable levels	Faster-growing viruses in a mixture overgrow slower ones, which are not isolated
Sensitivity	Greater sensitivity (especially than monolayers)	Sensitivity improved using suspended cell assay
Sub-culture	Sub-culturing easy	Sub-culturing difficult (impossible without c.p.e.)
Virus separation	Impossible to separate virus types	Separation of viruses possible by plaque picking
Statistical precision	Bad precision, large bias where few replicates used (as is usual)	Good, especially where all concentrate tested in one assay

monolayers were the least sensitive system, tube cultures were of intermediate sensitivity (for MPN determination, though only four tubes were set up per dilution) and the suspended-cell assay was the most sensitive. BGM cells gave the best recoveries and RD cells were variable. RD cells have been reported susceptible to Coxsackie virus A strains (Block and Schwartzbrod, 1989) though they are less sensitive than suckling mice, which is the only other system that supports growth of this group of viruses.

Virus infectivity may also be determined by immunofluorescence or immunoperoxidase techniques, which are particularly useful where limited replication occurs and a distinct c.p.e. is not produced. It may also be determined by molecular biology techniques such as the detection of virus-specific mRNA.

Identification

Viruses may be identified by the serum neutralisation test (SNT), immunoassay (Payment et al., 1982; Pandya et al., 1988), immunoperoxidase (Payment and Trudel 1985, 1987) or by genome-sequence analysis.

Flow cytometry has been used by Abad et al. (1998), Baradi et al. (1998) and Bosch et al. (2004) to sort rotavirus-infected $CaCO_2$ and MA-104 cells automatically.

Detection of viruses by molecular biology

The use of molecular biological detection techniques has permitted faster detection times and, in many cases, increases in sensitivity. It is particularly useful in the detection of viruses which do not multiply in cell culture and, since most of the gastroenteritis viruses fall into this category, this is an important development.

Techniques were first validated against cell-culture methods, which led to the development of molecular biology-based detection methods for enteroviruses in environmental concentrates, which were then taken forward in the development of methods for the detection of enteric pathogens.

Gene probes were the first approach made in the molecular biological detection of enteric viruses, and have been widely used (Dubrou et al., 1991; Enriquez et al., 1993; Margolin et al., 1993; Moore and Margolin, 1993). However they lack sensitivity and they have largely been superseded. Richardson et al. (1991) reviewed the water industry application of gene probes.

The PCR reaction (Saiki et al., 1988) overcomes these problems. Ease of use and increased sensitivity has made the technique commonplace in many laboratories. Problems encountered with PCR include the possible presence of fulvic and humic acids in the concentrates which inhibit the RT and/or polymerase reactions, and different solutions have been found including adsorption of the extracted RNA to silica (e.g. Shieh et al., 1995). Pallin et al. (1997) devised a method for recovering all the virus in a concentrate into a single PCR tube, which allowed direct comparisons of sensitivity with cell-culture methods where the whole of the concentrate is tested at one time.

The polymerase chain reaction

Numerous investigations have been done using RT-PCR to detect enteroviruses in different environmental samples, including river and marine recreational waters (e.g. Kopecka et al., 1993; Gilgen et al., 1995; Wyn-Jones et al., 1995), ground waters (Abbaszadegan et al., 1993; Regan and Margolin, 1997) and sludge-amended field soils (Straub et al., 1995). Detection of enteroviruses is a practical proposition since the picornavirus group contains well-conserved nucleotide sequences at the 5′ end of the genome, which are used to prepare pan-enterovirus primers, which are the starting reagents in the PCR. The technique has been extended to cover other virus groups present in water including adenoviruses (Puig et al., 1994), HAVs (Graff et al., 1993), astroviruses (Marx et al., 1995) and rotaviruses (Gajardo et al., 1995). Van Heerden et al. (2005a) compared two nested PCR methods for detection of adenoviruses in river and treated drinking water; the same group investigated swimming pools for the presence of adenoviruses by nested PCR (van Heerden et al., 2005b) and Jiang and Chu (2004) investigated rivers and coastal waters.

Lodder and de Roda Husman (2005) investigated the incidence of noroviruses, rotaviruses, enteroviruses and reoviruses in source waters and sewage. They developed a quantitative approach by analysing 10-fold serial dilutions of the extracted RNA and found noroviruses between 4 and 4900 "PCR-detectable units" per litre of river water. Higher titres were found in sewage. The Lordsdale strain of norovirus GGII was the most prevalent. Other viruses were also found. This approach to quantitation was extended and supported by statistical estimation by Westrell et al. (2006) who found norovirus titres up to 1700 (mean 12) PCR detectable units per litre in source water samples from the River Meuse. Borchardt et al. (2004) used a similar approach to estimate viruses transported by river water infiltrating municipal wells. Half the well water samples tested were positive for one or more of a range of enteric viruses, though no infectious virus was found.

Refinement of the RT-PCR and restriction enzyme analysis of amplicons has permitted the differentiation of virus types within the enterovirus group. Hughes et al. (1993) compared the nucleotide sequences of six Coxsackie virus B4 (CB4) isolates from the aquatic environment with those of four CB4 isolates from clinical specimens and found that the isolates fell into two distinct groups not related to their origin, and Sellwood et al. (1995) reported a system using restriction fragment length polymorphism analysis to discriminate between wild and vaccine-like strains of poliovirus.

Many (RT-)PCR-based analyses relate to outbreak investigations. Yeats et al. (2002) described an outbreak of illness in about 90 children followed by their attendance at a summer camp. Analysis by RT-PCR of stool specimens and drinking and swimming pool water samples revealed the presence of an enterovirus, later typed as echovirus 3 (EV3), in several of each kind of sample. Parshionikar et al. (2003) investigated an outbreak of gastroenteritis at a tourist saloon in the US and by RT-PCR found norovirus GGI.3 in both stool specimens and well water samples; Hoebe et al. (2004) conducted an epidemiological and virological investigations into an outbreak of gastroenteritis in children who had played in a recreational fountain, and found the same norovirus sequences in the stool samples as in the water samples. During investigations into the outbreak of SARS, Wang et al. (2005) used semi-nested RT-PCR followed by sequencing to detect and identify SARS-CoV RNA in sewage.

The persistence of HAV in many communities (and therefore the local environment) led Morace et al. (2002) to develop a rapid method for monitoring its environmental presence at sewage treatment plants in Southern Italy. RT-PCR was used to detect HAV in sewage and effluent, and the sensitivity could be refined by the use of an antigen-capture stage. Grimm and Fout (2002) developed an RT-PCR method for the detection of HEV in spiked water samples.

Most (RT-)PCR methods focus on the polymerase-gene sequence of the virus. However, it is often necessary to refine the analyses to discern different strains of viruses (e.g. noroviruses). Bon et al. (2005), in a molecular epidemiological study of calicivirus cases and outbreaks over a 6-year period found it important to target the capsid gene region as well as the polymerase region in order to discriminate between strains in outbreaks where more than one strain was involved. It is likely that this approach, where capsid-gene sequence can be related to serological information, will become increasingly useful in molecular epidemiological studies.

In further modifications designed to reduce the analysis time, Papaventsis et al. (2005) developed a method for culturing enteroviruses directly on the filter following adsorption, then further analysed by RT-PCR, restriction fragment length polymorphism and sequencing. Coxsackie A, B, and polioviruses were found.

Multiplex PCR methods have been developed by several investigators, but must be employed with caution and the appropriate controls. Egger et al. (1995) devised a multiplex PCR for the differentiation of polioviruses from non-polioviruses, which made an important step in the accumulation of public health information, and multiplex (RT-)PCR reactions have been described by Fout et al. (2003), Formiga-Cruz et al. (2005), Denis-Mize et al. (2004) and Li et al. (2002) for a range of viruses in several different aquatic matrices.

There is an important use for RT-PCR in the screening of samples for enteroviruses; negative ones can be discarded and positives investigated further for presence of infectious virus.

Real-time-PCR

Real-time PCR provides the possibility to quantify the number of specific sequences in a sample and has been applied to a number of environmental virology investigations. Choi and Jiang (2005) used it to estimate human adenoviruses in 114 river water samples; 16% were positive, each containing between 10^2 and 10^4 adenovirus genomes per litre. Plaque assays on A549 and HEK-293 cultures were negative, suggesting that the viruses detected by quantitative PCR (QPCR) were non-infectious. The group went on to develop a TaqMan® assay for Ad40 in a variety of environmental samples (Jiang et al., 2005). Pusch et al. (2005) detected a wide range of viruses in samples taken downstream of a waste-water treatment plant. By QPCR they estimated the range of titres of astroviruses to be $3.7 \times 10^3 - 1.2 \times 10^8$ and of noroviruses to be $1.8 \times 10^4 - 9.7 \times 10^5$ "genome equivalents" per litre. Laverick et al. (2004) devised a QPCR for noroviruses and used it in an in-depth 14-months surveillance of sewage, marine and riverine recreational waters. Absolute quantitation of template was obtained from a standard curve constructed using quantitative standards produced by cloning a modified sequence of the norovirus forward primer. Le Cann et al. (2004) devised a real-time RT-PCR for astrovirus in sewage, and reported mean values of 4.1×10^6 "astrovirus genomes" per 100 ml inlet sewage and 1.04×10^4 genomes in the effluent. HAV in polluted seawater was estimated by Brooks et al. (2005) to contain 90–523 copies of HAV per litre at one location and 347–2656 copies per litre at another, the range at each site being attributed to the variation in rainfall.

Nucleic acid sequence-based amplification

Nucleic acid sequence-based amplification (NASBA) amplifies target RNA at a single temperature (usually 41°C) and provides an alternative approach to the amplification of DNA sequences at varying temperatures. One advantage of this is that thermal stressing of blocks or carousels is avoided, another is that the time of the overall process is reduced compared with PCR. The progress of the reaction may still be monitored in real-time. The technique and its application to food and environmental materials have been reviewed by Cook (2003). Jean et al. (2002) used a NASBA coupled to an ELISA reaction for the detection of rotavirus in seeded sewage effluent samples, and Abd el-Galil et al. (2005) developed a NASBA reaction coupled to a molecular beacon for real-time detection of HAV in seeded surface water samples. The technique was used by Rutjes et al. (2005) for the detection of enteroviruses in surface water samples, though it was slightly less sensitive in detecting target virus sequences than RT-PCR. Rutjes et al. (2006) also developed a broadly reactive NASBA reaction for the detection of waterborne noroviruses and found it to be more sensitive than RT-PCR and, further, that the reaction was unaffected by inhibitors in the sample.

Molecular biology and virus infectivity

The principal drawback of molecular detection methods is that in their native form they give no indication of infectivity. Although knowledge of the structure of the target virus and some knowledge of how it behaves in the environment can lead to inferences about its infectivity, there is no direct indication of this in the data obtained from examination of an agarose gel or thermal cycler printout. This has led to much (mostly inconclusive) debate about the relationship between infectivity assay and molecular data. Difficulties in interpretation have arisen since the two kinds of information are not really comparable, being based on different properties of the virus. A number of approaches have been made to overcome this.

Integrated cell culture– PCR

Combination of cell culture with PCR has permitted detection of infectious virus even where it normally fails to produce a c.p.e., or where the c.p.e. takes a long time to appear. This technique, integrated cell culture–PCR (ICC-PCR, or ICC/RT-PCR for RNA viruses) has been used by several groups. Reynolds et al. (1996) and Murrin and Slade (1997) inoculated BGM cultures with concentrates and tested the supernatants at intervals up to 10 days. Virus was detectable by RT-PCR as early as 1 day post-inoculation, instead of more than 3 days by normal visualisation of c.p.e. Lee and Jeong (2004) compared ICC–PCR with total culturable virus assay for detection of enteroviruses, adenoviruses and reoviruses in water and found the ICC–(RT)PCR applicable as long as the limitations of the primers used were recognised; Spinner and Di Giovanni (2001) applied the technique to reovirus detection in drinking water sources. Jiang et al. (2004), investigating HAV in water, refined the technique in developing an integrated cell culture/strand-specific RT-PCR procedure capable of distinguishing between infectious and non-infectious HAV in spiked water samples. This involved initial propagation of infectious virus in cell culture followed by detection of the negative-strand RNA of the replicative intermediate using strand-specific RT-PCR. Greening et al. (2002) were able to detect naturally-occurring infectious enteroviruses and infectious adenoviruses in three days and five days respectively by ICC-(RT)PCR, compared with five days and 10 days if plaque assays or immunofluorescence were used. Cromeans et al. (2004) used a similar approach for the detection of HAV in water. The detection of the double-stranded replicative form of RNA viruses in cultured cells permits the conclusion that the virus is actually replicating and that it is not the sample inoculum which is being detected.

Detection of virus-specific mRNA

DNA viruses that do not replicate well in cell culture may be detected by the detection of virus-specific mRNA. This is particularly a useful approach in the detection of adenoviruses in water sample concentrates, particularly Ad40 and 41, which do not produce a clear c.p.e. Adenoviruses have a high particle/infectious virion ratio in culture (Brown et al., 1992), which is important when estimating the infectious viruses in a sample. Ko et al. (2003) developed a method for detection of infectious Ads2 and 41 in culture by detecting virus-specific mRNA, which is only produced during virus replication. The mRNA of Ad2 was detected as soon as 6 h after infection, and of Ad41 as soon as 24 h after infection of A549 cell cultures. This is in contrast to the development of up to 10 days for environmental isolates of "culturable" adenoviruses and several weeks (if at all) for the growth of Ad41 in culture. The group went on to develop the technique for use in detecting Ads in water sample concentrates and found they could detect as little as two infectious units (IU) Ad2 and 10 IU Ad41 in sample concentrates inoculated into cell cultures (Cromeans et al., 2004).

The combination of real-time PCR and detection of components produced only by replicating virus has significant meaning for the progress of detection of enteropathogenic viruses in aquatic matrices and the understanding of the significance of enteric viruses in the environment.

References

Abad FX, Pintó RM, Bosch A. Flow cytometry detection of infectious rotaviruses in environment and clinical samples. Appl Environ Microbiol 1998; 64: 2392–2396.

Abbaszadegan M, Huber MS, Gerba CP, Pepper IL. Detection of enteroviruses in ground water by PCR. Appl Environ Microbiol 1993; 65: 444–449.

Abd el-Galil KH, el-Sokkary MA, Kheira SM, Salazar AM, Yates MV, Chen W, Mulchandani A. Real-time nucleic acid sequence-based amplification assay for detection of hepatitis A virus. Appl Environ Microbiol 2005; 71: 7113–7116.

AranhaCreado H, Oshima K, Jafari S, Brandwein H. Virus retention by a hydrophilic triple-layer PVDF microporous membrane filter. PDA J Pharm Sci Technol 1997; 51: 119–124.

Baggi F, Peduzzi R. Genotyping of rotaviruses in environmental water and stool samples in Southern Switzerland by nucleotide sequence analysis of 189 base pairs at the 5' end of the VP7 gene. J Clin Microbiol 2000; 38: 3681–3685.

Baradi CRM, Emslie KR, Vesey G, Williams KL. Development of a rapid and sensitive quantitative assay for rotavirus based on flow cytometry. J Virol Methods 1998; 74: 31–38.

Belfort G, Paluszek A, Sturman LS. Enterovirus concentration using automated hollow fiber ultrafiltration. Water Sci Technol 1982; 14: 257–272.

Bergh O, Borsheim KY, Bratbakg G, Heldal M. High abundance of viruses found in aquatic environment. Appl Environ Microbiol 1989; 34: 467–468.

Beytout D, Laveran H, Reynaud MP. Mèthode practique d'evaluation numerique applicable aux techniques miniaturisèes de titrage en plages. Ann Biol Clin 1975; 33: 379–384.

Bitton G, Pancorbo O, Gifford GE. Factors affecting the adsorption of poliovirus to magnetite in water and wastewater. Water Res 1976; 10: 973–980.

Block JC, Schwartzbrod L. Viruses in Water Systems. Detection and Identification. New York: VCH Publishers, Inc.; 1989.

Bon F, Ambert-Balay K, Giraudon H, Kaplon J, Le Guyader S, Pommepuy M, Gallay A, Vaillant V, de Valk H, Chikhi-Brachet R, Flahaut A, Pothier P, Kohli E. Molecular epidemiology of caliciviruses detected in sporadic and outbreak cases of gastroenteritis in France from December 1998 to February 2004. J Clin Microbiol 2005; 43: 4659–4664.

Borchardt MA, Haas NL, Hunt RJ. Vulnerability of drinking-water wells in La Crosse, Wisconsin, to enteric-virus contamination from surface water contributions. Appl Environ Microbiol 2004; 70: 5937–5946.

Bosch A, Pintó RM, Blanch AR, Jofre JT. Detection of human rotavirus in sewage through two concentration procedures. Water Res 1988; 22: 343–348.

Bosch A, Pinto RM, Comas J, Abad FX. Detection of infectious rotaviruses by flow cytometry. Methods Mol Biol 2004; 268: 61–68.

Brooks HA, Gersberg RM, Dhar AK. Detection and quantification of hepatitis A virus in seawater via real-time RT-PCR. J Virol Methods 2005; 127: 109–118.

Brown M, Wilson-Friesen HL, Doane F. A block in release of progeny virus and a high particle-to-infectious unit ratio contribute to poor growth of enteric adenovirus types 40 and 41 in cell culture. J Virol 1992; 66: 3198–3205.

Chaudhiri M, Sattar SA. Enteric virus removal from water by coal-based sorbents: development of low-cost water filters. Water Sci Technol 1986; 18: 77–82.

Choi S, Jiang SC. Real-time PCR quantification of human adenoviruses in urban rivers indicates genome prevalence but low infectivity. Appl Environ Microbiol 2005; 71: 7426–7433.

Cook N. The use of NASBA for the detection of microbial pathogens in food and environmental samples. J Microbiol Methods 2003; 53: 165–174.

Cooper PD. The plaque assay of animal viruses. Adv Virus Res 1967; 8: 319–378.

Cromeans TL, Jothikumar N, Jung K, Ko G, Wait D, Sobsey MD. Development of Molecular Methods to Detect Infectious Viruses in Water. Denver, CO: Awwa Research Foundation; 2004.

Dahling DR, Berg G, Berman D. BGM, a continuous cell line more sensitive than primary rhesus and African green kidney cells for the recovery of viruses from water. Health Lab Sci 1974; 11: 275–282.

Dahling DR, Phirke PM, Wright BA, Safferman RS. Use of bituminous coal as an alternative technique for the field concentration of waterborne viruses. Appl Environ Microbiol 1985; 49: 1222–1225.

Dahling DR, Wright BA. Optimization of the BGM cell line and viral assay procedures for monitoring viruses in the environment. Appl Environ Microbiol 1986; 51: 790–812.

Dahling DR, Wright BA. Optimisation of suspended cell method and comparison with cell monolayer technique for virus assays. J Virol Methods 1988; 20: 169–179.

Denis-Mize K, Fout GS, Dahling DR, Francy DS. Detection of human enteric viruses in stream water with RT-PCR and cell culture. J Water Health 2004; 31: 37–47.

Divizia M, de Filippis P, di Napoli A, Venuti A, Perez-Bercoff R, Pana A. Isolation of wild-type hepatitis A virus from the environment. Water Res 1989a; 23: 1155–1160.

Divizia M, Santi AL, Pana A. Ultrafiltration: an efficient second step for hepatitis A and poliovirus concentration. J Virol Methods 1989b; 23: 55–62.

Dubrou S, Kopecka H, Lopez-Pila JM, Maréchal J, Prèvot J. Detection of hepatitis A virus and other enteroviruses in wastewater and surface water samples by gene probe assay. Water Sci Technol 1991; 24: 267–272.

Egger D, Pasamontes L, Ostermayer M, Bienz K. Reverse transcription multiplex PCR for differentiation between polio- and enteroviruses from clinical and environmental samples. J Clin Microbiol 1995; 33: 1442–1447.

Ehlers MM, Grabow WOK, Pavlov DN. Detection of enteroviruses in untreated and treated drinking water supplies in South Africa. Water Res 2005; 39: 2253–2258.

Ehlers MM, Heim A, Grabow WOK. Prevalence, quantification and typing of adenoviruses detected in river and treated drinking water in South Africa. J Appl Microbiol 2005; 99: 234–242.

Enriquez CE, Abbaszadegan M, Pepper IL, Richardson KJ, Gerba CP. Poliovirus detection in water by cell culture and nucleic acid hybridization. Water Res 1993; 27: 1113–1118.

Farrah SR, Gerba CP, Wallis C, Melnick JL. Concentration of viruses from large volumes of tap water using pleated membrane filters. Appl Environ Microbiol 1976; 31: 221–226.

Fogh J, Wright WC, Loveless JD. Absence of HeLa cell contamination in 169 cell lines derived from human tumors. J Natl Cancer Inst 1977; 58: 209–214.

Formiga-Cruz M, Hundesa A, Clemente-Casares P, Albinana-Gimenez N, Allard A, Girones R. Nested multiplex PCR assay for detection of human enteric viruses in shellfish and sewage. J Virol Methods 2005; 125: 111–118.

Fout GS, Martinson BC, Moyer MW, Dahling DR. A multiplex reverse transcription-PCR method for detection of human enteric viruses in groundwater. Appl Environ Microbiol 2003; 69: 3158–3164.

Gajardo R, Bouchrit N, Pintó RM, Bosch A. Genotyping of rotaviruses isolated from sewage. Appl Environ Microbiol 1995; 61: 3460–3462.

Gerba CP. Recovering viruses from sewage, effluents and water. In: Methods for Recovering Viruses from the Environment (Berg G, editor). Boca Raton, FL: CRC Press; 1987; pp. 1–23.

Gilgen M, Germann D, Luethy J, Huebner P. Three-step isolation method for sensitive detection of enterovirus, rotavirus, hepatitis A virus and small round-structured viruses in water samples. Int J Food Microbiol 1997; 37: 189–199.

Gilgen M, Wegmuller B, Burkhalter P, Buhler HP, Muller U, Luethy J, Candrian U. Reverse transcription PCR to detect enteroviruses in surface water. Appl Environ Microbiol 1995; 61: 1226–1231.

Grabow WOK, Nupen EM, Bateman BW. South African research on enteric viruses in drinking water. Monogr Virol 1984; 15: 146–155.

Graff J, Ticehurst J, Flehmig B. Detection of hepatitis A virus in sewage by antigen capture polymerase chain reaction. Appl Environ Microbiol 1993; 59: 3165–3170.

Greening GE, Hewitt J, Lewis GD. Evaluation of integrated cell culture-PCR (C-PCR) for virological analysis of environmental samples. J Appl Microbiol 2002; 93: 745–750.

Grimm AC, Fout GS. Development of a molecular method to identify hepatitis E virus in water. J Virol Methods 2002; 101: 175–188.

Guttman-Bass N, Catalano-Sherman J. Humic acid interference with virus recovery by electropositive microporous filters. Appl Environ Microbiol 1986; 52: 556–561.

Haramoto E, Katayama H, Ohgaki S. Detection of noroviruses in tap water in Japan by means of a new method for concentrating enteric viruses in large volumes of freshwater. Appl Environ Microbiol 2004; 70: 2154–2160.

Hill VR, Polaczyk AL, Hahn D, Narayanan J, Cromeans TL, Roberts JM, Amburgey JE. Development of a rapid method for simultaneous recovery of diverse microbes in drinking water by ultrafiltration with sodium polyphosphate and surfactants. Appl Environ Microbiol 2005; 71: 6878–6884.

Hoebe CJ, Vennema H, de Roda Husman AM, van Duynhoven YT. Norovirus outbreak among primary schoolchildren who had played in a recreational water fountain. J Infect Dis 2004; 189: 699–705.

Hou K, Gerba CP, Goyal SM, Zerda KS. Capture of Latex Beads, Bacteria, Endotoxin, and Viruses by Charge-Modified Filters. Appl Environ Microbiol 1980; 40: 892–896.

Hovi T, Stenvik M, Partanen H, Kangas A. Poliovirus surveillance by examining sewage specimens. Quantitative recovery of virus after introduction into sewerage at remote upstream location. Epidemiol Infect 2001; 127: 101–106.

Huang PW, Laborde D, Land VR, Matosn DO, Smith AW, Jiang X. Concentration and detection of caliciviruses in water samples by reverse transcription-PCR. Appl Environ Microbiol 2000; 66: 4383–4388.

Hugues B, André M, Champsaur H. Virus concentration from waste water: glass wool versus glass powder adsorption methods. Biomed Lett 1991; 46: 103–107.

Hughes MS, Hoey EM, Coyle PV. A nucleotide sequence comparison of Coxsackievirus B4 isolates from aquatic samples and clinical specimens. Epidemiol Infect 1993; 110: 389–398.

Jakubowski W, Hoff JC, Anthony NC, Hill WF. Epoxy-fiberglass adsorbent for concentrating viruses from large volumes of potable water. Appl Microbiol 1974; 28: 501.

Jean J, Blais B, Darveau A, Fliss I. Rapid detection of human rotavirus using colorimetric nucleic acid sequence-based amplification (NASBA)-enzyme-linked immunosorbent assay in sewage treatment effluent. FEMS Microbiol Lett 2002; 23: 143–147.

Jiang SC, Chu W. PCR detection of pathogenic viruses in southern California urban rivers. J Appl Microbiol 2004; 97: 17–28.

Jiang S, Dezfulian H, Chu W. Real-time quantitative PCR for enteric adenovirus serotype 40 in environmental waters. Can J Microbiol 2005; 51: 393–398.

Jiang YJ, Liao GY, Zhao W, Sun MB, Qian Y, Bain CX, Jiang SD. Detection of infectious hepatitis A virus by integrated cell culture/strand-specific reverse transcriptase-polymerase chain reaction. J Appl Microbiol 2004; 97: 1105–1112.

Jothikumar N, Khanna P, Paulmurugan R, Kamatchiammal S, Padmanabhan P. A simple device for the concentration and detection of enterovirus, hepatitis E virus and rotavirus from water samples by reverse transcription-polymerase chain reaction. J Virol Methods 1995; 55: 410–415.

Katzenelson E, Fattal B, Hostovesky T. Organic flocculation: an efficient second step concentration method for the detection of viruses in tap water. Appl Environ Microbiol 1976; 32: 638–639.

Keswick BH, Pickering LK, DuPont HL, Woodward WE. Organic Flocculation: an efficient second step concentration method for the detection of viruses in tap water. Appl Environ Microbiol 1983; 46: 813–816.

Kittigul L, Khamoun P, Sujirarat D, Utrarachkij F, Chitpirom K, Chaichantanakit N, Vathanophas K. An improved method for concentrating rotavirus from water samples. Mem Inst Oswaldo Cruz 2001; 96: 815–821.

Ko G, Cromeans TL, Sobsey MD. Detection of infectious adenovirus in cell culture by mRNA reverse transcription-PCR. Appl Environ Microbiol 2003; 69: 7377–7384.

Kopecka H, Dubrou S, Prèvot J, Maréchal J, Lopez-Pila JM. Detection of naturally-occurring enteroviruses in waters by reverse transcription, polymerase chain reaction, and hybridization. Appl Environ Microbiol 1993; 59: 1213–1219.

Lahke SB, Parhad NM. Concentration of viruses from water on bituminous coal. Water Res 1988; 22: 635–640.

Laverick MA, Wyn-Jones AP, Carter MJ. Quantitative RT-PCR for the enumeration of noroviruses (Norwalk-like viruses) in water and sewage. Lett Appl Microbiol 2004; 39: 127–136.

Le Cann P, Ranarijaona S, Monpoeho S, Le Guyader F, Ferre V. Quantification of human astroviruses in sewage using real-time RT-PCR. Res Microbiol 2004; 155: 11–15.

Lee HK, Jeong YS. Comparison of total culturable virus assay and multiplex integrated cell culture-PCR for reliability of waterborne virus detection. Appl Environ Microbiol 2004; 70: 3632–3636.

Li JW, Wang XW, Yuan CQ, Zheng JL, Jin M, Song N, Shi XQ, Chao FH. Detection of enteroviruses and hepatitis A virus in water by consensus primer multiplex RT-PCR. World J Gastroenterol 2002; 8: 699–702.

Lodder WJ, de Roda Husman AM. Presence of noroviruses and other enteric viruses in sewage and surface waters in The Netherlands. Appl Environ Microbiol 2005; 71: 1453–1461.

Lund E, Hedstrom CE. The use of an aqueous polymer phase system for enterovirus isolations from sewage. Am J Epidemiol 1966; 84: 287–291.

Ma J-F, Naranjo J, Gerba CP. Evaluation of MK filters for recovery of enteroviruses from tap water. Appl Environ Microbiol 1994; 60: 1974–1977.

Mack WN, Yue-Shoung L, Coohon DB. Isolation of poliomyelitis virus from a contaminated well. Public Health Rep 1972; 87: 271–274.

Margolin AB, Gerba CP, Richardson KJ, Naranjo JE. Comparison of cell culture and a poliovirus gene probe assay for the detection of enteroviruses in environmental water samples. Water Sci Technol 1993; 27: 311–314.

Marx FE, Taylor MB, Grabow WOK. Optimization of a PCR method for the detection of astrovirus type 1 in environmental samples. Water Sci Technol 1995; 31: 359–362.

Moore N, Margolin AB. Evaluation of radioactive and non-radioactive gene probes and cell culture for detection of poliovirus in water samples. Appl Environ Microbiol 1993; 59: 3145–3146.

Morace G, Aulicino FA, Angelozzi C, Costanzo L, Donadio F, Rapicetta M. Microbial quality of wastewater: detection of hepatitis A virus by reverse transcriptase-polymerase chain reaction. J Appl Microbiol 2002; 92: 828–836.

Morris R. Detection of enteroviruses: an assessment of ten cell lines. Water Sci Technol 1985; 17: 81–88.

Morris R, Waite WM. Evaluation of procedures for the recovery of viruses from water—I Concentration systems. Water Res 1980; 14: 791–793.

Murphy AM, Grohmann GS, Sexton MFH. Infectious gastroenteritis in Norfolk Island and recovery of viruses from drinking water. J Hyg 1983; 91: 139–146.

Murrin K, Slade J. Rapid detection of viable enteroviruses in water by tissue culture and semi-nested polymerase chain reaction. Water Sci Technol 1997; 35: 429–432.

Muscillo M, Carducci A, la Rosa G, Cantiani L, Marianelli C. Enteric virus detection in Adriatic seawater by cell culture, polymerase chain reaction and polyacrylamide gel electrophoresis. Water Res 1997; 31: 1980–1984.

Myrmel M, Rimstad E, Wasteson Y. Immunomagnetic separation of a Norwalk-like virus (genogroup I) in artificially contaminated environmental water samples. Int J Food Microbiol 2000; 62: 17–26.

Nupen EM, Basson NC, Grabow WOK. Efficiency of ultrafiltration for the isolation of enteric viruses and coliphages from large volumes of water in studies on wastewater reclamation. Water Pollut Res 1981; 13: 851–863.

Pallin R, Place BM, Lightfoot NF, Wyn-Jones AP. The detection of enteroviruses in large volume concentrates of recreational waters by the polymerase chain reaction. J Virol Methods 1997; 57: 67–77.

Pandya G, Jana AM, Tuteja U, Rao KM. Identification of Group A Coxsackieviruses form sewage samples by indirect enzyme-linked immunosorbent assay. Water Res 1988; 22: 1055–1057.

Papaventsis D, Siafakas N, Markoulatos P, Papageorgiou GT, Kourtis C, Chatzichristou E, Economou C, Levidiotou S. Membrane adsorption with direct cell culture combined with reverse transcription-PCR as a fast method for identifying enteroviruses from sewage. Appl Environ Microbiol 2005; 71: 72–79.

Parshionikar SU, Willian-True S, Fout GS, Robbins DE, Seys SA, Cassady JD, Harris R. Waterborne outbreak of gastroenteritis associated with a norovirus. Appl Environ Microbiol 2003; 69: 5263–5268.

Patel JR, Daniel J, Mathan VI. A comparison of the susceptibility of three human gut tumour-derived differentiated epithelial cell lines, primary monkey kidney cells and human rhabdomyosarcoma cell line to 66-prototype strains of human enteroviruses. J Virol Methods 1985; 12: 209–216.

Payment P, Tremblay C, Trudel M. Rapid identification and serotyping of poliovirus isolates by an immunoassay. J Virol Methods 1982; 5: 301–308.

Payment P, Trudel M. Efficiency of several micro-fiber glass filters for recovery of poliovirus from tap water. Appl Environ Microbiol 1979; 38: 365–368.

Payment P, Trudel M. Immunoperoxidase method with human immune serum globulin for broad spectrum detection of cultivable viruses in environmental samples. Appl Environ Microbiol 1985; 50: 1308–1310.

Payment P, Trudel M. Detection and quantitation of human enteric viruses in wastewaters: increased sensitivity using a human immune serum immunoglobulin immunoperoxidase assay on MA-104 cells. Can J Microbiol 1987; 33: 568–570.

Percival S, Chalmers R, Embrey M, Hunter P, Sellwood J, Wyn-Jones P. Microbiology of Waterborne Diseases. Amsterdam: Elsevier Academic Press; ISBN 0-12-551570-7; 2004.

Pina S, Jofre J, Emerson SU, Purcell RH, Girones R. Characterization of a strain of infectious hepatitis E virus isolated from sewage in an area where hepatitis E is not endemic. Appl Environ Microbiol 1998; 64: 4485–4488.

Puig M, Jofre J, Lucena F, Allard A, Wadell G, Girones R. Detection of adenoviruses and enteroviruses in polluted water by nested PCR amplification. Appl Environ Microbiol 1994; 60: 2963–2970.

Pusch D, Oh DY, Wolf S, Dumke R, Schroter-Bobsin U, Hohne M, Roske I, Schreier E. Detection of enteric viruses and bacterial indicators in German environmental waters. Arch Virol 2005; 150: 929–947.

Ramia S, Sattar SA. Second-step concentration of viruses in drinking and surface waters using polyethylene glycol extraction. Can J Microbiol 1979; 25: 587.

Rao C, Sullivan R, Read RB, et al. A simple method for concentrating and detecting viruses. J Am Water Works Assoc 1968; 60: 1288–1294.

Reed LJ, Muench H. A simple method of estimating fifty percent endpoints. Am J Hyg 1938; 27: 493–495.

Regan PM, Margolin AB. Development of a nucleic acid capture probe with reverse transcriptase-polymerase chain reaction to detect poliovirus in groundwater. J Virol Methods 1997; 64: 65–72.

Reynolds CA, Gerba CP, Pepper IL. Detection of infectious enterovirus by an integrated cell culture-PCR procedure. Appl Environ Microbiol 1996; 62: 1424–1427.

Richardson KJ, Stewart MH, Wolfe RL. Application of gene probe technology to the water industry. J Am Water Works Assoc 1991; 83: 71–81.

Rotem Y, Katzenelson E, Belfort G. Virus concentration by capillary ultrafiltration. J Environ Eng-ASCE 1979; 5: 401–407.

Rutjes SA, Italiaander R, van den Berg HH, Lodder WJ, de Roda Husman AM. Isolation and detection of enterovirus RNA from large-volume water samples by using the NucliSens miniMAG system and real-time nucleic acid sequence-based amplification. Appl Environ Microbiol 2005; 71: 3734–3740.

Rutjes SA, van den Berg HHJL, Lodder WJ, de Roda Husman AM. Real-time detection of noroviruses in surface water by use of a broadly reactive nucleic acid sequence-based amplification assay. Appl Environ Microbiol 2006; 72: 5349–5358.

Saiki RK, Gelfand DH, Stoffel S, Scharf SJ, Higuchi R, Horn GT, Mullis KB, Erlich HA. Primer-directed enzymatic amplification of DNA with a thermostable DNA polymerase. Science 1988; 239: 487–491.

Sarrette B, Danglot B, Vilagines R. A new and simple method for the recuperation of enteroviruses from water. Water Res 1977; 11: 355–358.

Sattar SA, Ramia S. Use of talc-selite layers in the concentration of enteroviruses from large volumes of potable waters. Water Res 1979; 13: 1351–1353.

Sattar SA, Westwood JCN. Viral pollution of surface waters due to chlorinated primary effluents. Appl Environ Microbiol 1978; 36: 427–431.

Schwab KJ, Leon R, Sobsey MD. Immunoaffinity concentration and purification of waterborne enteric viruses for detection by reverse transcriptase PCR. Appl Environ Microbiol 1996; 62: 2086–2094.

Schwartzbrod L, Lucena-Gutierrez F. Concentration des enterovirus dans les eaux par adsorption sur poudre de verre: proposition d'un appareillage simplifiè. Microbia 1978; 4: 55–58.

Sedmak G, Bina D, MacDonald J, Couillard L. Nine-year study of the occurrence of culturable viruses in source water for two drinking water treatment plants and the influent and effluent of a wastewater treatment plant in Milwaukee, Wisconsin (August 1994 through July 2003). Appl Environ Microbiol 2005; 71: 1042–1050.

Sellwood J, Litton PA, McDermott J, Clewley JP. Studies on wild and vaccine strains of poliovirus isolated from water and sewage. Water Sci Technol 1995; 31: 317–321.

Shieh Y-SC, Wait D, Tai L, Sobsey MD. Methods to remove inhibitors in sewage and other faecal wastes for enterovirus detection by the polymerase chain reaction. J Virol Methods 1995; 54: 51–66.

Sobsey MD, Glass JS. Poliovirus concentration from tap water with electropositive adsorbent filters. Appl Environ Microbiol 1980; 40: 201–210.

Sobsey MD, Hickey AR. Effects of humic and fulvic acids on poliovirus concentration from water by microporous filtration. Appl Environ Microbiol 1985; 49: 259–264.

Sobsey MD, Jones BL. Concentration of poliovirus from tapwater using positively-charged microporous filters. Appl Environ Microbiol 1979; 37: 588–595.

Spinner ML, Di Giovanni GD. Detection and identification of mammalian reoviruses in surface water by combined cell culture and reverse transcription-PCR. Appl Environ Microbiol 2001; 67: 3016–3020.

Standing Committee of Analysts (SCA). Methods for the isolation and identification of human enteric viruses from waters and associated materials. Methods for the Examination of Waters and Associated Materials. London: HMSO; 1995.

Straub TM, Pepper IL, Gerba CP. Comparison of PCR and cell culture for detection of enteroviruses in sludge-amended field soils and determination of their transport. Appl Environ Microbiol 1995; 61: 2066–2068.

Tsai YL, Sobsey MD, Sangermano LR, Palmer CJ. Simple method of concentrating enteroviruses and hepatitis A virus from sewage and ocean water for rapid detection by reverse transcriptase-polymerase chain reaction. Appl Environ Microbiol 1993; 59: 3488–3491.

United States Environmental Protection Agency (USEPA). USEPA/APHA Standard methods for the examination of water and wastewater. 2007.

Urase T, Yamamoto K, Ohgaki, S. Evaluation of virus removal in membrane separation processes using coliphage Q beta. In: Development and Water Pollution Control in Asia

(Liu S, Bhamidimarri R, Li X, editors). Pergamon Press; 1994; pp. 9–15; ISBN 0080424910.

Vaidya SR, Chitambar SD, Arankalle VA. Polymerase chain reaction-based prevalence of hepatitis A, hepatitis E and TT viruses in sewage from an endemic area. J Hepatol 2002; 37: 131–136.

van Heerden J, Ehlers MM, Heim A, Grabow WOK. Prevalence, quantification and typing of adenoviruses detected in river and treated drinking water in South Africa. J Appl Microbiol 2005a; 99: 234–242.

van Heerden J, Ehlers MM, Heim A, Grabow WOK. Detection and risk assessment of adenoviruses in swimming pool water. J Appl Microbiol 2005b; 99: 1256–1264.

Vilaginès P, Sarrette B, Husson G, Vilaginès R. Glass wool for virus concentration at ambient water pH level. Water Sci Technol 1993; 27: 299–306.

Vilaginès Ph, Sarrette B, Vilaginès R. Détection en continu du poliovirus dans des eaux de distribution publique. C R Acad Sci (Paris) 1988; t307(serie III): 171–176.

Wallis C, Melnick JL. Concentration of viruses from sewage by adsorption on to Millipore membranes. Bull World Health Organ 1967a; 36: 219–225.

Wallis C, Melnick JL. Concentration of viruses on aluminium and calcium salts. Am J Epidemiol 1967b; 85: 459–468.

Wallis C, Melnick JL. Concentration of viruses on membrane filters. J Virol 1967c; 1: 472–477.

Wang XW, Li J, Guo T, Zhen B, Kong Q, Yi B, Li Z, Song N, Jin M, Xiao W, Zhu X, Gu C, Yin J, Wei W, Yao W, Liu C, Li J, Ou G, Wang M, Fang T, Wang G, Qiu Y, Wu H, Chao F, Li J. Concentration and detection of SARS coronavirus in sewage from Xiao Tang Shan Hospital and the 309th Hospital of the Chinese People's Liberation Army. Water Sci Technol 2005; 52: 213–221.

Watt PM, Johnson DC, Gerba CP. Improved method for concentration of *Giardia, Cryptosporidium* and poliovirus from water. J Environ Sci Health 2002; 37: 321–330.

Wellings FM, Lewis AL, Mountain CW. Demonstration of solids-associated virus in wastewater and sludge. Appl Environ Microbiol 1976; 31: 354–360.

Westrell T, Teunis P, van den Berg H, Lodder W, Ketelaarse H, Stenstrom TA, de Roda Husman AM. Short- and long-term variations of norovirus concentrations in the Meuse river during a 2-year study period. Water Res 2006; 40: 2613–2620.

Wolfaardt M, Moe CL, Grabow WOK. Detection of small round-structured viruses in clinical and environmental samples by polymerase chain reaction. Water Sci Technol 1995; 31: 375–382.

Wyn-Jones AP, Pallin R, Sellwood J, Tougianidou D. Use of the polymerase chain reaction for the detection of enteroviruses in river and marine recreational waters. Water Sci Technol 1995; 31: 337–344.

Wyn-Jones AP, Sellwood J. Review of methods for the isolation, concentration, identification and enumeration of enteroviruses. Project WW-11B Research and Development—Bathing Water Policy, UK Water Industry Research Ltd.; 1998.

Wyn-Jones AP, Sellwood J. Enteric viruses in the aquatic environment. J Appl Microbiol 2001; 91: 945–962 Invited Review.

Yeats J, Smuts H, Serfontein CJ, Kannemeyer J. Investigation into a school enterovirus outbreak using PCR detection and serotype identification based on the 5′ non-coding region. Epidemiol Infect 2002; 133: 1123–1130.

© 2007 Elsevier B.V. All rights reserved
DOI 10.1016/S0168-7069(07)17010-5

Chapter 10

Viruses in Shellfish

Françoise S. Le Guyader[a], Robert L. Atmar[b]

[a]*Laboratoire de Microbiologie, IFREMER, BP 21105, 44311 Nantes cedex 03, France*
[b]*Departments of Medicine and Molecular Virology & Microbiology, Baylor College of Medicine, 1 Baylor Plaza, MS BCM280, Houston, TX 77030*

Background

Shellfish have been identified as a vector for human enteric pathogens for more than 150 years. During the 1800s, outbreaks of typhoid fever and cholera were associated with shellfish consumption (Richards, 1985). Contamination of shellfish-growing waters with human sewage was recognized as a contributing cause of the outbreaks and this recognition led to the development of bacteriologic criteria to assess the impact of sewage on shellfish and shellfish-growing waters. Shellfish filter large volumes of water during their feeding, and in the process they concentrate small particles containing microalgae and microorganisms. The practice of consuming shellfish either raw or undercooked prevents inactivation or killing of any enteric pathogens that are present and can lead to transmission of disease.

The identification of enteric bacteria in shellfish or their growing waters was recognized to be an indicator of exposure of the shellfish to fecal flora, and thus potentially to human enteric pathogens. The institution of regulations to specify acceptable levels of bacterial enteric pathogens in shellfish tissues or in shellfish growing waters in Europe (European regulation, 91/492/EC) and the United States (National Shellfish Sanitation Program) and improvements in sewage waste treatment procedures were followed by the virtual elimination of shellfish-associated outbreaks of typhoid fever and cholera in the United States (Richards, 1985; Rippey, 1994).

As outbreaks of shellfish-associated bacterial infection declined, a new problem emerged. Outbreaks of nonbacterial gastroenteritis and infectious hepatitis were described in association with shellfish consumption (Richards, 1987). These outbreaks were determined to be caused by viruses that either grow poorly in cell culture or cannot be cultivated at all, including hepatitis A virus (HAV) and noroviruses (NoVs) (Portnoy et al., 1975; Murphy et al., 1979). In many outbreaks, the shellfish and shellfish-growing waters met regulatory criteria for fecal-bacterial levels. Furthermore, the practice of depuration, a process by which shellfish "purify" themselves of enteric bacteria by filtering clean waters, failed to eliminate the risk of viral mediated disease (Mackowiak et al., 1976; Grohmann et al., 1981).

The failure of bacterial indicators (e.g. fecal coliform counts) to identify shellfish contaminated with enteric viruses led to the exploration of other possible indicators of fecal contamination. Human feces contain not only bacterial species but also bacteriophages that infect the fecal-bacterial organisms. Because routine detection of viral contamination has not been readily available, phages were proposed to be a more accurate indicator of the viral risk associated with shellfish sold for consumption (Chung et al., 1998; Doré et al., 2000). Phage properties that made them an attractive candidate indicator included the following: they are abundant in fecal samples; their genetic and physical properties are similar to those of the important human enteric viruses; and methods to measure their presence in shellfish and water samples are readily available (Doré et al., 2000). Three different phage species have been proposed as potential indicators (somatic, F-specific RNA (F-RNA) or *Bacteroides fragilis* phages), but most of the available data are on F-RNA phages.

Several studies have examined the correlation between detection of phage and human enteric viruses. Lee et al. (1999) found no relationship between HAV contamination of shellfish collected from three different areas in Hong Kong and the detection of phages. Chung et al. (1998) found that all the samples collected from a contaminated area in the United States were found positive for F-RNA phages, and half of the samples were also positive for enteroviruses. In England, a relationship between the presence of NoV and F-RNA phages in shellfish from polluted sites was noted (Doré et al., 2000). Hernroth et al. (2002) evaluated several different phage indicators in mussels harvested in Sweden and found that only F-RNA phages were significantly associated with the presence of enteroviruses in these shellfish; on the other hand, the F-RNA phages were not associated with the presence of either NoVs or adenoviruses. They concluded that phages were no better an indicator than *E. coli* as evidence for fecal contamination of an area and that improved methods of detection of enteric pathogens are needed (Hernroth et al., 2002). Myrmel et al. (2004) found a correlation between the presence of NoV and F-RNA phage contamination of mussels harvested from Norwegian waters. However, more than half of the NoV positive samples were negative for F-RNA phages and a positive result for F-RNA phages was less than twice as common in samples with NoV as in those without NoV, raising the question about the value of F-RNA as an indicator (Myrmel et al., 2004). Mussels collected from the Adriatic

Sea were found to be contaminated with HAV but were not contaminated with F-RNA phages or *E. coli* (Croci et al., 2000). In contrast, a 1-year study in the Netherlands identified the presence of phages (F-RNA or somatic) in 67% of oyster samples analyzed without detecting pathogenic viruses such as NoV or HAV (Lodder-Verschoor et al., 2005). A survey of shellfish collected from several European countries showed geographic variations in the presence of F-RNA phages and the presence of human enteric viruses (Formiga-Cruz et al., 2003). These studies demonstrate the difficulty in selecting an indicator for monitoring contamination of shellfish with human viral pathogens.

Enteric viruses causing shellfish-associated disease

Many enteric viruses can be detected in shellfish (Table 1), but only a few have been associated with outbreaks of human disease following consumption of contaminated shellfish. A brief description of the viruses causing shellfish-associated disease is given below, while a more extensive discussion of these viruses is provided in other chapters of this issue.

Hepatitis A virus (HAV)

Infectious hepatitis, caused by the HAV, is one of the most serious illnesses transmitted by shellfish. The HAV belongs to the genus *Hepatovirus* of the family *Picornaviridae*, and it is very stable in the environment, remaining viable for up to several weeks in water or on fomites (Abad et al., 1994; Arnal et al., 1998). Hepatitis A has an average incubation period of 28 days (range 15–50 days). It is generally asymptomatic or associated with a mild illness in young children, while in older children and adults the illness is characterized by jaundice in more than 70% of individuals (CDC, 2006). There is only a single serotype, and an effective vaccine is available for prevention of infection (CDC, 2006).

Noroviruses (NoV)

NoVs are the most common infection currently associated with shellfish consumption. *Norovirus* is a genus in the family *Caliciviridae*, and the genus is divided into five genogroups (Zheng et al., 2006). Genogroups I, II, and IV contain human strains, and the genogroups are further subdivided into genotypes based upon analyses of the amino acid sequence of the major capsid protein, VP1. NoV infection causes gastroenteritis characterized by the symptoms of vomiting and diarrhea (Atmar and Estes, 2006). The prevalence of vomiting along with the short incubation period (1–2 days) and short clinical illness (1–3 days) has been used epidemiologically to identify probable outbreaks of NoV-associated gastroenteritis (Kaplan et al., 1982; Turcios et al., 2006). The virus is stable in the environment, and the infectious dose is relatively low. Cooking of shellfish (e.g. through

Table 1

Detection of human enteric viruses in shellfish collected in different countries

Type of shellfish	Country	Commercial area				Noncommercial area				References
		AdV	EV	HAV	NoV	AdV	EV	HAV	NoV	
Oyster	Brazil		19	0	23			49		Sincero et al. (2006)
	France[a]						38	8	25	Le Guyader et al. (2000)
	France		13	0			49[b]			Dubois et al. (2004)
	Imported				9					Beuret et al. (2003)
	Imported				10.5					Cheng et al. (2005)
	Japan				9					Nishida et al. (2003)
	Japan								60	Ueki et al. (2005)
	Korea	89	11							Choo and Kim (2006)
	The Netherlands		27	0	0		14	0	0	Lodder-Verschoor et al. (2005)
	UK				37					Henshilwood et al. (1998)
	US				18[b]				56	Costantini et al. (2006)
Mussels	China		14	12						Lee et al. (1999)
	Italy			36						Croci et al. (2000)
	Italy			23				20		Chironna et al. (2002)
	Italy			34						De Medici et al. (2001)
	Italy				19					De Medici et al. (2004)
	Morocco	0				20				Karamoto et al. (2005)
	Norway	19			7					Myrmel et al. (2004)
	Spain		13		20	50	17		20	Muniain-Mujika et al. (2003)
	Sweden	17	13	nr	20	57	31		30	Hernroth et al. (2002)
	Sweden	17				28	15		16.5	Formiga-Cruz et al. (2002)
	Tunisia[c]						0	0	35	Elamri et al. (2006)
Oyster/Mussels	Greece	28	10	11	6	37	21	26	5.5	Formiga-Cruz et al. (2002)
	The Netherlands	nr	5		16			2%		Boxman et al. (2006)
	Spain	31	0	0	12	35	32	4	30	Formiga-Cruz et al. (2002)
	UK	28	6	0	8	52	17	1	14	Formiga-Cruz et al. (2002)
Clams	Spain							53		Sunen et al. (2004)
Cockles	Italy			36						De Medici et al. (2001)

Abbreviations: AdV, adenovirus; EV, enterovirus; HAV, hepatitis A virus; NoV, norovirus.

Note: Number represent % of samples positive for the designated enteric virus; blanks were used when samples were not evaluated for the designated virus.
[a] In this study astroviruses and rotaviruses were detected in 17 and 27% of sample in commercial area and in 47 and 44% in noncommercial area, respectively.
[b] Human and animal strains detected.
[c] Astroviruses were detected in 61% of the sample (collected in a noncommercial area).

steaming) may fail to inactivate the virus, and outbreaks of infection have occurred after consumption of cooked shellfish (MacDonnell et al., 1997).

Other enteric viruses

Many other enteric viruses have been detected in shellfish tissues (Table 1), including rotaviruses, astroviruses, enteroviruses, and adenoviruses (Le Guyader et al., 2000; Hernroth et al., 2002; Elamri et al., 2006). However, human outbreaks of disease caused by these viruses are rarely recognized in association with consumption of contaminated shellfish based upon the lack of reports in the scientific literature. Reasons for the lack of shellfish-associated disease caused by these viruses may include relative lack of susceptibility of the persons consuming the shellfish (i.e. pre-existing immunity), higher infectious doses needed to establish infection, and milder symptomatic infection leading to lack of recognition (under-reporting). Shellfish consumption has been reported as a risk factor for hepatitis E virus (HEV) infection, but further studies are needed to establish this link (Cacopardo et al., 1997; Koizumi et al., 2004).

Economic importance of shellfish and impact of viral disease

The worldwide consumption of shellfish has increased considerably during the past 3 decades, and infectious outbreaks have been increasingly reported from almost all continents (Potasman et al., 2002). Oysters are the most commonly implicated, and most of the symptoms are self-limiting gastrointestinal symptoms. However, severe infections have been reported due to HAV and *Vibrio vulnificus*, especially in immunosuppressed patients (Potasman et al., 2002).

The European shellfish industry generates 460 M€ per year, and it has been increasing by approximately 7% each year. The European market produces 1,232,183 tons of mollusks per year, with oysters accounting for 137,187 tons of the total (OFIMER 2003). More than a third of the worldwide shellfish production comes from European production countries (e.g. in 1991, 180,000 tons of live weight; 72% from farmed bivalves; and 28% from the wild) (Eurostat data). Twenty-three thousand workers are employed by 8500 European companies, and this activity is one of the major sources of employment in many coastal areas (e.g. Ireland, France, Spain, the Netherlands). France produces almost 80% of the European oysters, while Spain is the second largest producer of mussels worldwide after China.

The costs of outbreaks of shellfish-associated viral disease have not been clearly defined, but they are likely to be substantial. In the USA, food-borne diseases are a major cause of morbidity and hospitalization, with about 325,000 hospitalizations and 5000 deaths per year (Butt et al., 2004). Among the 76 million estimated cases of food-borne disease, 10–19% of those for which a vehicle of transmission is identified involve seafood; half of these are caused by viruses, and half of the illnesses are associated with shellfish consumption (Mead et al., 1999; Butt et al.,

2004). In countries with higher seafood consumption, or where seafood is traditionally eaten raw, a larger percentage of food-borne illnesses are due to seafood
consumption. For example, in Japan as much as 70% of food-borne illness is
associated with seafood consumption (Butt et al., 2004).

The costs associated with some outbreaks have been assessed. A series of outbreaks associated with NoV infection in New York over a 5-month period in 1982
led to estimated market losses of 1.8 million dollars from price decreases and
additional costs of more than 600 thousand dollars (Brown and Folsom, 1983). An
outbreak of hepatitis A occurred in Italy in 1996 and was strongly associated with
consumption of uncooked shellfish (Lopalco et al., 1997). This outbreak spread to
other parts of Italy and a cost analysis determined that the average cost to the 5889
affected individuals was 662 dollars while the overall costs to society were more
than 24 million dollars, representing 0.4% of public health expenditures in the
region during this time (Lucioni et al., 1998). Thus, costs involve both direct market
losses (e.g. through closure of producing areas and loss of consumer confidence in
food safety leading to decreased demand) and costs of illness (e.g. direct medical
care, lost time from work). Many shellfish-producing areas have active tourism,
and negative publicity associated with illness outbreaks may also adversely impact
local economies.

Biology of shellfish and interactions with ingested enteric viruses

Feeding method of the oyster (filter feeder)

The oyster digestive system is adapted to process particulate food (microalgae) and
will be described as a model for other shellfish. Organs involved in the ingestion
and digestion of food and the elimination of feces include the mouth, a short
esophagus, stomach, crystalline style sac, digestive diverticula, midgut, rectum, and
anus. With the exception of a short section of the rectum, the entire alimentary
canal lies within the visceral mass and is completely immobilized by the surrounding connective tissue. Food is moved from the mouth toward the anus by the strong
ciliary activity from epithelial cells that line the alimentary tract. The stomach is a
complex pouch, connected to both the digestive diverticulum and to the intestinal
tract (via the midgut and style sac). The digestive gland surrounds the stomach
entirely and also part of the intestine. It comprises a series of branched ducts that
open into the stomach, and the duct branches serially to terminate in blind-ending
tubules. The digestive tubules undergo cyclical cytological changes, possibly corresponding to tidal cycles: holding phase (flat digestive cells and wide lumen),
absorption and intracellular digestion (begins with the arrival of food), disinte-
grating phase (apical part of the digestive cells swells and protrudes into the lumen
releasing numerous free spherules), and reconstituting phase (crypt cells regenerate
new digestive cells) (Gosling, 2003).

Shellfish pump water over their gills and suspended particles are captured and
passed on to the alimentary tract. However, some sorting of particles occurs prior

to ingestion to help regulate what is presented to the digestive tract. Food particles enter the stomach through the short esophagus, and particles are sorted further according to size, density, and digestibility. The ciliary action of epithelial cells sorts the particles in the stomach as follows: small and heavy (or excess) particles are immediately rejected through the intestinal groove to the midgut while larger or lighter particles are recirculated for further degradation. Food particles are embedded in mucous strings from the esophagus and are carried forward by the rotation of the crystalline style and subjected to mechanical and chemical (mainly glucanases) degradation. Small particles and probably soluble molecules enter the digestive gland via the brush border of the ducts. A second phase of extracellular digestion may occur in the lumen of the tubules, where extracellular enzymes are present. However, intracellular digestion is the main digestive process in this part of the alimentary tract, and then nutrients are transported to the hemolymph, amoebocytes, and periglandular connective tissue. Undigested remnants accumulate in residual bodies. In the final phase of the digestive process, the digestive cells break up to release their apical pole filled with residual bodies and lysosomes and expelled into the lumen of tubules, thereafter reaching the stomach via the ciliated duct section. Waste products are passed on to the rectum via the intestine, where digestion and absorption of some nutrients may also occur (Shumway et al., 1985; Gosling, 2003).

Uptake, distribution, and clearance of enteric viruses from shellfish

Virus uptake by shellfish is a fast and dynamic process, and high titers of virus can be accumulated by shellfish in a short period of time (Di Girolamo et al., 1977). Bioaccumulation experiments have shown that both Norwalk virus (NV) and HAV can be detected in shellfish tissues after as little as 12–16 h exposure (Schwab et al., 1998a; Kingsley and Richards, 2003; Le Guyader et al., 2006a). There is variability in the bioaccumulative capabilities of individual shellfish, especially following exposure to low virus concentrations (Metcalf et al., 1979).

It is generally thought that oysters act as mere filters or ionic traps, passively concentrating particles. From this, a simple depuration process should be sufficient to rid oysters of virus, as observed for bacteria (Richards, 1988). The inefficiency of depuration or relaying leads to the persistence of viruses in shellfish for extended periods of time, and it is this characteristic that is suspected to have an important impact on public health (Metcalf, 1982; Richards, 1988; Loisy et al., 2005a). For example, only 7% of bioaccumulated NV is removed by depuration after 48 h compared to a 95% reduction in levels of *E. coli* (Schwab et al., 1998a). Bioaccumulated virus is located primarily in the digestive diverticula, although low levels are also detected outside the digestive tract. A variety of mechanisms have been proposed to explain observed differences between oyster species or virus accumulation, such as mechanical entrapment or ionic bonding (Metcalf, 1982; Schwab et al., 1998a; Burkhardt and Calci, 2000). Virus accumulation in oysters may also be affected by other factors such as water temperature, mucus production, glycogen

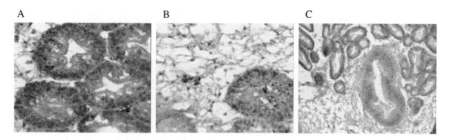

Fig. 1 Immunohistochemical detection of Norwalk virus virus-like particles (NV VLPs) in oyster digestive tissue. Three oysters were allowed to bioaccumulate 10^9 NV VLPs per 1l for 12h in clean
seawater. VLPs were detected by immunohistochemistry in intraepithelial cells or in the lumen of a
digestive tubule (A), or in the connective tissue (B). Oyster negative control (C).

content of the connective tissue, or gonadal development (Burkhardt and Calci, 2000).

We recently demonstrated specific binding of viral particles from a genogroup I (GI) NoV to the oyster digestive tract, suggesting an additional specific mechanism for virus particle concentration to occur. In this study, we used recombinant NV virus-like particles (NV VLPs). We, and others, previously have shown that VLPs are a good surrogate for native particles in that their antigenic and morphologic appearance is similar to native virus and VLPs have similar stability in the marine environment process (Caballero et al., 2004; Loisy et al., 2004, 2005a). The VLPs performed in a similar fashion to native NV in bioaccumulation experiments as well as in tissue binding experiments. NV VLPs bind specifically to carbohydrate structures in the midgut and digestive diverticulum (Fig. 1) (Le Guyader et al., 2006a). The carbohydrates are similar to, but distinct from, the histo-blood group antigens that NV has been shown to bind in human tissues (Harrington et al., 2002; Hutson et al., 2004).

Virus detection methods

Virus concentration

A number of molecular methods have been described over the last 15 years to detect viruses in bivalve molluscs. Many of these methods were adapted from methods described earlier to detect virus using cell culture (Metcalf et al., 1980; Lewis and Metcalf, 1988). The initial steps consist of virus elution from shellfish tissues, recovery of viral particles, and then virus concentration. These approaches can be further subdivided based upon whether the whole shellfish or dissected tissues are analyzed (Table 2). When whole shellfish are analyzed, generally 10–50 g of tissue are sampled, representing different numbers of individual shellfish depending on the species. This approach is particularly useful for small species, such as clams or cockles, because dissection may be technically difficult. Some methods have used acidic adsorption prior to virus elution, but most methods have

Table 2

Overview of methods used for virus detection in shellfish

Shellfish	Mass analyzed	Final %	Elution	Concentration	NA extraction	Sensitivity	References
Oysters	25 g	2 (0.5 g)	Glycine[a]	PEG	QIAamp	EV 1.2 PFU/g	Shieh et al. (1999)
	50 g	1 (0.5 g)	Water[a]	PEG, precipitate	Boiling		Chung et al. (1996)
	18 g	5 (1 g)	Glycine	Ultracentri.	GuSCN		Muniain-Mujika et al. (2003)
	50 g		Sonication	PEG	GuSCN		Green et al. (1998)
	25 g	2 (0.4 g)	Glycine	PEG	Tri-reagent	NoV 22 RT-PCR$_{50}$U HAV 27 PFU	Kingsley and Richards (2001)
	1.5 g DT	20 (0.3 g)	Chloroform-but, CatFloc	PEG	Prot. K, CTAB	NoV 100 RT-PCRs HAV 100 PFU	Atmar et al. (1995)
	1.5 g DT	8 (0.12 g)	Glycine-threonine	PEG	GuSCN + QIAamp		Beuret et al. (2003)
	1.5 g DT		Chloroform-but, CatFloc	PEG	Prot. K, CTAB	?	Schwab et al. (2001)
	1.5 g DT	6 (0.1 g)	Chloroform-but, CatFloc	Ultracentri.	QIAamp kit		Nishida et al. (2003)
	1.5 g DT	6 (0.09 g)	Zirconia beads		RNeasy kit		Lodder-Verschoor et al. (2005)
	2 g DT	0.5 (0.01 g)	Proteinase K		GuSCN		Jothikumar et al. (2005)
	10 g DT	0.8 (0.08 g)	Trizol		GuSCN		Boxman et al. (2006)

Table 2 (*continued*)

Shellfish	Mass analyzed	Final %	Elution	Concentration	NA extraction	Sensitivity	References
Mussels	20 g	0.2 (0.04 g)	Glycine, cat-Floc	Ag capture	QIAamp		Lee et al. (1999)
	50 g		Glycine	PEG	Guanidium, CsCl		Croci et al. (2000)
	100 g	1 (1 g)	Glycine	Ultracentri.	GuSCN		Pina et al. (1998)
	10 g	8 (0.8 g)	Glycine	PEG	RNeasy kit		Chironma et al. (2002)
	25 g		Threonine[a]	PEG	GuSCN		Mullendore et al. (2001)
	25 g DT	20 (1.5 g)	Glycine	Ultracentri.	Tryzol+Boom		Myrmel et al. (2004)
	2 g DT	5 (0.1 g)	Glycine	Ultracentri.	GuSCN		Hernroth et al. (2002)
Clams	50 g	1 (2.5 g)			Poly A	HAV 1.2 TCID$_{50}$/g	Goswami et al. (2002)
	25 g	5 (1.25 g)	Glycine, chlorof.	Ultracentri.	NucleospinRNA kit	HAV 600 PCRU/g	Sunen et al. (2004)
	1.5 g DT	5 (0.07 g)	Chlorof-but, CatFloc	PEG	RNeasy kit		Costafreda et al. (2006)

Abbreviations: Glycine, glycinebuffer; Chlorof, chloroform; but, butanol; PEG, polyethylene glycol precipitation; ultracentri, ultracentrifugation; prot. K, proteinase K; GuSCN, guanidinium thiocyanate; EV, enterovirus; NoV, norovirus; HAV, hepatitis A virus.

Note: Gray, only digestive tissues analyzed.

[a]Elution performed after acidic adsorption.

performed virus elution using a large volume of buffer (2–5 times the shellfish weight). Glycine buffer at basic pH (pH range 9.5–10) has been the most common buffer used for this purpose. Eluted viruses are then concentrated using either polyethylene glycol precipitation or ultracentrifugation. Direct analysis of eluted samples without virus concentration has also been described. The capture of HAV particles using virus-specific antibodies has also been used as a successful concentration method (Desenclos et al., 1991).

Early in the 1980s, the hypothesis that viruses are concentrated in digestive diverticulum tissues was proposed (Metcalf et al., 1980), and this was confirmed in the 1990s by detection of HAV using *in situ* transcription in oysters following virus bioaccumulation (Romalde et al., 1994). Removal of digestive tissues provides several advantages, including increased sensitivity, decreased processing time, and decreased interference with reverse transcription-polymerase chain reaction (RT-PCR) (Atmar et al., 1995). Since the initial description of analyzing only digestive tissues, a number of variations have been published, and most have analyzed the same weight (1.5–2 g) of digestive tissues. Viruses are eluted using various buffers (e.g. glycine or glycine-threonine) before concentration by polyethylene glycol or ultracentrifugation. These approaches are similar to those used for analysis of whole shellfish. However for digestive tissues, direct lysis of virus particles is used more frequently. For example proteinase K, or Trizol and lysis of shellfish tissues using Zirconia beads and a denaturing buffer have all been used for virus elution. A disadvantage of this direct approach is that a lower quantity of shellfish tissues is analyzed in RT-PCR assay (Table 2).

Nucleic acid purification

The methods used for nucleic acid extraction are dependent on those used for virus elution and concentration. Most methods are based on guanidium extraction either using the methods described by Boom et al. (1990) or using a kit, based on similar chemistry (QIAamp or RNeasy kit by Qiagen®). Capsid lysis by proteinase K and then purification of nucleic acid using phenol-chloroform and cetyl trimethyl ammonium bromide (CTAB) precipitation is more labor-intensive but was one of the first successful methods described (Atmar et al., 1993). When whole virus is concentrated, such as with antigen capture methods, nucleic acids are released by heat denaturation of the virus (Desenclos et al., 1991).

RT-PCR

Complementary DNA (cDNA) synthesis is accomplished by RT. Most assays utilize a virus-specific primer in the RT reaction (Atmar et al., 1995; Le Guyader et al., 2000; Formiga-Cruz et al., 2002; Kingsley et al., 2002; de Medici et al., 2004; Myrmel et al., 2004; Boxman et al., 2006) but random hexamers are used in some assays (Chung et al., 1996; Green et al., 1998; Cheng et al., 2005). PCR amplification is usually performed for at least 40 cycles; some methods use a nested PCR format and may utilize fewer than 40 cycles in the first amplification reaction. Probe hybridization is then performed as a confirmation step and enhances both assay sensitivity and specificity (Atmar et al., 1995; Chung et al., 1996; Shieh et al., 1999; Le Guyader et al. 2000; Costantini et al., 2006). Virus-specific amplicons may also be sequenced to obtain additional information about the virus(es) present in the sample. Real-time RT-PCR assays, in which the RT, PCR, and hybridization assays are combined in a single reaction, are being developed and used successfully to detect enteric viruses in shellfish (Nishida et al., 2003; Jothikumar et al., 2005; Loisy et al., 2005b; Costafreda et al. 2006).

One of the limitations in developing RT-PCR assays for the detection of enteric viruses has been the selection of primer and probe combinations that allow the detection of most or all strains of concern. For example, the development of broadly reactive primers for the detection of NoV has been problematic. No single assay has been able to detect all virus strains (Atmar and Estes, 2001; Vinje et al., 2003). Broadly reactive primers are required, but no single assay stands out as the best based upon evaluation of sensitivity, detection limit, and assay format for detection of viruses in stool samples (Vinje et al., 2003). In the absence of such a universal primer set, multiple sets need to be used to be able to detect all strains (Le Guyader et al., 1996a; Atmar and Estes, 2001). The use of multiple primer sets enhances the chance to detect a greater number of strains, and the homology of the primers with the NoV strain is important in terms of sensitivity (Le Guyader et al., 1996a; Atmar and Estes, 2001). For example, in three outbreak reports, primer sets targeting different area of the NoV genome needed to be used to be able to amplify the strain in both clinical and environmental samples (Shieh et al., 2000; Le Guyader et al., 2003, 2006b).

One of the goals of extraction methods is to remove inhibitors of the RT and PCR sufficiently to allow detection of viral nucleic acids. Polysaccharides present in shellfish tissue are at least one substance that can inhibit the PCR (Atmar et al., 1993). The different methods described in Table 2 accomplish inhibitor removal to varying degrees, although no systematic evaluation of the efficiency of inhibitor removal has been performed. Internal standards have been used to detect the presence of significant sample inhibition, and the amount of sample inhibition varies depending upon the shellfish tissue being analyzed (Atmar et al., 1995; Schwab et al., 1998b). Dilution of the extracted sample is another approach for circumventing the problem of persistent inhibitors. Thus, a smaller quantity of shellfish tissues is analyzed. For the methods described in Table 2, the weight of digestive tissues analyzed in each RT reaction varies between 0.01 and 2.5 g. The method analyzing the smallest shellfish tissue weight (0.01 g) is based on direct lysis of virus without a concentration step (Jothikumar et al., 2005), while the method analyzing the largest tissue weight (2.5 g) is based upon direct extraction of all nucleic acids followed by purification of nucleic acid using a poly A capture (Goswami et al., 2002). Focusing analysis of shellfish tissues on the digestive tissues, where the majority of viruses in an individual animal are concentrated, enhances assay performance by eliminating tissues (e.g. adductor muscle) that are rich in inhibitors but contain relatively little virus. For example, the digestive tissues represent about one tenth of the total animal weight for oysters.

Although numerous methods for detection of viruses in shellfish have been published, it is difficult to compare assay performance between methods. Reasons for this include the use of different reference virus strains, different species of shellfish, different RT-PCR methods, and the paucity of reports of direct comparisons of different methods. Because of the increasing importance given to direct detection of viruses in shellfish, it is important to develop a universal standard to facilitate comparison of methods. For example, Costafreda et al. (2006) proposed use of Mengo virus to evaluate virus concentration and nucleic acid extraction efficiency. Known quantities of RNA transcripts corresponding to a portion of the genome to be amplified have also been spiked into nucleic acid extracts to allow quantitative (Costafreda et al., 2006) or qualitative (Schwab et al., 1998b; Le Guyader et al., 2003) assessment of amplification efficiency. The more widespread applicability of these approaches will be dependent upon recognition of the need for such controls in the analysis of shellfish for enteric viruses.

Culture (when used)

Cell culture can be used to detect those enteric viruses that are cultivable, but to do so the virus concentration methods must maintain virus viability. Methods such as those proposed by Lewis and Metcalf (1988) have been successful for detection of enteroviruses. More recently, cell culture has been used to initially amplify viral nucleic acids, and potentially remove inhibitors, prior to detection of viral RNA by RT-PCR. This approach is called integrated cell culture (ICC) RT-PCR, and has been used for detection of adenoviruses, enteroviruses, and HAV (Chung et al., 1996; De Medici et al., 2001; Choo and Kim, 2006). The method also allows detection of viruses that may not cause cytopathic changes in cell culture (e.g. HAV) and an analysis of the infectivity of viral RNA found in shellfish samples (Chironna et al. 2002; Croci et al., 2005). In some instances, the number of samples that are positive by ICC-RT-PCR is lower than that obtained by direct PCR methods; this finding may be due to the elimination of inactivated virus that is still detectable using molecular methods (De Medici et al., 2001). Alternatively, the lower sensitivity may reflect the inability of the cell line used to support the growth of some virus strains. The relative value of this approach for screening shellfish remains to be determined for cultivable viruses, but it cannot be used for NoV at this time because of the lack of availability of a cell culture system.

Detection of viral contamination in shellfish

Many studies from different countries are now available examining the prevalence of enteric viruses in shellfish (Table 1). These include imported shellfish and those collected from producing areas and from the market. Reported prevalence of NoV detection varies from 6 to 37% in commercially distributed shellfish (Table 1).

Factors explaining the observed variability include the analysis of shellfish collected in different years, the use of different concentration/extraction methods, and the use of different RT-PCR assays. It is also possible that prevalence surveys with positive findings may be over-represented based upon publication bias.

NoV strains identified in shellfish have been examined in some studies. Oysters harvested in England over a 14-month period showed that GI strains were more prevalent than GII strains, although 42% of samples contained a mixture of both genogroups (Henshilwood et al., 1998). Studies from other countries have also seen a relatively high prevalence of GI strains: Spain, 12% GI vs. 14% GII; Sweden 17% GI vs. 24% GII; and United Kingdom, 5% GI vs. 5% GII (Formiga-Cruz et al., 2002). In France, GI strains are detected as frequently as GII in contaminated shellfish (data not published). In contrast, GII strains are overwhelmingly the most commonly recognized strains circulating in the community (Atmar and Estes, 2006). Few quantitative data are available but one study demonstrates that the number of NoV copies has been estimated to be about 10^2–10^4 per oyster (Nishida et al., 2003).

HAV contamination of shellfish is much less common. Only a few countries (Table 1), mainly in southern Europe, Brazil, and China, report finding shellfish contaminated with this virus (Chironna et al., 2002; Elamri et al., 2006; Sincero et al., 2006). This observation likely reflects the circulation of HAV in the population of these countries. In shellfish collected from more contaminated areas (noncommercial or European class B area) HAV prevalence is higher.

Studies done on shellfish contamination also demonstrate that animal virus strains can be detected (Dubois et al., 2004; Costantini et al., 2006). This finding suggests a possible route for zoonotic transmission and for generation of novel variants. Molecular assays must also be designed to allow the differentiation of human strains from animal strains, particularly in circumstances where zoonotic transmission is known not to occur. Failure to use an assay that distinguishes animal strains from human strains could lead to an overestimate of the exposure of shellfish to human sewage.

Evaluation of outbreaks: lessons learned and remaining questions

The association of virus outbreaks with consumption of contaminated shellfish has been recognized for decades. Initial studies described epidemiologic links between illness and shellfish consumption (Mackowiak et al., 1976; Grohmann et al., 1981; Richards, 1987; Lees, 2000). The link between specific pathogens (i.e. HAV and NoV) and shellfish consumption strengthened in the 1970s and 1980s as diagnostic methods were developed for identification of the etiology of specific clinical syndromes. In one instance, a possible viral pathogen (small round viruses) was observed by electron microscopy in cockles implicated in a gastroenteritis outbreak (Appleton and Pereiras, 1977). However, it has only been possible to directly detect viruses in shellfish implicated in transmission of disease since 1990 (Table 3). The application of these methods is beginning to provide information for risk assessments, but it is also raising new questions.

Table 3

Evaluation of shellfish from different outbreaks linked to shellfish consumption worldwide

Shellfish	Country	Imported	Category	No. of cases	Stool analysis	Shellfish analysis	Comments	References
Oysters	Japan	No	-	-	5/5 + NoV (GI)	1/2 + NoV GII + GI	Multiple strains	Sugieda et al. (1996)
Oysters	Denmark	Yes	A	356	3/11stool + NoV, 4/11 + EV	NoV +, EV +	No sequence	Christensen et al. (1998)
Oysters	France	No	A	19	13/25 + NoV, 8 strains	+ NoV, GI	Same sequence	Gilles et al. (2003)
Oysters	France	No	A	14	4/4 + Nov (3 strains-GI)	2/2 + NoV GI	Same sequence in 2 patients, 100 RNA copies/oyster	Le Guyader et al. (2003)
Oysters	USA (CA)	No	A&B	171	1/2 + NoV	2/3 + NoV	Same sequence, GII	Shieh et al. (2000)
Oysters	USA (LA)	No	A	132	12/12 + GI.1, 5/12 + SMA	9/10 + NoV (GI)	7/81 nucleotides different between shellfish and stool samples	Kohn et al. (1995)
Oysters	Singapore	Yes	B	305	4/5 + NoV	8 + NoV	No sequence	Le Guyader et al. (1996b) Ng et al. (2005)
Oysters	The Netherlands	Yes	A	-	8/9 NoV GI, GII, GIV	5/5 NoV GI (1 + GII)	Same sequence for GI	Boxman et al. (2006)
Oysters	France	No	B°	127	7/12 GII.4, GII.b GI.4	3/3 + NoV GI.4, GII.4, GII.8	Flooding and STP failure	Le Guyader et al. (2006b)
Oysters	Italy	Yes		202	22/41 GII.4, GII.8, GI.6, GI.4		125 RNA copies GII/oyster	
Mussels	Italy	No	-	103	24/24 + No	5/11 from market + NoV GII & GI	19 cases + 3 mussels: GII.8 2 cases + 1 mussel: GI	Prato et al. (2004)
Clams	US	Yes	B°°	5	5/5 NoV	1 + NoV GII, + HAV	No case of hepatitis A	Kingsley et al. (2002)
Oyster	US	No	A	61	61 stool: 59+ for HAV	HAV Ag + PCR	No sequence	Desenclos et al. (1991)
Clam	China	No	A	638*	638 cases (initial): + HAV	HAV +(CC)	No sequence	Halliday et al. (1991)
Clams	Spain	Yes	-	184	34/57 + HAV	15/20 + HAV	Several antigenic variants found in shellfish	Sanchez et al. (2002)

Note: -, data not specified; B°, depuration specified; B°°, based on *E. coli* counts detected in the samples analyzed; *, cases directly linked to clam consumption.

Category (A or B): sanitary classification of the sample based on *E. coli* detection in shellfish meat (EEC regulation) or in water (US regulation).

Oysters have been the predominant shellfish implicated in gastroenteritis outbreaks, but disease associated with consumption of clams and mussels have also been reported (Kingsley et al., 2002; Sanchez et al., 2002; Prato et al., 2004). The reasons for the predominance of outbreaks associated with oysters are unclear. Application of molecular methods to both fecal samples and shellfish have shown that multiple virus strains can be detected in the samples at the same time; in fact, consumption of shellfish appears to be a risk factor for co-infection with multiple strains of NoV (Kageyama et al., 2004; Le Guyader et al., 2006b). Sequencing of amplicons has provided direct links between virus detected in shellfish and that found in human samples (Shieh et al., 2000; Gilles et al., 2003; Le Guyader et al., 2003, 2006b; Boxman et al., 2006). However, unexplained differences in sequences obtained from shellfish and stool samples have also been reported despite strong epidemiologic links (Le Guyader et al., 1996b). Although NoVs are antigenically and genetically diverse, it is also possible to obtain identical sequences between epidemiologically unrelated shellfish and stool samples (unpublished observation); circulation of a common (or global) strain, such as the GII.4 NoVs, can complicate interpretation of results (Noel et al., 1999).

It appears that GI NoV strains are present at least as frequently as GII strains in oysters and that they are responsible for a large proportion of oysters-related outbreaks; in contrast, GII strains are generally predominant in other food-related outbreaks (Blanton et al., 2006). Possible reasons for the apparent predominance of GI strains in oysters include a greater stability of GI strains in the environment; specific binding of GI strains to oyster tissues leading to increased bioaccumulation (Le Guyader et al., 2006a), or other biophysical properties of the virus that have not yet been identified. Future studies should provide additional insight into these observations. Information on the quantity of virus in contaminated oysters associated with disease transmission is beginning to be generated (Le Guyader et al., 2003, 2006b; Costafreda et al., 2006) and this information along with ongoing studies examining the human infectious dose of NoV will provide data to allow the performance of risk assessments. Part of this calculation will need to include an assessment of the significance of a positive result of a molecular assay since inactivated virus can still be detected in RT-PCR assays (Richards 1999). However, the recent development of improved molecular methods, a better understanding of virus epidemiology, and standardization of methods across laboratories will lead to improved safety of shellfish for the consumer.

References

Abad FX, Pinto RM, Diez JM, Bosch A. Disinfection of human enteric viruses in water by copper and silver in combination with low levels of chlorine. Appl Environ Microbiol 1994; 60: 2377–2383.

Appleton H, Pereiras MS. A possible virus aetiology in outbreaks of food-poisoning from cockles. Lancet 1977; 9: 780–781.

Arnal C, Crance JM, Gantzer C, Schwartzbrod L, Deloince R, Billaudel S. Persistence of infectious hepatitis A virus and its genome in artificial sewater. Zentralbl Hyg Umweltmed 1998; 201: 279–284.

Atmar RL, Estes MK. Diagnosis of nonculturable gastroenteritis viruses, the human caliciviruses. Clin Microbiol Rev 2001; 14: 15–37.

Atmar RL, Estes MK. The epidemiologic and clinical importance of norovirus infection. Gastroenterol Clin North Am 2006; 35: 275–290.

Atmar RL, Metcalf TG, Neill FH, Estes MK. Detection of enteric viruses in oysters by using the polymerase chain reaction. Appl Environ Microbiol 1993; 59: 631–635.

Atmar RL, Neill FH, Romalde JL, Le Guyader F, Woodley CM, Metcalf TG, Estes MK. Detection of Norwalk virus and hepatitis A virus in shellfish tissues with the PCR. Appl Environ Microbiol 1995; 61: 3014–3018.

Beuret C, Baumgartner A, Schluep J. Virus-contaminated oysters: a three-month monitoring of oysters imported to Switzerland. Appl Environ Microbiol 2003; 69: 2292–2297.

Blanton LH, Adams SM, Beard RS, Wei G, Bulens SN, Widdowson M-A, Glass RI, Monroe SS. Molecular and epidemiologic trends of caliciviruses associated with outbreaks of acute gastroenteritis in the United States, 2000–2004. J Infect Dis 2006; 193: 413–421.

Boom R, Sol CJA, Salimans MMM, Jansen CL, Wertheim-Van Dillen PME, Ven Der Noordaa J. Rapid and simple method for purification of nucleic acids. J Clin Microbiol 1990; 28: 495–503.

Boxman ILA, Tilburg JJHC, te Loeke NAJM, Vennema H, Jonker K, de Boer E, Koopmans M. Detection of noroviruses in shellfish in the Netherlands. Int J Food Microbiol 2006; 108: 391–396.

Brown JW, Folsom WD. Economic impact of hard clam associated outbreaks of gastroenteritis in New York state. NOAA Technical Memorandum NMFS-SEFC-121. Charleston, SC: U.S. Department of Commerce; 1983.

Burkhardt W, Calci KR. Selective accumulation may account for shellfish-associated viral illness. Appl Environ Microbiol 2000; 66: 1375–1378.

Butt AA, Aldridge KE, Sanders CV. Infections related to the ingestion of seafood. Part I: viral and bacterial infections. Lancet Infect Dis 2004; 4: 201–212.

Caballero S, Abad FX, Loisy F, Le Guyader FS, Cohen J, Pinto RM, Bosch A. Rotavirus virus-like particles as surrogates in environmental persistence and inactivation studies. Appl Environ Microbiol 2004; 70: 3904–3909.

Cacopardo B, Russo R, Preiser W, Benanti F, Brancati G, Nunnari A. Acute hepatitis E in Catania (eastern Sicily) 1980–1994. The role of hepatitis E virus. Infection 1997; 25: 313–316.

CDC. Prevention of hepatitis A through active or passive immunization. MMWR 2006; 55(RR7): 1–23.

Cheng PKC, Wong DKK, Chung TWH, Lim WWL. Norovirus contamination found in oysters worldwide. J Med Virol 2005; 76: 593–597.

Chironna M, Germinario C, De Medici D, Fiore A, Di Pasquale S, Quarto M, Barbuti S. Detection of hepatitis A virus in mussels from different sources marketed in Puglia region (South Italy). Int J Food Microbiol 2002; 75: 11–18.

Choo Y-J, Kim SJ. Detection of human adenoviruses and enteroviruses in Korean oysters using cell culture, integrated cell culture-PCR, and direct PCR. J Microbiol 2006; 44: 162–170.

Christensen BF, Lees D, Henshilwood K, Bjergskov T, Green J. Human enteric viruses in oysters causing a large outbreak of human food borne infection in 1996/97. J Shellfish Res 1998; 17: 1633–1635.

Chung H, Jaykus LA, Lovelace G, Sobsey MD. Bacteriophages and bacteria as indicators of enteric viruses in oysters and their harvest waters. Water Sci Technol 1998; 38: 37–44.

Chung H, Jaykus LA, Sobsey MD. Detection of human enteric viruses in oysters by *in vivo* and *in vitro* amplification of nucleic acids. Appl Environ Microbiol 1996; 62: 3772–3778.

Costafreda MI, Bosch A, Pinto RM. Development, evaluation, and standardization of a real-time TaqMan reverse transcription-PCR assay for quantification of hepatitis A virus in clinical and shellfish samples. Appl Environ Microbiol 2006; 72: 3846–3855.

Costantini V, Loisy F, Joens L, Le Guyader FS, Saif LJ. Human and animal enteric caliciviruses in oysters from different coastal regions of the United States. Appl Environ Microbiol 2006; 72: 1800–1809.

Croci L, De Medici D, Di Pasquale S, Toti L. Resistance of hepatitis A virus in mussels subjected to different domestic cooking. Int J Food Microbiol 2005; 105: 139–144.

Croci L, De Medici D, Scalfaro C, Fiore A, Divizia M, Donia D, Cosentino AM, Moretti P, Costanti G. Determination of enteroviruses, hepatitis A virus, bacteriophages and *Escherichia coli* in adriatic sea mussels. J Appl Microbiol 2000; 88: 293–298.

De Medici D, Croci L, Di Pasquale S, Fiore A, Toti L. Detecting the presence of infectious hepatitis A virus in molluscs positive to RT-nested-PCR. Lett Appl Microbiol 2001; 33: 362–366.

De Medici D, Croci L, Suffredini E, Toti L. Reverse transcription booster PCR for detection of noroviruses in shellfish. Appl Environ Microbiol 2004; 70: 6329–6332.

Desenclos JCA, Klontz KC, Wilder MH, Naiman OV, Margolis HS, Gunn RA. A multistate outbreak of hepatitis A caused by the consumption of raw oysters. Am J Public Health 1991; 81: 1268–1272.

Di Girolamo R, Liston J, Matches J. Ionic binding, the mechanism of viral uptake by shellfish mucus. Appl Environ Microbiol 1977; 33: 19–25.

Doré W, Henshilwood K, Lees DN. Evaluation of F-specific RNA bacteriophage as a candidate human enteric virus indicator for bivalves molluscan shellfish. Appl Environ Microbiol 2000; 66: 1280–1285.

Dubois E, Merle G, Roquier C, Trompette A, Le Guyader F, Cruciere C, Chomel J-J. Diversity of enterovirus sequences detected in oysters by RT-heminested PCR. Int J Food Microbiol 2004; 92: 35–43.

Elamri DE, Aouni M, Parnaudeau S, Le Guyader FS. Detection of human enteric viruses in shellfish collected in Tunisia. Lett Appl Microbiol 2006; 43: 399–404.

Formiga-Cruz M, Allard AK, Conden Hansson AC, Henshilwood K, Hernroth BE, Joffre J, Lees DN, et al. Evaluation of potential indicators of viral contamination in shellfish and their applicability to diverses geographical areas. Appl Environ Microbiol 2003; 69: 1556–1563.

Formiga-Cruz M, Tofino-Quesada G, Bofill-Mas S, Lees DN, Henshilwood K, Allard AK, et al. Distribution of human virus contamination in shellfish from different growing areas in Greece, Spain, Sweden, and the United Kingdom. Appl Environ Microbiol 2002; 68: 5990–5998.

Gilles C, de Casanove J-N, Dubois E, Bon E, Pothier P, Kohli E, Vaillant V. Epidemie de gastro-enterites à Norovirus liées à la consommation d'huîtres, Somme, janvier 2001. Bull Epidemiol Hebd 2003; 8: 47–48.

Gosling E. How bivalves feed. In: Bivalves Mollucsc, Biology, Ecology and Culture. Oxford, UK: Blackwell Publishing; 2003; pp. 87–123.

Goswami BB, Kulka M, Ngo D, Istafanos P, Cebula TA. A polymerase chain reaction based method for the detection of hepatitis A virus in produce and shellfish. J Food Prot 2002; 65: 393–402.

Green J, Henshilwood K, Gallimore CI, Brown DWG, Lees DN. A nested reverse transcriptase PCR assay for detection of small round structured viruses in environmentally contaminated molluscan shellfish. Appl Environ Microbiol 1998; 64: 858–863.

Grohmann GS, Murphy AM, Christopher PJ, Auty E, Greenberg HB. Norwalk virus gastroenteritis in volunteers consuming depurated oysters. Aust J Exp Biol Med Sci 1981; 59: 219–228.

Halliday ML, Kang LY, Zhou TR, Hu MD, Pan QC, Fu TY, Huang YS, Hu SL. An epidemic of hepatitis A attributable to the ingestion of raw clams in Shangai, China. J Infect Dis 1991; 164: 852–859.

Harrington PR, Lindesmith L, Yount B, Moe CL, Baric RS. Binding of Norwalk virus like particles to ABH histo blood group antigens is blocked by antisera from infected human, volunteers or experimentally vaccinated mice. J Virol 2002; 76: 12335–12343.

Henshilwood K, Green J, Lees DN. Monitoring the marine environment for small round structured viruses (SRSVs): a new approach to combating the transmission of these viruses by molluscan shellfish. Water Sci Technol 1998; 38: 51–56.

Hernroth BE, Conden-Hansson A-C, Rehnstam-Holm A-S, Girones R, Allard AK. Environmental factors influencing human viral pathogens and their potential indicator organisms in the blue mussel, *Mytilus edulis*: the first Scandinavian report. Appl Environ Microbiol 2002; 68: 4523–4533.

Hutson AM, Atmar RL, Estes MK. Norovirus disease: changing epidemiology and host susceptibility factors. Trends Microbiol 2004; 12: 279–287.

Jothikumar N, Lowther JA, Henshilwwod K, Lees DN, Hill VR, Vinje J. Rapid and sensitive detection of noroviruses by using TaqMan based one-step reverse transcription-PCR assays and application to naturally contaminated shellfish samples. Appl Environ Microbiol 2005; 71: 1870–1875.

Kageyama T, Shinohara M, Uchida K, Fukhusi S, Hoshino FB, Kojima S, Takai R, Oka T, Takeda N, Katayama K. Coexistence of multiple genotypes, including newly identified genotypes, in outbreaks of gastroenteritis due to norovirus in Japan. J Clin Microbiol 2004; 42: 2988–2995.

Kaplan JE, Feldman R, Campbell DS, et al. The frequency of a Norwalk-like pattern of illness in outbreaks of acute gastroenteritis. Am J Public Health 1982; 72: 1329–1332.

Karamoto Y, Ibenyassine K, Aitmhand R, Idaomar M, Ennaji MM. Adenovirus detection in shellfish and urban sewage in Morocco (Casablanca region) by the polymerase chain reaction. J Virol Meth 2005; 126: 135–137.

Kingsley DH, Meade GK, Richards GP. Detection of both hepatitis A virus and Norwalk-like virus in imported clams associated with food-borne illness. Appl Environ Microbiol 2002; 68: 3914–3918.

Kingsley DH, Richards GP. Rapid and efficient extraction method for reverse transcription-PCR detection of hepatitis A and Norwalk-like viruses in shellfish. Appl Environ Microbiol 2001; 67: 4152–4157.

Kingsley DH, Richards GP. Persistence of hepatitis A virus in oysters. J Food Prot 2003; 66: 331–334.

Kohn MA, Farley TA, Ando T, Curtis M, Willson SA, Monroe SS, Baron RC, Macfarland LM, Glass RI. An outbreak of Norwalk virus gastroenteritis associated with eating raw oysters. JAMA 1995; 273: 466–471.

Koizumi Y, Isoda N, Sato Y, Iwaki T, Ono K, Ido K, et al. Infection of a Japanese patient by genotype 4 hepatitis E virus while traveling in Vietnam. J Clin Microbiol 2004; 42: 3883–3885.

Lee T, Yam WC, Tam TY, Ho BSW, Ng MH, Broom MJ. Occurence of hepatitis A virus in green-lipped mussels (Perna viridis). Water Res 1999; 33: 885–889.

Lees D. Viruses and bivalve shellfish. Int J Food Microbiol 2000; 59: 81–116.

Le Guyader F, Estes MK, Hardy ME, Neill FH, Green J, Brown DWG, Atmar RL. Evaluation of a degenerate primer for the PCR detection of human caliciviruses. Arch Virol 1996a; 141: 2225–2235.

Le Guyader F, Haugarreau L, Miossec L, Dubois E, Pommepuy M. Three-year study to assess human enteric viruses in shellfish. Appl Environ Microbiol 2000; 66: 3241–3248.

Le Guyader F, Neill FH, Estes MK, Monroe SS, Ando T, Atmar RL. Detection and analysis of a small round-structured virus strain in oysters implicated in an outbreak of acute gastroenteritis. Appl Environ Microbiol 1996b; 62: 4268–4272.

Le Guyader FS, Bon F, DeMedici D, Parnaudeau S, Bertome A, et al. Detection of multiple noroviruses associated with an international gastroenteritis outbreak linked to oyster consumption. J Clin Microbiol 2006b; 44: 3878–3882.

Le Guyader FS, Loisy F, Atmar RL, Hutson AM, Estes MK, Ruvoen-Clouet N, Pommepuy M, Le Pendu J. Norwalk virus specific binding to oyster digestive tissues. Emerg. Infect Dis 2006a; 12: 931–936.

Le Guyader FS, Neill FH, Dubois E, Bon F, Loisy F, Kohli E, Pommepuy M, Atmar RL. A semi-quantitative approach to estimate Norwalk-like virus contamination of oysters implicated in an outbreak. Int J Food Microbiol 2003; 87: 107–112.

Lewis GD, Metcalf TG. Polyethylene glycol precipitation for recovery of pathogenic viruses, including hepatitis A virus and human rotavirus, from oyster, water, and sediment samples. Appl Environ Microbiol 1988; 54: 1983–1988.

Lodder-Verschoor F, de Roda Husman AM, van der Berg HHJL, et al. Year-round screening of non commercial and commercial oysters for the presence of human pathogenic viruses. J Food Prot 2005; 68: 1853–1959.

Loisy F, Atmar RL, Cohen J, Bosch A, Le Guyader FS. Rotavirus VLP2/6: a new tool for tracking rotavirus in the environment. Res Microbiol 2004; 155: 575–578.

Loisy F, Atmar RL, Guillon P, Le Cann P, Pommepuy M, Le Guyader SF. Real-time RT-PCR for norovirus screening in shellfish. J Virol Meth 2005b; 123: 1–7.

Loisy F, Atmar RL, Le Saux J-C, Cohen J, Caprais M-P, Pommepuy M, Le Guyader FS. Rotavirus virus like particles as surrogates to evaluate virus persistence in shellfish. Appl Environ Microbiol 2005a; 71: 6049–6053.

Lopalco PL, Malfait P, Salmaso S, Germinario C, Quarto M, Barbuti S, Cipriani R, Mundo A, Pesole G. A persisting outbreak of hepatitis A in Puglia, Italy, 1996: epidemiological follow-up. Euro Surveill 1997; 2: 31–32.

Lucioni C, Cipriani V, Mazzi S, Panunzio M. Cost of an outbreak of hepatitis A in Puglia, Italy. Pharmacoeconomics 1998; 13: 257–266.

MacDonnell S, Kirkland KB, Haldy WG, Aristeguita C, Hopkins RS, Monroe SS, Glass RI. Failure of cooking to prevent shellfish associated viral gastroenteritis. Arch Intern Med 1997; 157: 111–116.

Mackowiak PA, Caraway CT, Portnoy BL. Oyster-associated hepatitis: lessons from the Louisiana experience. Am J Epidemiol 1976; 103: 181–191.

Mead PS, Slutsker L, Dietz V, MacCaig LF, Bresee JS, Shapiro C, Griffin PM, Tauxe RV. Food-related illness and death in the United States. Emerg Infect Dis 1999; 5: 607–625.

Metcalf TG. Viruses in shellfish growing waters. Environ Int 1982; 7: 21–27.

Metcalf TG, Moulton E, Eckerson D. Improved method and test strategy for recovery of enteric viruses from shellfish. Appl Environ Microbiol 1980; 39: 141–152.

Metcalf TG, Mullin B, Eckerson D, Moulton E, Larkin EP. Bioaccumulation and depuration of enteroviruses by the soft-shelled clam, mya arenaria. Appl Environ Microbiol 1979; 38: 275–282.

Mullendore J, Sobsey MD, Shieh YSC. Improved method for the recovery of hepatitis A virus from oysters. J Virol Meth 2001; 94: 25–35.

Muniain-Mujika I, Calvo M, Lucena F, Girones R. Comparative analysis of viral pathogens and potential indicators in shellfish. Int J Food Microbiol 2003; 83: 75–85.

Murphy AM, Grohmann GS, Christopher PJ, Lopez WA, Davey GR, Millson RH. An Australian wide outbreak of gastroenteritis from oysters caused by Norwalk virus. Med J Aust 1979; 2: 329–333.

Myrmel M, Berg EMM, Rimstad E, Grinde B. Detection of enteric viruses in shellfish from the Norwegian coast. Appl Environ Microbiol 2004; 70: 2678–2684.

Ng TL, Chan PP, Phua TH, Loh JP, et al. Oyster-associated outbreaks of norovirus gastroenteritis in Singapore. J Infect 2005; 51: 413–418.

Nishida T, Kimura H, Saitoh M, Shinohara M, Kato M, et al. Detection, quantitation, and phylogenic analysis of noroviruses in Japanese oysters. Appl Environ Microbiol 2003; 69: 5782–5786.

Noel JS, Fankhauser RL, Ando T, Monroe SS, Glass RI. Identification of a distinct common strain of 'Norwalk-like viruses' having a global distribution. J Infect Dis 1999; 179: 1334–1344.

OFIMER. Annual report. French Ministry of Agriculture and Fish; 2003. Available on line www.ofimer.fr

Pina S, Puig M, Lucena F, Jofre J, Girones R. Viral pollution in the environment and in shellfish: human adenovirus detection by PCR as an index of human viruses. Appl Environ Microbiol 1998; 64: 3376–3382.

Portnoy BL, Mackowiak PA, Caraway CT, Walker JA, McKinley TW, Klein CA. Oyster-associated hepatitis. Failure of shellfish certification programs to prevent outbreaks. JAMA 1975; 233: 1065–1068.

Potasman I, Paz A, Odeh M. Infectious outbreaks associated with bivalve shellfish consumption: a worldwide perspective. Clin Infect Dis 2002; 35: 921–928.

Prato R, Lopalco PL, Chironna M, Barbuti G, Germinario C, Quarto M. Norovirus gastroenteritis general outbreak associated with raw shellfish consumption in South Italy. BMC Infect Dis 2004; 4: 1–6.

Richards GP. Outbreaks of shellfish-associated enteric virus illness in the United States: requisite for development of viral guidelines. J Food Prot 1985; 48: 815–823.

Richards GP. Shellfish associated enteric virus illness in the United States, 1934–1984. Estuaries 1987; 10: 84–85.

Richards GP. Microbial purification of shellfish: a review of depuration and relaying. J Food Prot 1988; 51: 218–251.

Richards GP. Limitations of molecular biological techniques for assessing the virological safety of foods. J Food Prot 1999; 62: 691–697.

Rippey SR. Infectious diseases associated with molluscan shellfish consumption. Clin Microbiol Rev. 1994; 7: 419–425.

Romalde JL, Estes MK, Szucs G, Atmar RL, Woodley CM, Metcalf TG. In situ detection of hepatitis A virus in cell cultures and shellfish tissues. Appl Environ Microbiol 1994; 60: 1921–1926.

Sanchez G, Pinto R, Vanaclocha H, Bosch A. Molecular characterization of hepatitis A virus isolates from a transcontinental shellfish-borne outbreak. J Clin Microbiol 2002; 40: 4148–4155.

Schwab KJ, Neill FH, Estes MK, et al. Distribution of Norwalk virus within shellfish following bioaccumulation and subsequent depuration by detection using RT-PCR. J Food Prot 1998a; 61: 1674–1680.

Schwab KJ, Neill FH, Estes MK, Atmar RL. Improvements in the RT-PCR detection of enteric viruses in environmental samples. Water Sci Technol 1998b; 38: 83–86.

Schwab KJ, Neill FH, Le Guyader F, Estes MK, Atmar RL. Development of a reverse transcription-PCR -DNA enzyme immunoassay for detection of Norwalk-like viruses and hepatitis A virus in stool and shellfish. Appl Environ Microbiol 2001; 67: 742–749.

Shieh YSC, Calci KR, Baric RS. A method to detect low levels of enteric viruses in contaminated oysters. Appl Environ Microbiol 1999; 65: 4709–4714.

Shieh YSC, Monroe SS, Fankhauser RL, Langlois GW, Burkhardt W, Baric RS. Detection of Norwalk-like virus in shellfish implicated in illness. J Infect Dis 2000; 181: 360–366.

Shumway SE, Cucci TL, Newell RC, Yentsch CM. Particles selection, ingestion and absorption in filter-feeding bivalves. J Exp Mar Biol Ecol 1985; 91: 77–92.

Sincero TCM, Levin DB, Simoes CMO, Barardi CRM. Detection of hepatitis A virus (HAV) in oysters (*Crassostrea gigas*). Water Res. 2006; 40: 895–902.

Sugieda M, Nakajima K, Nakajima S. Outbreaks of Norwalk-like virus associated gastroenteritis traced to shellfish: coexistence of two genotypes in one specimen. Epidemiol Infect 1996; 116: 339–346.

Sunen E, Casa N, Moreno B, Zigorraga C. Comparison of two methods for the detection of hepatitis A in clam samples (Tapes spp.) by reverse transcription-nested PCR. Int J Food Microbiol 2004; 91: 147–154.

Turcios RM, Widdowson MA, Sulka AC, Mead PS, Glass RI. Reevaluation of epidemiological criteria for identifying outbreaks of acute gastroenteritis due to norovirus: United States, 1998–2000. Clin Infect Dis 2006; 42: 964–969.

Ueki Y, Sano D, Watanabe T, Akiyama K, Omura T. Norovirus pathway in water environment estimated by genetic analysis of strains from patients of gastroenteritis, sewage, treated wastewater, river water and oysters. Water Res 2005; 39: 4271–4280.

Vinje J, Vennema H, Maunula L, Von Bonsdorff C-H, Hoehne M, Shreier E, Richards A, Green J, Brown D, Beard S, Monroe S, De Bruin E, Svensson L, Koopmans MPG. International collaborative study to compare reverse transcriptase PCR assays for detection and genotyping of noroviruses. J Clin Microbiol 2003; 41: 1423–1433.

Zheng D-P, Ando T, Fankhauser RL, Beard RS, Glass R, Monroe SS. Norovirus classification and proposed strain nomenclature. Virology 2006; 346: 312–323.

Human Viruses in Water
Albert Bosch (Editor)
DOI 10.1016/S0168-7069(07)17011-7

Chapter 11

Indicators of Waterborne Enteric Viruses

Juan Jofre

Department of Microbiology, School of Biology, University of Barcelona,
Diagonal 645, 08028-Barcelona, Spain

Introduction

Amid the numerous microorganisms responsible for waterborne infections, enteric viruses play an important role. Up to recently it has not been possible to detect most of human waterborne viruses in water samples since they were not cultivable. At present, with the advent of genomic techniques, it is possible to detect the presence of genomes of many viruses in water samples. These techniques do not allow for the moment to distinguish whether the detected viruses are infectious or not, without introducing additional steps that impair the efficiency of the detection method. Moreover, repeatedly the human enteric viruses are present in such low densities that bothersome concentration procedures are needed to detect them. Thus, the detection of viruses, though possible, does not seem realistic from the routine point of view. Therefore, for many aspects related to water quality control surrogate indicators are still needed.

Indicator concept

The term indicator organism includes different concepts, which frequently lead confusion. The term marker to substitute the conventional indicator concept was introduced by Mossel (1982) in food microbiology and the term model, with the same meaning, was preferred in water microbiology (IAWPRC Study Group on Health Related Water Microbiology, 1991; Armon and Kott, 1996). This chapter will deal with model organisms of waterborne enteric viruses. Therefore, the term model organisms, substituting the conventional indicator organisms, will be used, following recommended reviews on bacteriophages (IAWPRC Study Group on

Health Related Water Microbiology, 1991; Armon and Kott, 1996). According to this usage, the index and the indicator functions can be attributed to the model organisms. Considering these two functions frequently facilitates the discussion about the suitability of a given potential indicator.

An index organism is related to the occurrence of the surrogated microorganisms. The relationship may be direct, such as indexes of human viruses, or indirect, such as an index of faecal pollution. The criteria for an index organism are very similar to those presently used for bacterial faecal indicators.

In contrast, the model organisms with an indicator function have a much broader definition. An indicator is basically a model that has behavioural characteristics similar to those of the surrogated microorganisms, and has the same or greater resistance to environmental stresses, but that does not necessarily originate from the same source as the pathogen (e.g. faeces). There is no doubt that the best model organisms have the same source as the surrogate organisms. This criterion, and others mentioned previously, define an indicator as a marker for the efficiency of treatments or for the variation in numbers of the surrogated microorganisms in specific environments. In general, criteria for indicator organisms are less restrictive than the criteria for index organisms.

Ideally, a perfect model organism will fulfil the index and indicator functions. However, the more it is known about faecal pathogens and potential model microorganisms, the more unlikely it seems that we will find a single faecal microorganism fulfilling both functions.

Conventional bacterial indicators

Bacterial groups that include *Escherichia coli* (total, faecal and thermotolerant coliforms), *Enterococcus* spp. (faecal streptococci) and *Clostridium perfringens* (sulphite-reducing clostridia or spores of sulphite-reducing clostridia) have been the conventional indicators. More recently, with the advent of methods that allow their direct detection, these bacterial species or genera have been used as indicators of water pollution. However, the scientific community involved in water microbiology has raised doubts about the capability of these indicators to measure water quality and predict waterborne viral diseases. This concern has risen as a consequence of two types of observations. Firstly, there is often a lack of correlation between these indicators and viruses in water samples (Gerba, 1979; Lucena et al., 1988; Wyer et al., 1995; Borchardt et al., 2004). Secondly, viruses are more resistant to natural stressors and disinfection processes than conventional bacterial indicators (Miescier and Cabelli. 1982; Payment et al., 1985a,b; Harwood, et al., 2005). These differences have led many researchers in the subject to advocate the need of alternative indicators of waterborne viruses.

Alternative indicators for viruses

Human enteric virus and bacteriophages of enteric bacteria have both been advocated as potential viral indicators.

Probably because they are the most easily detectable by cell culture of all enteric viruses, enteroviruses have been considered as potential indicators of the human enteric viruses. Most frequently, in water virology, the term enterovirus is used to define viruses replicating in BGM (Buffalo green monkey) cells. The scarce regulations on water quality in which criteria based on viruses appear are based on enteroviruses. Examples are the still ruling European Union Guidelines on bathing waters (EEC, 1976) or the United States Environmental Protection Agency (USEPA) regulations about the quality of sludges with unrestricted use in agriculture (USEPA, 1992).

After the advent of the genomic techniques, the genome of viruses excreted by humans had also been proposed as potential viral indicator. Because of the occurrence and levels in sewage of both infectious viruses (Irving and Smith, 1981; Hurst et al., 1988) and viral genomes (Pina et al., 1998), their resistance to some environmental stressors and disinfectants (Enriquez et al., 1995; Gerba et al., 2002) and because being DNA virus the nucleic acid amplification does not require the RT step, adenoviruses genomes are gaining some ground as potential indicators of waterborne viruses. However, genomic methods are inconvenient as they do not allow, without introducing additional steps, (Nuanualsuwan and Cliver. 2002), to distinguish between infectious and non-infectious viruses. This distinction is essential when evaluating water treatments.

However, both infectious viruses and viral genomes present some other drawbacks as potential indicators of waterborne enteric viruses. Firstly, their concentrations are usually low and require cumbersome concentration procedures. Secondly, different studies show that there is no correlation between the densities of various viruses in wastewater effluents (Irving and Smith, 1981), surface (Chapron et al., 2000) and marine water samples (Jiang et al., 2001). This lack of coincidence is explainable by several factors. Firstly, many viral waterborne viral infections are seasonal and seasonality varies even according to the climate of the area (Payne et al., 1986; Trabelsi et al., 2000; da Rosa et al., 2001; Pang et al., 2001; Chau et al., 1977). Consequently, it can be expected that the densities of the different viruses in sewage depend on the epidemic situation and seasonal distribution in the area. This seasonality has indeed been observed in the occurrence of human viruses in water (Irving and Smith, 1981; Hejkal et al., 1984; Krikelis et al., 1985; Tani et al., 1995). Also, increased densities of viruses have been observed in wastewaters during a local outbreak of norovirus gastroenteritis (Lodder et al., 1999). Secondly, important differences among different viruses have been reported regarding their resistance to environmental stressors or to disinfectants used in water treatments (Gironés et al., 1989; Abad et al., 1994; Callahan et al., 1995; Shin and Sobsey, 1998; Sobsey et al., 1995; Abad et al., 1997; Gerba et al., 2002). These differences have even been detected amid different viruses of the same genera, and even between different isolates of the same species (Engelbrecht et al., 1980; Payment et al., 1985a).

Bacteriophages infecting enteric bacteria seem good candidates. As early as 1948, Guelin advocated phages as potential indicators of enteric microorganisms (Guelin, 1948). He found that phages infecting *E. coli* correlated with coliform bacteria numbers in fresh and marine waters. Later on, the potential value of phages infecting enteric bacteria has been vastly studied. At present we can say that phages are not perfect as viral indicators since there is no unequivocal correlation between the presence of phages and viruses in many water samples. However, available information indicates that the presence of a given number of certain phages in water samples will indicate the likely presence of viruses.

Characteristics of potential indicator bacteriophages

General characteristics of bacteriophages

Bacteriophages or phages are viruses that infect bacteria. They basically consist of a nucleic acid molecule (genome) surrounded by a protein coat called capsid. Many phages also contain additional structures such as tails and spikes and, much less frequently, lipids. These features imply that in terms of composition, structure, morphology and capsid size, phages share many properties with human viruses. Bacteriophages are extremely abundant in nature and probably the most abundant life form on Earth. Attending to electron microscope observations, phages are one order of magnitude more abundant than bacteria in the oceans (Bergh et al., 1989; Fuhrman, 1999). Whitman et al. (1998) estimated that in the biosphere there are around 5×10^{30} bacteria. Extrapolating the oceanic data, the estimated number of bacteriophages on Earth may approach 10^{31}.

Bacteriophages only can replicate inside susceptible and metabolising host bacteria. Phages can only infect certain bacteria. The host specificity of phages is mainly determined by receptor molecules on the surface of the bacteria. Phage receptors have been described in different parts of bacteria (capsule, cell wall, flagella and pili). Phages attached to the receptors located in the cell wall are typically known as somatic phages. Bacteriophages may follow two different life-cycle pathways: lytic cycle and lysogenic cycle. Those phages that only can undergo the lytic cycle are known as virulent phages and those that can undergo both cycles are named temperate bacteriophages. Lytic cycle usually proceeds with replication immediately after phages infect the host cell, and new viruses are released in large numbers by the rupture of the host cell wall (lysis) after a short period of time; as little as 25–30 min are needed for some bacteriophages infecting E. coli. Both the number of phages released from an infected bacteria and the time needed for replication vary according to the host bacteria and the bacteriophage, but also depend on the physiological status of the host cell. As a consequence of host lysis of the first infected cell and then of the surrounding ones, most virulent phages produce clear plaques on a law of susceptible host bacteria (Adams, 1959).

Temperate phages do not necessarily start lytic replication immediately after cell infection. Depending on a number of conditions, the genome of these phages replicates as a part of the genome of the host strain, until induced to replicate by the lytic-cycle pathway. Bacterial cells with genomes of phages integrated in their genome are known as lysogens and the integrated phages are known as prophages.

Various agents or conditions such as UV light irradiation and a number of chemical agents induce the genome of lysogenic phages to undergo the lytic cycle. Generally, these phages are released at low frequency with limited impact on host populations (Adams, 1959). The contribution of induction to lytic cycle of prophages to the occurrence of free phages in water environments is not well known and needs to be further investigated.

Phages may readily be grouped on the basis of a few gross characteristics including host range, morphology, nucleic acid, strategies of infection, morphogenesis, phylogeny, serology, sensitivity to physical and chemical agents, and dependence on properties of hosts and environment. The present classification, which has been adopted by the International Committee on Taxonomy of Viruses, is mostly based on phage morphology and characteristics of the nucleic acid (Murphy et al., 1995). Phages of particular interest in water quality assessment are included in the seven families whose characteristics are summarised in Table 1.

Groups of bacteriophages proposed as potential indicators of waterborne enteric viruses

Somatic coliphages (Kott et al., 1974; IAWPRC Study Group on Health Related Water Microbiology, 1991), F-specific RNA bacteriophages (Havelaar and Pot Hogeboom, 1988; IAWPRC Study Group on Health Related Water Microbiology, 1991) and bacteriophages infecting *Bacteroides fragilis* (Tartera and Jofre, 1987; IAWPRC Study Group on Health Related Water Microbiology, 1991) have so far been advocated as potential model microorganisms for various aspects of water quality control.

Somatic coliphages

The term somatic coliphages is used to define the bacteriophages infecting *E. coli* through the cell wall. The great majority of known somatic coliphages belong to the following families: *Myoviridae*, *Siphoviridae*, *Podoviridae* and *Microviridae*. All types are found in sewage, although the most abundant are *Myoviridae* and *Siphoviridae* (Rajala-Mustonen and Heinonen-Tanski, 1994; Muniesa et al., 1999). Somatic coliphages attach to the bacterial cell wall and may lyse the host cell in 20–30 min under optimal physiological conditions. They produce plaques of widely different size and morphology. The methodology to detect them is very simple and results may be obtained in 4 h.

Natural host strains of somatic coliphages include, besides *E. coli*, other related bacterial species as *Shigella* spp and *Klebsiella* spp. Some of these may occur in pristine waters, so exceptionally somatic coliphages may also multiply in these environments. In addition, they have been reported to replicate in *E. coli* under environmental conditions (Vaughn and Metcalf, 1975; Seeley and Primrose, 1980; Borrego et al., 1990). Therefore, one of the potential drawbacks of somatic coliphages has been their potential replication outside the gut. However, recent

Table 1

Bacteriophage families holding phages of particular interest in water quality assessment

Group (family)	Characteristics	Representative phages (host)
Myoviridae	dsDNA Long contractile tail. Capsids up to 100 nm Isometric or elongated capsids	T2, T4 (*E. coli*)
Siphoviridae	dsDNA Long non-contractile tails Capsids up to 60 nm Isometric capsids	λ (*E. coli*) B40-8 (*Bacteroides fragilis*)
Podoviridae	dsDNA Short tails Capsids up to 65 nm Isometric capsids	T7, P22 (*E. coli*)
Microviridae	ssDNA Without tail Capsids about 25–30 nm Isometric capsids	φX174 (*E. coli*)
Leviviridae	ssRNA Without tail Capsids about 25 nm Isometric capsid	f2, MS2, GA, Qβ, F1 (F-plasmid bearing bacteria)
Inoviridae	dsDNA Capsids about 800 × 6 nm Long and flexible rods	fd, M13 (F-plasmid bearing bacteria)
Tectiviridae	dDNA Without tail Capsids up to 60 nm Isometric capsids Lipid membrane below the capsid	PRD1, PR722 (enterobacteria)

reports on the host range of phages infecting the hosts strains used in standardised methods (Muniesa et al., 2003), the densities of phages and bacteria needed for replication to occur and the factors that affect phage replication (Wigins and Alexander, 1985; Muniesa and Jofre, 2004) suggest that the contribution of replication outside the gut to the number of somatic coliphages in water environments is negligible in most aquatic environments.

The term somatic coliphages covers many types of phages with a wide range of characteristics, including differential resistance to inactivating factors. Moreover, different strains of *E. coli* as well as different assay media count different numbers and types of somatic coliphages (Havelaar and Hogeboom, 1983; Muniesa et al., 1999). Consequently, the information available on both the presence and the behaviour of somatic coliphages in water environments has to be interpreted cautiously, since the data reported in the literature had been obtained with different host strains of *E. coli*, media and assay conditions. Recently, methods for the detection and enumeration of somatic coliphages have been standardised. To avoid confusion, from now on in this report, the term somatic coliphages, unless otherwise indicated, will be restricted to phages infecting *E. coli* C strains (*E. coli* ATCC 13706, or their nalidixic acid-resistant mutants *E. coli* CN13 or *E. coli* WG5) as established by Standard Methods (APHA, 1998) or ISO (2000) or USEPA (2000a,b). Bacteriophages frequently used to study somatic coliphage resistance to environmental stressors or disinfection or behaviour in concentration methods are T-4, T-7 and φX147.

F-specific RNA bacteriophages

F-specific bacteriophages are those that infect bacteria through the sex pili, which are coded by the F plasmid first detected in *E. coli* K12. The F plasmid is transferable to a wide range of Gram-negative bacteria and consequently F-specific phages may have several hosts; however, it is thought that their main host is *E. coli*.

F-specific RNA bacteriophages consist of a simple capsid of cubic symmetry of 21–30 nm in diameter and contain single-stranded RNA as the genome. They belong to the family *Leviviridae* (Murphy et al., 1995). The F plasmid is not synthesised below 32°C. Therefore, the probability of these phages replicating in the environment is very small. Their infectious process is inhibited by the presence of RNase in the assay medium, which can be used to distinguish between the F-specific RNA bacteriophages and the rod-shaped F-specific DNA bacteriophages of the family *Inoviridae*, which also infect the host cell, through the sex pili.

Strains *Salmonella typhimurium* WG49 (Havelaar and Hogeboom, 1984) and *E. coli* HS (DeBartolomeis and Cabelli, 1991) tailored to detect F-specific bacteriophages also detect small percentages of somatic phages. Nevertheless, all phages detected by the tailored strains are usually referred to as F-specific bacteriophages. The number of F-specific RNA bacteriophages is the difference between the number of phages counted in the presence and in the absence of RNase in the assay medium. More than 90–95% of the phages detected in sewage by tailored strains are F-specific RNA bacteriophages (Havelaar and Hogeboom, 1984; DeBartolomeis and Cabelli, 1991). This percentage may be lower in receiving waters and treated wastewaters. Strain WG49 counts numbers of F-specific phages slightly greater than strain HS (Chung et al., 1998; Schaper and Jofre 2000).

F-specific RNA bacteriophages group (*Leviviridae*) contains two genera (*Levivirus* and *Allolevirus*) and three minor unclassified groups (Murphy et al., 1995). *Levivirus* contains group I and group II phages, whereas *Allolevirus* contains group III and group IV phages (Hsu et al., 1995). These four groups coincide with the serotypes I (e.g. phages f2 and MS2), II (e.g., GA), III (e.g., Qβ) and IV (e.g., F1) first described by Furuse (1987). Genomic characterisation has allowed establishing genogroups that roughly correspond to those serotypes (Beekwilder et al., 1995; Hsu et al., 1995). Genogrouping by nucleic acid hybridisation or

RT-PCR has been used to differentiate subgroups in waters receiving faecal wastes (Schaper and Jofre 2000; Schaper et al., 2002, Vinjé et al., 2004). Groups I and IV are the predominant types found in waters contaminated with animal faecal residues whereas II and III are mostly associated with human pollution (Furuse 1987; Havelaar et al., 1990; Schaper et al., 2002; Cole et al., 2003). The study of genotypes distribution may contribute to source tracking faecal pollution (Scott et al., 2002; Blanch et al., 2004).

In spite of the temperature-conditioned production of sexual pili, necessary for phage replication, conflicting reports on replication of F-specific bacteriophages in wastewater and groundwater exist (Havelaar and Pot Hogeboom, 1988). But a study on the effect of host cell and phage numbers, nutrition and competition from insusceptible cells indicates that replication outside the gut is very improbable and therefore the influence of replication outside the gut on the number of these phages in the environment is negligible (Woody and Cliver, 1997).

Standardised methods for the detection and enumeration of both F-specific RNA bacteriophages (ISO, 1995) and F-specific bacteriophages (USEPA, 2000a,b) are now available. Results can be obtained in 12 h. Bacteriophages frequently used to study F-RNA bacteriophage resistance to environmental stressors or disinfection or behaviour in concentration methods are MS2 and, to a lesser extent, f2 and Qβ.

Bacteriophages infecting Bacteroides fragilis

The predominant phages infecting *B. fragilis* described so far are *Siphoviridae*, which attach to the cell wall of the host bacteria and may lyse the host cell in 30–40 min under optimal conditions. The great majority of *B. fragilis* phages detected in sewage are also *Siphoviridae* with flexible and frequently curved and curly tails (Queralt et al., 2003).

Most *B. fragilis* phages have a narrow host range (Tartera and Jofre, 1987). Concurrently, strains of *B. fragilis* differ in the numbers of phages that they recover from sewage or sewage-polluted waters (Puig et al., 1999). They also differ in their capability to discriminate human from animal faecal contamination. Thus, strain HSP40 preferably detects phages in human faecal wastes (Tartera et al., 1989), whereas strain RYC2056 detects phages in both human and non-human faecal wastes (Puig et al., 1999; Blanch et al., 2004). RYC2056 detects similar numbers of phages in sewage around the globe (Puig et al., 1999; Lucena et al., 2003) whereas strain HSP40 shows geographic differences. Despite being able to detect good numbers, from 10^3–10^4 plaque forming units (PFU) per 100 ml, in the Mediterranean area (Tartera et al., 1989; Armon, 1993) and South Africa (Grabow et al., 1993), strain HSP40 fails to detect significant numbers in Northern Europe (Puig et al., 1999) and the USA (Chung et al., 1998). Strain VPI3625 tested in the USA seems to behave similarly to strain RYC2056 (Chung et al., 1998).

A similar behaviour seems applicable to other species of *Bacteroides*. A strain of *Bacteroides tethaiaomicron* named GA17 has recently been reported to detect nearly 10^5 PFU/100 ml of municipal and hospital wastewater in the Mediterranean area, but it shows geographic differences (Payán et al., 2005). However, an easy method to isolate *Bacteroides* hosts, convenient for a given geographical area, has been described (Payán et al., 2005).

Bacteriophages infecting *B. fragilis* have not been reported to replicate outside the gut likely because of the special requirements, anaerobiosis and nutrients, of the host strain to support phage replication (Tartera and Jofre, 1987).

A standardised method for the detection and enumeration of *B. fragilis* phages is now available (ISO, 2001). It is slightly more complex and costly than methods for detecting the other groups. Results can be obtained in 18 h. The phage mostly used to study phage resistance to environmental stressors or disinfection or behaviour in concentration methods is B40-8.

Detection and enumeration methods

Typically bacteriophages are detected by their effects on the host bacteria that they infect. Easy methods to quantify bacteriophages and to determine their presence or absence in a given volume of sample are available.

Numbers of phages are generally determined by direct quantitative plaque assays, the principles of which were designed as early as 1936 by Gratia (Adams, 1959). Melted agar medium is mixed with a suitable volume of the sample under investigation and host bacterium at a temperature just above the solidification temperature of the agar. This mixture is poured on top of a bottom agar layer in a conventional petri dish. The plates are incubated and plaques scored as PFU, or plaque forming particles (PFP), after incubation.

The presence of phages in a given volume of sample can be determined by the qualitative presence–absence enrichment test. Enrichment is accomplished by adding host bacterium and nutrients to a sample, and then incubating under conditions that permit infection of the bacteria and replication of the phages present in the sample. The number of phages increases to the point where they are readily detectable by plaque assay or spot test on a lawn of the host strain (Adams, 1959). Enrichment of multiple tube serial dilutions allows estimating numbers of phages by "quantal" methods, as for example the most probable number procedure.

The most important factor in defining a method for the detection of a given group of bacteriophages is the choice of the host strain. Standardised plaque assay and enrichment methods has been settled for somatic coliphages (APHA, 1998; USEPA, 2000a,b; ISO, 2000), F-specific RNA phages (USEPA, 2000a,b, ISO, 1995) and bacteriophages infecting *B. fragilis* (ISO, 2001). The American Public Health Association (APHA) Standard Method has been reported to perform poorly (Green et al., 2000). These methods are easily implemented in routine laboratories (Mooijman et al., 2005). The cost, considering material and labour, of testing for somatic coliphages is about the same as that of testing for a conventional bacterial indicator. F-specific RNA bacteriophage and *B. fragilis* phage testing will cost about the double.

Molecular methods for the detection of F-specific RNA bacteriophages (Rose et al., 1997) and bacteriophages infecting *B. fragilis* (Puig et al., 2000) has been described.

Quality assurance of testing phages in water samples requires the availability of reference suspensions of bacteriophages. At present certified reference suspensions of phage are not available, but it is quite easy to prepare in-lab bacteriophage suspensions of high quality and stability. These suspensions can be prepared not only with the reference phages indicated in the standardised methods (Mooijman et al., 2005) but also with suspensions of phages naturally occurring in sewage (Mendez et al., 2002).

Though the enrichment method allows testing relatively large volumes (up to 1 l is feasible), sometimes concentration may be required either because greater volumes need to be tested or because quantification is needed. Many methods have been described which are based on the same principles as the methods used for animal viruses. Not all methods, mostly those based on adsorption–elution, probed to be adequate for a given human virus are necessarily helpful for concentrating all kinds of bacteriophages (Grabow, 2001). ISO has launched a standardised procedure for validation of concentration methods for bacteriophages (ISO, 2002). For volumes ranging from 100 to 1000 ml two methods arise as the most recommendable. For water with low turbidity, Sobsey et al. (1990) developed a simple, inexpensive and practical procedure for the recovery and detection of F-specific RNA phages using mixed cellulose and acetate membrane filters with a diameter of 47 mm and a pore size of 0.45 µm; this method has been slightly modified by Mendez et al. (2004) and has an excellent performance for up to 1 litre for somatic coliphages, F-specific RNA phages and bacteriophages of *B. fragilis*. For water samples with high turbidity, a method based on flocculation with magnesium hydroxide (Schulze and Lenk, 1983) is recommended. Both methods are feasible in non-specialised laboratories (Contreras-Coll et al., 2002; Mendez et al., 2004). For volumes from 1 to 100 l, the adsorption–elution method using electropositive filters described by Logan et al. (1980) seems to be the one of choice.

Conservation of samples before testing has been studied (Mendez et al., 2002). Samples can be kept at 4°C for at least 48 h without any significant change in the number of infectious bacteriophages. In the case of small volumes, for example phage concentrates, they can be kept without loss in phage viability at −20 or −80°C for months after the addition of 10% v/v glycerol.

Bacteriophages in faeces

Somatic coliphages have been isolated in variable percentages (up to 60%) of human stool samples. Values as high as 10^9 PFU/g have been reported. The percentages of isolation reported in mammals and birds range from 100% in pigs to 38% in rabbits (Havelaar et al., 1986, Grabow et al., 1995) with counts up to 10^7 PFU/g.

A maximum incidence of 30% positive human samples for F-specific phages has been reported (Havelaar et al., 1986; Grabow et al., 1995; Schaper et al., 2002). Such phages have been isolated inconsistently from domestic and feral animal faeces, though incidences are in general higher than in humans (Grabow et al., 1995; Calci et al. (1998); Schaper et al., 2002), with densities up to 10^4 PFU/g.

B. fragilis RYC2056 has been reported to recover phages from 28% of human stool samples, although it also recovers phages from animal faeces (Puig et al., 1999), with the maximum incidence, 30%, in pigs. *B. fragilis* HSP40 phages have been isolated from 10–13% of human stool samples with maximum values above 10^8 PFU/g, but never from animal faeces (Grabow et al., 1995; Tartera and Jofre, 1987).

Bacteriophages in faecal wastes

Municipal and hospital sewage and abattoir wastewater

Somatic coliphages are the most abundant indicator phages in raw municipal and hospital sewage with values ranging from 10^6 to 10^7 PFU/100 ml, usually less than one order of magnitude lower than the number of faecal coliforms wherever counted (Nieuwstad et al., 1988; Grabow et al., 1993; Chung et al., 1998; Contreras-Coll et al., 2002; Lucena et al., 2003; Lodder and de Roda, 2005). They are also the most abundant in abattoir wastewater, with values that keep them in proportion to faecal coliforms as in municipal wastewater (Havelaar and Hogeboom, 1984; Tartera et al., 1989; Grabow et al., 1993; Blanch et al., 2004).

F-specific RNA bacteriophages rank second in abundance in both municipal and hospital raw sewage and raw wastewater from abattoirs, with values ranging from 5×10^5 to 5×10^6 PFU/100 ml, usually about one order of magnitude lower than the values of somatic coliphages (Havelaar and Hogeboom, 1984; Nieuwstad et al., 1988; Grabow et al., 1993; Chung et al., 1998; Contreras-Coll et al., 2002; Lucena et al., 2003; Blanch et al., 2004) Lodder and de Roda, 2005).

Bacteriophages infecting *B. fragilis* RYC2056 are found in municipal sewage (Europe, South Africa and America) and hospital sewage (Europe), with average values ranging from 10^4 to 10^5 PFU/100 ml, about one order of magnitude lower than the values of F-specific RNA phages. Their ratio with respect to somatic coliphages and F-RNA phages is remarkably constant (Puig et al., 1999; Contreras-Coll et al., 2002; Lucena et al., 2003; Blanch et al., 2004). Strain VPI3625 has been described to recover approximately 10^4 PFU/100 ml in the USA (Chung et al., 1998). Strain HSP40 has been reported to detect approximately 10^4 PFU/100 ml in some geographic areas, but much lower values in other areas (Chung et al., 1998; Puig et al., 1999). Strain GA17 of *B. tethaiaomicron* detects between 5×10^4 and 10^5 PFU/100 ml in both municipal and hospital sewage in Southern Europe (Payán et al., 2005).

Bacteriophages detected by strain RYC2056 are usually present in wastewaters from abattoirs, but their ratios to somatic coliphages are significantly lower than the ratios in municipal or hospital sewage (Puig et al., 1999; Blanch et al., 2004). Densities of phages detected by the other strains of *Bacteroides* mentioned above are either null or very low (Tartera et al., 1989; Puig et al., 1999; Payán et al., 2005).

Numbers of the three groups of bacteriophages, when a geographically suitable host strain of *Bacteroides* is used, are fairly constant in raw sewage throughout the world, as are the numbers of bacterial indicators.

Septage

Calci et al. (1998) detected F-specific RNA phages in 11 of 17 samples of USA household septic tank with maximum values of 8.7×10^5 PFU/100 ml. In septic effluents of a high school in the USA, Deborde et al. (1998) detected somatic and F-specific RNA phages in all 43 samples tested with average values of 2.3×10^4 PFU of somatic coliphages and 7.4×10^4 PFU of F-specific RNA phages per 100 ml, though in this case the host strains used were not those recommended in the standardised methods. In Argentina, Lucena et al. (2003) reported the detection of somatic coliphages and F-specific RNA phages in all samples tested with values similar to those reported for sewage in the area; in contrast, phages infecting *B. fragilis* RYC2056 were detected only in 7 of 28 septage samples tested.

Slurries from animal wastes

Data available about bacteriophages in slurries are difficult to compare because of the vagueness of the term slurry. Occurrence and densities are in-between prevalence in faeces and in abattoir wastewaters (Hill and Sobsey, 1998; Schaper et al., 2002; Cole et al., 2003; Blanch et al., 2004)

Raw sludges

Numbers of the three groups of bacteriophages to be considered as potential indicators in raw sludge vary in the reports available, probably as a consequence of the lack of standardised methods for both phage extraction from sludges and phage counting when the investigations were done (Chauret et al., 1999; Lasobras et al., 1999; Mignotte-Cadiergues et al., 1999). In the available reports in which the three groups of phages were counted (Lasobras et al., 1999; Mignotte-Cadiergues et al., 1999), the highest counts correspond to somatic coliphages, followed by F-specific RNA bacteriophages and bacteriophages infecting *Bacteroides*. The accumulation of bacteriophages in raw primary and secondary sludges is similar to that of bacterial indicators, and the ratios of bacterial indicators to phages do not differ significantly from those in sewage (Lasobras et al., 1999; Williams and Hurst, 1988).

Bacteriophages as indexes of human viruses

The most frequently reported situation is that correlation between viruses and phages is not found in many sites or sets of samples (Griffin et al., 1999; Jiang et al., 2001; Abbaszadegan et al., 2003; Borchardt et al., 2004). However, certain degree of correlation has been reported in a few studies between: (i) enteroviruses and somatic coliphages in water at the different stages of a water treatment plant (Stetler, 1984); (ii) F-specific RNA bacteriophages and enteroviruses in fresh water (Havelaar et al., 1993); (iii) F-specific RNA bacteriophages and enteroviruses in shellfish (Chung et al., 1998); (iv) F-specific RNA bacteriophages and genomes of calicivirus in shellfish (Doré et al., 1998); (v) enteroviruses and rotaviruses and phages infecting *B. fragilis* HSP40 in sea sediments (Jofre et al.,1989); (vi) somatic coliphages and enteroviruses in marine coastal water (Mocé-Llivina et al., 2005); (vii) different phages and the genomes of adenovirus, hepatitis A and norovirus in shellfish (Brion et al., 2005).

The absence of phages might not guarantee the absence of virus. However, taking into consideration that bacteriophages infecting enteric bacteria are always present in raw municipal sewage, that at least in industrialised countries sewage constitutes the main input of faecal pollution to receiving waters and that some phages behave similarly to human viruses, the presence of certain numbers of indicator phages in waters may indicate the presence of virus.

Bacteriophages as indicators

The indicator purpose requires the removal and inactivation of the surrogate indicator to be similar or lesser to that of the surrogated pathogens both in nature and in water treatments.

Viruses are removed from water by adsorption to particles, sedimentation and filtration through porous matrices, either soil or man-constructed filters. These processes do not inactivate viruses, but transfer them from one compartment to another.

Viruses and phages in water and the solid compartments (sediments and bio-solids) undergo inactivation by both natural and man-introduced physical, chemical and biological factors. Amid the natural factors, temperature, sunlight, pH and ionic environment play a major role in virus and phage inactivation. Man-introduced physical and chemical treatments as irradiation (UV, gamma and electron beam radiation), thermal treatment and chemical disinfection (lime, halogens, ozone and others) play a major role in the elimination of viruses in water and sludge treatments.

In the evaluation of phages as viral indicators, the question is whether their outcome in all these natural and man-driven processes resembles viral outcome and whether after all these processes they still persist in numbers higher than those of viruses in recreational waters, drinking water sources, drinking water, reclaimed water, water used for shellfish growing, shellfish, etc. But the effect of each of these natural and man-driven processes to the removal and inactivation of viruses and bacteriophages from water is difficult to determine. Comparing numbers of viruses and phages before and after the processes and finding out the fate of single-seeded laboratory-grown phages or viruses in laboratory or pilot experiments are the habitual procedures to evaluate and compare removal and inactivation. However, both procedures have important shortcomings and comparisons are difficult to interpret because the detection methods and the sample volumes analysed for phages and viruses are very different. Moreover, separate studies show significant differences in the removal and inactivation of seeded laboratory-grown and naturally occurring viruses and phages. This difficulty can be partially overcome with phages since naturally occurring phages can be partially purified from waste-water at densities high enough to make some laboratory or pilot inactivation studies that surely give a better approximation to the real world than the one given by the laboratory-grown bacteriophages.

There is a vast literature about removal and inactivation of viruses and bacteriophages in nature and in water treatments as well as about occurrence and levels of viruses and bacteriophages in water environments. A significant fraction of this information has been compiled in several reviews (IAWPRC Study Group on Health Related Water Microbiology, 1991; Armon and Kott, 1996, Grabow, 2001, Jofre, 2002). Information available suggests that the fate of phages in natural water environments and their outcome in water and sludge treatments resembles those of human pathogenic viruses.

Like viruses, bacteriophages adsorb to solids (Payment et al., 1988). This feature has a major contribution in their removal from waters. Indeed, adsorption of phages to suspended solids facilitates their sedimentation in both natural environments and water-treatment processes as indicated by increased phage densities in sediments (Araujo et al., 1997) and in sewage sludges (Lasobras et al., 1999; Mignotte-Cadiergues et al., 1999). Though most of the studies done on adsorption of phages in soil columns or in pilot fields have been done with seeded laboratory-cultured phages, results suggest that phage behaviour resembles that of viruses (Yates et al., 1985; Sobsey et al., 1995; Schijven et al., 2000). Adsorption also influences the effective retention of viruses and bacteriophages by microfilters used in water treatments, whose pore size is greater than viruses and phages (Herath et al., 1998; Farahbakhsh and Smith, 2004). In conclusion, removal of phages and viruses in water environments and in water treatments follow similar trends. Moreover, the accumulation and the poor depuration of phages and viruses in shellfish are comparable and are probably influenced by the great tendency of viruses and phages to adsorb to solids (Lucena et al., 1994, Doré and Lees, 1995; Muniain-Mujika et al., 2002).

Viruses and bacteriophages persist longer than conventional bacterial indicators in water environments and their persistence resembles that of viruses as shown by model experiments (Gironés et al., 1989; Sinton et al., 1999; Duran et al., 2002; Mocé-Llivina et al., 2005) and by the decrease in the ratios between the numbers of conventional bacterial indicators and viruses and phages of the three groups in water environments with aged pollutants (Lucena et al., 1994, Duran et al., 2002; Contreras-Coll et al., 2002; Skraber et al., 2004; Lodder and de Roda, 2005; Mocé-Llivina et al., 2005).

In wastewater treatments, such as primary sedimentation, flocculation-aided sedimentation, activated sludge digestion, activated sludge digestion plus precipitation and the upflow anaerobic sludge blanket (UASB) processes, all bacterial indicators, bacteriophages and infectious enteroviruses undergo similar reduction in numbers (Grabow et al., 1984; Nieuwstad et al., 1988; Fleisher et al., 2000; Lucena et al., 2004; Lodder and de Roda, 2005). In contrast, wastewater and sludge treatments that include disinfection processes result in substantial differences between the elimination of conventional bacterial indicators and viruses and bacteriophages. Thus, treatments, such as UV disinfection (Jancangelo et al., 2003; Harwood et al., 2005), chemical disinfection (Tyrrell et al., 1995; Duran et al., 2003; Harwood et al., 2005), thermal treatment of sludges (Mignotte-Cadiergues et al., 2002; Mocé-Llivina et al., 2003), liming (Grabow et al., 1978; Mignotte-Cadiergues et al., 2002) or lagooning (Campos et al., 2002, Lucena et al., 2004), result in reduction of densities of conventional bacterial indicators clearly higher than those of naturally occurring phages and viruses. The effects on phages and human viruses of most of these processes are comparable, though with different degrees of likeness, which depends on the virus and the phages group studied.

The situation is similar in drinking water treatments, though in this case the information on the elimination of naturally occurring phages and viruses in the treatments is still scarce. However, a few descriptions on the reduction in numbers of naturally occurring phages and enteroviruses (Payment et al., 1985b) and bacteriophages (Jofre et al., 1995) indicate similar outcomes to different treatments for phages and viruses, regarding both the extent of reduction in numbers, which is lesser than the reduction in numbers of bacteria, and the extent of the response to treatments of different complexity.

Which group of bacteriophages better surrogates human viruses?

As to which of the three groups better fulfils the model function, each of the groups of phages advocated as model organisms has advantages and weaknesses.

Somatic coliphages are the most abundant, and the methods for their detection and enumeration is the simplest and fastest, with results available in one working day. The most important weakness attributed to this group was their potential replication outside the gut, but in the view of available information, it seems that the contribution of replication outside the gut to the numbers found in water samples is negligible in the great majority of situations. Perhaps the main feebleness of this group is their resistance to some disinfectants, which is low as compared to that of other bacteriophages, although high compared to that of bacterial indicators.

F-specific RNA bacteriophages rank second in abundance. The method for detecting them is simple and fast, although not as much so for the somatic coliphages. They offer good perspectives as index organisms for viruses in groundwater and to monitor some water treatments, for example, UV disinfection. In contrast, their persistence in surface waters, mainly in warm climates, is low, as is their resistance to some inactivating treatments, mostly treatments based on heat and high pH.

Bacteriophages infecting *Bacteroides* rank third in abundance. Some strains detect phages with an unequivocal origin in the human gut. They are more resistant than the other groups to most inactivating factors and treatments and they do not replicate outside the gut. The method for detection requires anaerobiosis. Their low numbers in waters and the need of different hosts for different geographical areas are at present the main drawbacks.

Now that standardised and feasible methods are available, more research should be done to guarantee the best choice for the various potential applications of bacteriophages as model organisms.

References

Abad FX, Pinto RM, Bosch A. Survival of enteric viruses on environmental fomites. Appl Environ Microbiol 1994; 60: 3704–3710.

Abad FX, Pintó RM, Bosch A. Disinfection of human enteric viruses on fomites. FEMS Microbiol Lett 1997; 156: 107–111.

Abbaszadegan M, LeChevalier M, Gerba C. Occurrence of viruses in US groundwaters. J Am Water Works Assoc 2003; 95: 107–120.

Adams MH. Bacteriophages. New York, NY: Interscience Publishers, Inc.; 1959.

APHA. Standard methods for the examination of water and wastewater. Washington DC: American Public Health Association; 1998.

Araujo R, Puig A, Lasobras J, Lucena, Jofre J. Phages of enteric bacteria in fresh water with different levels of faecal pollution. J Appl Microbiol 1997; 82: 281–286.

Armon R. Bacteriophages monitoring in drinking water: do they fulfil the index or indicator function. Water Sci Technol 1993; 27: 463–467.

Armon R, Kott Y. Bacteriophages as indicators of pollution. Crit Rev Env Sci Technol 1996; 26: 299–335.

Beekwilder J, Niewenhuizen R, Havelaar AH, van Duin J. An oligonucleotide hybridisation assay for the identification and enumeration of F-specific RNA phages in surface water. J Appl Bacteriol 1995; 80: 179–186.

Bergh O, Borsheim Y, Bratbak G, Hendal M. High abundance of viruses found in aquatic environment. Nature 1989; 340: 467–468.

Blanch AR, Belanche-Muñoz L, Bonjoch X, Ebdon J, Gantzer C, Lucena F, Ottoson J, Kourtis C, Iversen A, Kühn I, Mocé L, Muniesa M, Schwartzbrod J, Skraber S, Papageorgiou G, Taylor HD, Wallis J, Jofre J. Tracking the origin of faecal pollution in surface water: an ongoing project within the European Union research programme. J Water Health 2004; 2: 249–260.

Borchardt MA, Haas NL, Hunt RJ. Vulnerability of drinking-water wells in La Crosse, Wisconsin, to enteric-virus contamination from surface water contributions. Appl Environ Microbiol 2004; 70: 5937–5946.

Borrego JJ, Cornax R, Moriñigo A, Martinez-Manzanares P, Romero P. Coliphages as an indicator of faecal pollution in water. Their survival and productive infectivity in natural aquatic environment. Water Res 1990; 24: 111–116.

Brion G, Viswanathan C, Neelakantan TR, Lingireddy S, Girones R, Lees D, Allard A, Vantarakis A. Artificial neural network prediction of viruses in shellfish. Appl Environ Microbiol 2005; 71: 5244–5253.

Calci KR, Burkhardt III W, Watkins WD, Rippey SR. Occurrence of male-specific bacteriophages in feral and domestic animal wastes, human faeces and human associated wastewaters. Appl Environ Microbiol 1998; 64: 5027–5029.

Callahan KM, Taylor DJ, Sobsey MD. Comparative survival of hepatitis A virus, poliovirus and indicator viruses in geographically diverse seawaters. Water Sci Technol 1995; 31: 189–193.

Campos C, Guerrero A, Cárdenas M. Removal of bacterial and viral fecal indicator organisms in a waste stabilization pond system in Choconta, Cundinamarca (Colombia). Water Sci Technol 2002; 45: 61–66.

Chapron CD, Ballester NA, Fontaine JH, Frades CN, Margolin AB. Detection of astrovirus, enterovirus and adenovirus types 40 and 41 in surface waters collected and evaluated by the information collection rule and an integrated cell culture-nested PCR procedure. Appl Environ Microbiol 2000; 66: 2520–2525.

Chau TN, Lai ST, Lai JY, Yuen H. Acute viral hepatitis in Hong Kong: a study of recent incidences. Hong Kong Med J 1977; 3: 261–266.

Chauret C, Springthorpe S, Sattar S. Fate of *Cryptosporidium* oocysts, *Giardia* cyst, and microbial indicators during wastewater treatment and anaerobic sludge digestion. Can J Microbiol 1999; 45: 257–262.

Chung H, Jaykus LA, Lovelance G, Sobsey MD. Bacteriophages and bacteria as indicators of enteric viruses in oysters and their harvest waters. Water Sci Technol 1998; 38: 37–44.

Cole D, Long S, Sobsey M. Evaluation of F + RNA and DNA coliphages as source-specific indicators of faecal contamination in surface waters. Appl Environ Microbiol 2003; 69: 6507–6514.

Contreras-Coll N, Lucena F, Mooijman K, Havelaar A, Pierzo V, Boqué M, Gawler A, Höller C, Lambiri M, Mirolo G, Moreno B, Niemi M, Sommer R, Valentin B, Wiedenmann A, Young V, Jofre J. Occurrence and levels of indicator bacteriophages in bathing waters through Europe. Water Res 2002; 36: 4963–4974.

Da Rosa ML, Gomes F, Pires I. Epidemiological aspects of rotavirus infections in Minas Gerais, Brazil. Braz J Infect Dis 2001; 5: 215–222.

DeBartolomeis J, Cabelli VJ. Evaluation of an *E. coli* host strain for enumeration of F male-specific bacteriophages. Appl Environ Microbiol 1991; 57: 1301–1305.

Deborde DC, Woessner WW, Lauerman B, Ball P. Coliphage prevalence in high school septic effluent and associated ground water. Water Res 1998; 32: 3781–3785.

Doré WJ, Hensilwood K, Lees DN. The development of management strategies for control of virological quality in oysters. Water Sci Technol 1998; 38: 29–35.

Doré WJ, Lees DN. Behaviour of *E. coli* and male specific bacteriophages in environmentally contaminated bivalve. Appl Environ Microbiol 1995; 61: 2830–2834.

Duran AE, Muniesa M, Méndez X, Valero F, Lucena FJ, Jofre J. Removal and inactivation of indicator bacteriophages in fresh waters. J Appl Microbiol 2002; 92: 338–347.

Duran AE, Muniesa M, Mocé-Llivina L, Campos C, Jofre J, Lucena F. Usefulness of different groups of bacteriophages as model micro-organisms for evaluating chlorination. J Appl Microbiol 2003; 95: 29–37.

EEC. The Council of European Economic Communities Directive of 8 December, 1975 concerning the quality of bathing waters. Official Journal of the European Communities; Directive no 76/160/EEC. Brussels: European Communities; 1976.

Engelbrecht RS, Weber MJ, Salter BL, Smith CA. Comparative inactivation of viruses by chlorine. Appl Environ Microbiol 1980; 40: 249–256.

Enriquez CE, Hurst CJ, Gerba CP. Survival of enteric adenoviruses 40 and 41 in tap, sea and waste water. Water Res 1995; 29: 2548–2553.

Farahbakhsh K, Smith DW. Removal of coliphages in secondary effluent by microfiltration–mechanisms of removal and impact of operating parameters. Water Res 2004; 38: 585–592.

Fleisher J, Schlafmann K, Otchwemah R, Botzenhart K. Elimination of enteroviruses, F-specific coliphages, somatic coliphages and *E. coli* in four sewage treatment plants of southern Germany. J Water Supply Res Technol 2000; 49: 127–138.

Fuhrman JA. Marine viruses and their biogeochemical and ecological effects. Nature 1999; 399: 541–548.

Furuse K. Distribution of coliphages in the environment: general considerations. In: Phage ecology (Goyal SM, Gerba CP, Bitton G, editors). New York, NY: Wiley; 1987; pp. 87–124.

Gerba CP. Failure of indictor bacteria to reflect the occurrence of enteroviruses in marine waters. Am J Public Health 1979; 69: 1116–1119.

Gerba CP, Gramos DM, Nwachuku N. Comparative inactivation of enteroviruses and adenoviruses 2 by UV light. Appl Environ Microbiol 2002; 68: 5167–5169.

Gironés R, Jofre J, Bosch A. Natural Inactivation of enteric viruses in seawater. J Environ Qual 1989; 18: 34–39.

Grabow WO, Middendorff IG, Basson NC. Role of lime treatment in the removal of bacteria, enteric viruses, and coliphages in a wastewater reclamation plant. Appl Environ Microbiol 1978; 35: 663–669.

Grabow WOK. Bacteriophages: update on application as models for viruses in water. Water S Afr 2001; 27: 251–268.

Grabow WOK, Coubrough P, Nupen EM, Bateman BW. Evaluation of coliphages as indicators of the virological quality of sewage-polluted water. Water S Afr 1984; 10: 7–13.

Grabow WOK, Holtzhausen CS, De Villiers CJ. Research on bacteriophages as indicators of water quality 1990–1992. Water Research Commission Report no. 321/1/93. Pretoria: Water Research Commission; 1993.

Grabow WOK, Neubrech TE, Holtzhausen CS, Jofre J. *Bacteroides fragilis* and *E. coli* bacteriophages. Excretion by humans and animals. Water Sci Technol 1995; 31: 223–230.

Green J, Brice K, Conner C, Mutesi R. The APHA standard method for the enumeration of somatic coliphages in water has low efficiency of platting. Water Res 2000; 34: 759–762.

Griffin DW, Gibson III CJ, Lipp EK, Riley K, Paul III JH, Rose JB. Detection of viral pathogens by reverse transcriptase PCR and of microbial indicators by standard methods in the canals of the Florida Keys. Appl Environ Microbiol 1999; 65: 4118–4125.

Guelin A. Etude quantitative de bacteriophages de la mer. Ann Inst Pasteur 1948; 74: 104–112.

Harwood VJ, Levine AD, Scott TM, Chivukula V, Lukasik J, Farrah SR, Rose JB. Validity of the indicator organism paradigm for pathogen reduction in reclaimed water and public health protection. Appl Environ Microbiol 2005; 71: 3163–3170.

Havelaar AH, Furuse K, Hogeboom WH. Bacteriophages and indicator bacteria in human and animal faeces. J Appl Bacteriol 1986; 60: 255–262.

Havelaar AH, Hogeboom WH. A method for the enumeration of male specific bacteriophages in water. J Appl Bacteriol 1984; 56: 439–447.

Havelaar AH, Hogeboom WM. Factors affecting the enumeration of coliphages in sewage and sewage polluted waters. Antonie van Leeuwenhoek 1983; 49: 387–397.

Havelaar AH, Pot Hogeboom WM. F-specific RNA bacteriophages as model viruses in water hygiene: ecological aspects. Water Sci Technol 1988; 20: 399–407.

Havelaar AH, Pot-Hogeboom WM, Furuse K, Pot R, Hormann MP. F-specific RNA bacteriophages and sensitive host strains in wastewater of human and animal origin. J Appl Bacteriol 1990; 69: 30–37.

Havelaar AH, van Olphen M, Drost Y. F-specific RNA bacteriophages are adequate model organisms for enteric viruses in fresh water. Appl Environ Microbiol 1993; 59: 2956–2962.

Hejkal TW, Smith EM, Gerba CP. Seasonal occurrence of rotavirus in water. Appl Environ Microbiol 1984; 47: 588–592.

Herath G, Yamamoto K, Urase T. Mechanism of bacterial and viral transport through microfiltration membranes. Water Sci Technol 1998; 38: 489–496.

Hill VR, Sobsey MD. Microbial indicator reductions in alternative treatment systems for swine wastewater. Water Sci Technol 1998; 38: 119–122.

Hsu FC, Shieh YS, van Duin J, Beekwilder MJ, Sobsey MD. Genotypic male specific RNA coliphages by hybridisation with oligonucleotide probes. Appl Environ Microbiol 1995; 61: 3960–3966.

Hurst CJ, McClellan KA, Benton WH. Comparison of cytopathogenicity, immunoflourescence and in situ hybridization as methods for the detection of adenoviruses. Water Res 1988; 22: 1547–1552.

IAWPRC Study Group on Health Related Water Microbiology. Bacteriophages as model viruses in water quality control. Water Res 1991; 25: 529–545.

Irving LG, Smith FA. One-year survey of enteroviruses, adenoviruses, and reoviruses isolated from effluent at an activated-sludge purification plant. Appl Environ Microbiol 1981; 4: 51–59.

ISO. ISO 10705-1: Water quality. Detection and enumeration of bacteriophages—part 1: Enumeration of F-specific RNA bacteriophages. Geneva: International Organisation for Standardisation; 1995.

ISO. ISO 10705-2: Water quality. Detection and enumeration of bacteriophages—part 2: Enumeration of somatic coliphages. Geneva: International Organisation for Standardisation; 2000.

ISO. ISO 10705-4: Water quality. Detection and enumeration of bacteriophages—part 4: Enumeration of bacteriophages infecting *Bacteroides fragilis*. Geneva: International Organisation for Standardisation; 2001.

ISO. ISO 10705-3. Water quality. Enumeration and detection of bacteriophages. Part 3. Validation of methods for concentration of bacteriophages from water. Geneve: International Standardisation Organization; 2002.

Jancangelo JG, Loughram P, Petric B, Simpson D, McIlroy C. Removal of enteric viruses and selected microbial indicators by UV irradiation of secondary effluent. Water Sci Technol 2003; 47: 193–198.

Jiang S, Noble R, Chu W. Human adenoviruses and coliphages in urban runoff-impacted coastal waters of Southern California. Appl Environ Microbiol 2001; 67: 179–184.

Jofre J. Bacteriophages as indicators. In: Encyclopedia of Environmental Microbiology (Gabriel Bitton, editor). Vol. 1. New York, NY: Wiley; 2002; pp. 354–363.

Jofre J, Blasi M, Bosch A, Lucena F. Occurrence of bacteriophages infecting Bacteroides fragilis and other viruses in polluted marine sediments. Water Sci Technol 1989; 21: 15–19.

Jofre J, Ollé E, Ribas F, Vidal A, Lucena F. Potential usefulness of bacteriophages that infect *Bacteroides fragilis* as model organisms for monitoring virus removal in drinking water treatment plants. Appl Environ Microbiol 1995; 61: 3227–3231.

Kott Y, Roze N, Sperber S, Betzer N. Bacteriophages as viral pollution indicators. Water Res 1974; 8: 165–171.

Krikelis V, Spyrou N, Markoulatos P, Serie C. Seasonal distribution of enterovirus and adenovirus in domestic sewage. Can J Microbiol 1985; 31: 24–25.

Lasobras J, Dellundé J, Jofre J, Lucena F. Occurrence and levels of phages proposed as surrogate indicators of enteric viruses in different types of sludges. J Appl Microbiol 1999; 86: 723–729.

Lodder WJ, de Roda AM. Presence of noroviruses and other enteric viruses in sewage and surface water in the Netherlands. Appl Environ Microbiol 2005; 71: 1453–1461.

Lodder WJ, Vinjé J, van de Heide R, de Roda Husman AM, Leenen EJTM, Koopmans MPG. Molecular detection of Norwalk-like calicivirus in sewage. Appl Environ Microbiol 1999; 65: 5624–5627.

Logan KB, Rees GE, Primrose SB. Rapid concentration of bacteriophages from large volumes of freshwater: evaluation of positively charged, microporous filters. J Virol Methods 1980; 1: 87–97.

Lucena F, Bosch A, Ripoll J, Jofre J. Faecal pollution in Llobregat river: interrelationship of viral, bacterial and physico-chemical parameters. Water Air Soil Pollut 1988; 39: 15–25.

Lucena F, Duran AE, Morón A, Calderón E, Campos C, Gantzer C, Skraber S, Jofre J. Reduction of bacterial indicators and bacteriophages infecting faecal bacteria in primary and secondary wastewater treatments. J Appl Microbiol 2004; 97: 1069–1076.

Lucena F, Lasobras J, Mcintosh D, Forcadell M, Jofre J. Effect of distance from the polluting focus on relative concentrations of *Bacteroides fragilis* phages and coliphages in mussels. Appl Environ Microbiol 1994; 60: 2272–2277.

Lucena F, Mendez X, Morón A, Calderón E, Campos C, Guerrero A, Cárdenas M, Gantzer C, Schwartzbrod L, Skraber S, Jofre J. Occurrence and densities of bacteriophages proposed as indicators and bacterial indicators in river waters from Europe and South America. J Appl Microbiol 2003; 94: 808–815.

Mendez J, Audicana A, Isern A, Llaneza J, Tarancon ML, Jofre J, Lucena F. Standardised evaluation of the performance of a simple membrane filtration-elution method to concentrate bacteriophages from drinking water. J Virol Methods 2004; 117: 19–25.

Mendez J, Jofre J, Lucena F, Contreras N, Mooijman K, Araujo R. Conservation of phage reference materials and water samples containing bacteriophages of enteric bacteria. J Virol Methods 2002; 106: 215–224.

Miescier JJ, Cabelli VJ. Enterococci and other microbial indicators in municipal wastewater effluents. J Water Pollut Control Fed 1982; 54: 1599–1606.

Mignotte-Cadiergues B, Gantzer C, Schwartzbrod L. Evaluation of bacteriophages during the treatment of sludge. Water Sci Technol 2002; 46: 189–194.

Mignotte-Cadiergues B, Maul A, Schwartzbrod L. Comparative study of techniques used to recover viruses from residual urban sludge. J Virol Methods 1999; 78: 71–80.

Mocé-Llivina L, Lucena F, Jofre J. Enteroviruses and bacteriophages in bathing waters. Appl Environ Microbiol 2005; 71: 6838–6844.

Mocé-Llivina L, Muniesa M, Pimenta-Vale H, Lucena F, Jofre J. Survival of bacterial indicator species and bacteriophages after thermal treatment of sludge and sewage. Appl Environ Microbiol 2003; 69: 1452–1456.

Mooijman KA, Ghameshlou Z, Bahar M, Jofre J, Havelaar A. Enumeration of bacteriophages in water by different laboratories of the European Union in two interlaboratory comparison studies. J Virol Methods 2005; 127: 60–68.

Mossel DAA. Marker (index and indicator organisms) in food and drinking water. Semantics, ecology, taxonomy and enumeration. Antonie van Leeuwenhoek 1982; 48: 609–611.

Muniain-Mujika I, Girones R, Tofiño-Quesada G, Calvo M, Lucena F. Depuration dynamics of viruses in shellfish. Int J Food Microbiol 2002; 77: 125–133.

Muniesa M, Jofre J. Factors influencing the replication of somatic coliphages in the water environment. Antonie van Leeuwenhoek 2004; 86: 65–76.

Muniesa M, Lucena F, Jofre J. Study of the potential relationship between the morphology of infectious somatic coliphages and their persistence in the environment. J Appl Microbiol 1999; 87: 402–409.

Muniesa M, Mocé-Llivina L, Katayama H, Jofre J. Bacterial host strains that support replication of somatic coliphages. Antonie van Leeuwenhoek 2003; 83: 305–315.

Murphy FA, Fauquet CM, Bishop DHL, Ghabrial SA, Jarvis AW, Martelli GP, Mayo MA, Summers MD. Virus taxonomy: the classification and nomenclature of viruses. Sixth Report of the Internacional Committe on Taxonomy of Viruses. Arch Virol 1995; 10: 268–274.

Nieuwstad TJ, Mulder EP, Havelaar AH, van Olphen M. Elimination of microorganisms from wastewater by tertiary precipitation followed by filtration. Water Res 1988; 22: 1389–1397.

Nuanualsuwan S, Cliver DO. Pretreatment to avoid positive RT-PCR results in inactivated viruses. J Virol Methods 2002; 104: 217–225.

Pang XL, Zeng SQ, Honma S, Nakata S, Vesikari T. Effect of rotavirus vaccine on Sapporo virus gastroenteritis in Finnish infants. Paediatr Infect Dis J 2001; 20: 295–300.

Payán A, Ebdon J, Taylor H, Gantzer C, Ottoson J, Papageorgiou GT, Blanch AR, Lucena F, Jofre J, Muniesa M. Method for isolation of *Bacteroides* bacteriophages host strains suitable for tracking sources of faecal pollution in water. Appl Environ Microbiol 2005; 71: 5659–5662.

Payment P, Morin E, Trudel M. Coliphages and enteric viruses in the particulate phase of river water. Can J Microbiol 1988; 34: 907–910.

Payment P, Tremblay M, Trudel M. Relative resistance to chlorine of poliovirus and coxsackievirus isolates from environmental sources and drinking water. Appl Environ Microbiol 1985a; 49: 981–983.

Payment P, Trudel M, Plante R. Elimination of viruses and indicator bacteria at each step of treatment during preparation of drinking water at seven water treatment plants. Appl Environ Microbiol 1985b; 49: 1418–1482.

Payne CM, Ray CG, Borduin V, Minnich LL, Lebowitz MD. An eight-year study of the viral agents of acute gastroenteritis in humans: ultrastructural observations and seasonal distribution with major emphasis on coronavirus-like particles. Diagn Microbiol Infect Dis 1986; 5: 39–54.

Pina S, Puig M, Lucena F, Jofre J, Girones R. Viral pollution in the environment and in shellfish: human adenovirus detection by PCR as an index of human viruses. Appl Environ Microbiol 1998; 64: 3376–3382.

Puig M, Jofre J, Gironés R. Detection of phage infecting *Bacteroides fragilis* HSP40 using a specific DNA probe. J Virol Methods 2000; 88: 163–173.

Puig A, Queralt N, Jofre J, Araujo R. Diversity of *Bacteroides fragilis* strains in their capacity to recover phages from human and animal wastes and from fecally polluted wastewater. Appl Environ Microbiol 1999; 65: 1772–1776.

Queralt N, Jofre J, Araujo R, Muniesa M. Homogeneity of the morphological groups of bacteriophages infecting bacteroides fragilis strain HSP40 and strain RYC2056. Curr Microbiol 2003; 46: 163–168.

Rajala-Mustonen RL, Heinonen-Tanski H. Sensitivity of host strains and host range of coliphages isolated from Finnish and Nicaraguan wastewater. Water Res 1994; 2: 1811–1815.

Rose JB, Zhou X, Griffin DW, Paul JH. Comparison of PCR and plaque assay for the detection and enumeration of coliphage in polluted marine waters. Appl Environ Microbiol 1997; 63: 4564–4566.

Schaper M, Jofre J. Comparison of methods for detecting F-RNA bacteriophages and fingerprinting the origin of faecal pollution in water samples. J Virol Methods 2000; 89: 1–10.

Schaper M, Jofre J, Uys M, Grabow W. Distribution of genotypes of F-specific RNA bacteriophages in human and non-human sources of faecal pollution in South Africa and Spain. J Appl Microbiol 2002; 92: 657–667.

Schijven JF, Medema G, Vogelaar AJ, Hassanizadeh SM. Removal of microorganisms by deep well injection. J Contam Hydrol 2000; 44: 301–327.

Schulze E, Lenk J. Concentration of coliphages from drinking water by $Mg(OH)_2$ flocculation. Naturwissenschaften 1983; 70: S162.

Scott TM, Rose JB, Jemkims TM, Farrah SR, Lukasik J. Microbial source tracking: current methodology and future directions. Appl Environ Microbiol 2002; 68: 5796–5803.

Seeley ND, Primrose SB. The effect of temperature on the ecology of aquatic bacteriophages. J Gen Virol 1980; 46: 87–95.

Shin G-A, Sobsey MD. Reduction of Norwalk virus, poliovirus and coliphage MS2 by monochloramine disinfection of water. Water Sci Technol 1998; 38: 151–154.

Sinton LW, Finley RK, Lynch PA. Sunlight inactivation of fecal bacteriophages and bacteria in sewage-polluted seawater. Appl Environ Microbiol 1999; 65: 3605–3613.

Skraber S, Gassilloud B, Gantzer C. Comparison of coliforms and coliphages as tools for assessment of viral contamination in river water. Appl Environ Microbiol 2004; 70: 3644–3649.

Sobsey MD, Hall RM, Hazard RL. Comparative reductions of hepatitis A virus, enteroviruses and coliphage MS2 in miniature soil columns. Water Sci Technol 1995; 31: 203–209.

Sobsey MD, Schwab KJ, Handzel TR. A simple membrane filter method to concentrate and enumerate male-specific RNA coliphages. J Am Water Works Assoc 1990; 82: 52–59.

Stetler R. Coliphages as indicators of enteroviruses. Appl Environ Microbiol 1984; 48: 668–670.

Tani N, Kurumatani N, Yosemasu K. Seasonal distribution of adenoviruses, enteroviruses and reoviruses in urban river water. Microbiol Immunol 1995; 39: 557–580.

Tartera C, Jofre J. Bacteriophages active against *Bacteroides fragilis* in sewage polluted waters. Appl Environ Microbiol 1987; 53: 1632–1637.

Tartera C, Lucena F, Jofre J. Human origin of *Bacteroides fragilis* bacteriophages present in the environment. Appl Environ Microbiol 1989; 55: 2696–2701.

Trabelsi A, Peenze I, Pager C, Jeddi M, Steele D. Distribution of rotavirus VP7 serotypes and VP4 genotypes circulating in Sousse, Tunisia, from 1995 to 1999: Emergence of Natural Human Reassortants. J Clin Microbiol 2000; 38: 3415–3419.

Tyrrell SA, Rippey SR, Watkins WD. Inactivation of bacterial and viral indicators in secondary sewage effluents, using chlorine and ozone. Water Res 1995; 29: 2483–2490.

USEPA. Technical support document for land application of sewage sludge, VI. I.EPA/822/ R-093900/9. Washington, DC: Environmental Protection Agency; 1992.

USEPA. Method 1601. Male-specific (F+) and somatic coliphage in water by two-step enrichment procedure. EPA 821.R-00-009. Washington, DC: Environmental Protection Agency; 2000a.

USEPA. Method 1602. Male-specific (F+) and somatic coliphage in water by single agar layer procedure. EPA 821.R-00-010. Washington, DC: Environmental Protection Agency; 2000b.

Vaughn JM, Metcalf TC. Coliphages as indicators of enteric viruses in shellfish and shellfish raising estuarine waters. Water Res 1975; 9: 613–616.

Vinjé J, Oudejans SJG, Stewart JR, Sobsey MD, Long SC. Molecular detection and genotyping of male-specific coliphages by reverse transcription-PCR and reverse line blot hybridization. Appl Environ Microbiol 2004; 70: 5996–6004.

Whitman WB, Coleman DC, Wiebe WJ. Prokaryotes: the unseen majority. Proc Natl Acad Sci USA 1998; 95: 6578–6583.

Wigins BA, Alexander M. Minimum bacterial density for bacteriophages replication: implications for significance of bacteriophages in natural ecosystems. Appl Environ Microbiol 1985; 49: 19–23.

Williams FP, Hurst CJ. Detection of environmental viruses in sludge. Enhancement of enterovirus plaque assay titer with 5-iodo2′-deoxiuridine and comparison to adenoviruses and coliphage titers. Water Res 1988; 22: 847–851.

Woody MA, Cliver DO. Replication of coliphages Qβ as affected by host cell number, nutrition, competition from insusceptible cells and non-FRNA coliphages. J Appl Microbiol 1997; 82: 431–440.

Wyer MD, Fleisher JM, Gough J, Kay D, Merrett H. An investigation into parametric relationship between enteroviruses and faecal indicator organisms in the coastal waters of England and Wales. Water Res 1995; 29: 1863–1868.

Yates MV, Gerba CP, Kelly LM. Virus persistence in groundwater. Appl Environ Microbiol 1985; 49: 778–781.

Human Viruses in Water   251
Albert Bosch (Editor)
© 2007 Elsevier B.V. All rights reserved
DOI 10.1016/S0168-7069(07)17012-9

Chapter 12

Quality Control, Environmental Monitoring and Regulations

Jane Sellwood

Health Protection Agency, Environmental Virology Unit, Reading, UK

Quality control

Every laboratory, including those that undertake research on surveillance or monitoring of human viruses in water, should work according to internationally accepted standards of competency. The ISO and European Standard, 17025:2005 (ISO, 2005) sets out the requirements to be met by the testing laboratories to confirm their competence. The same principles are described for water microbiology laboratories in the UK document *Microbiology of Drinking Water* (Standing Committee of Analysts, 2002) and the United States Environmental Protection Agency (USEPA) Manual for certification of laboratories (USEPA, 1997). These principles should also be used as the basis of good practice in research laboratories, which would ensure confidence in the quality of outcomes. All types of laboratories should work according to a defined quality system, which sets out the management structure of the laboratory, the laboratory test and technological processes, equipment suitability and staff competency thereby optimising the production of reliable results. The quality system should include a documentation system that is comprehensive and robust and thus able to provide accurate information on the work done and is also available for audit purposes.

The management structure of the laboratory should be well defined and staff should have authority and resources to carry out their functions. A quality control manager should be identified with responsibility to ensure that the organisation, staff and the processes meet the required standards. A quality manual should be maintained which documents the objectives of the quality system and how they will be implemented. The manual should describe all the documents that the laboratory should use and arrangements for internal and external quality assurance. The safety policy of the laboratory may be a part of the manual. Procedures should be described for corrective action in case of departures from the quality system. All staff should have access to the manual and it should be kept up-to-date.

Management review and audits of all aspects of the quality system will encourage best practice within the laboratory and identify areas needing attention. A set programme of audits will ensure that regular reviews are undertaken. The audit programme should include aspects such as a review of the quality manual, accommodation, staff training, staff competency, outside suppliers, equipment monitoring, media and reagents. Safety issues should be paid particular attention with safety audits covering equipment, accommodation, staff training, staff vaccinations, and disposal procedures for laboratory waste. It is common practice to have a laboratory safety manual. National safety requirements and standards must be adhered to (Anonymous, 1995, 2005). Risk and hazard assessments for each process done in a laboratory and for all chemicals and reagents used should be undertaken and documented.

Laboratory accommodation should be suitable for its use. Animal housing, cell culture, molecular techniques and work with specified pathogens all require specialised and dedicated facilities. Environmental monitoring of the laboratory may be appropriate for safety reasons and to assess potential contamination of work procedures. The safe storage of chemicals and reagents is essential. Comprehensive lists of these will provide accurate information on which to base adequate safety procedures. Good fire precautions and effective fire-fighting provisions are also essential.

Staff should be trained adequately and must have documented competency records. The training should cover the procedures to be used, including appropriate background information. The safety aspects of working in a microbiological or virological laboratory must be a part of training and an essential aspect of working. Staff should be offered appropriate vaccinations.

The method and techniques undertaken by a laboratory must be fit for the intended use and documented adequately in standard operating procedures (SOPs). An example of a SOP-based organisation used for European Standards is given below. A system to amend and update these documents should be in place. Records of samples tested or experiments done should be kept encompassing an audit trail of staff involved and material used. Validation studies may be required to compare new methods with those previously in use. International standards for comparison and validation of methods are available (ISO, 2000, 2004). All methods and techniques should be assessed for performance by the use of reference material, comparison of results with other methods, use in a range of laboratories and determinations of the aspects that affect the outcome. Laboratory records of work done should be kept for every assay/experiment and include information on reagents used, equipment used, conditions of the assay (such as pH, temperature, timings), date and staff involved.

Aspects to be included in a standard operating procedure

- Introduction—background discussion
- Scope—how the test or technique can be used
- Definitions—what the terms that are used mean in context of the SOP
- Principle—summary of the procedure
- Limitations—what the procedure is not suitable for
- Health and Safety—hazards involved
- Equipment—details of all equipment required to do the procedure
- Media and reagents—details listed including preparation
- Procedure—details of method/technique including use of equipment
- Calculations—processing results including examples if appropriate
- Expression of results—how results are to be recorded
- Quality Assurance—what reference materials and controls are needed
- References—general and technical references appropriate to the procedure

Methods for the concentration and detection of viruses in water produced and recommended by national bodies has primarily been for enteroviruses and has later been adapted for other enteric viruses. The current UK method is under review (SCA, 1995) and the USEPA manual was last updated in 2001 (USEPA, 2001). Only one European Standard method has been produced and that is for an enterovirus plaque assay (BS EN, 2005). No ISO methods exist for human viruses in water.

Equipment and reagents should be purchased from recognised suppliers and of a quality that produces reliable results. All equipment, whether it is a standard laboratory autoclave or specialised molecular platform, should be monitored or tested to ensure that it is working correctly and accurately. Equipment must be used according to the manufacturer's instructions, cleaned, calibrated, maintained and serviced regularly over time. Thermometers and temperature recorders should be calibrated against a certificated thermometer. Records should be kept for each item of equipment, which contains information on date obtained, calibration, servicing, faults and repair. Reagents that are prepared in the laboratory should have records of (as appropriate) ingredients used, sterilisation procedures, sterility checks, performance checks and staff involved. Risk assessments of chemicals and reagents should be undertaken and procedures amended when appropriate.

Results of tests and experiments must be recorded accurately, clearly and objectively with sufficient information to provide a reliable audit trail. Paper and electronic records must be easy to follow, straightforward to access and storable for reasonable time periods. Conclusions and interpretation should be included where appropriate. Results should be reviewed in the context of statistical considerations such as the 'uncertainty of measurement' of the technique. Bacteriological methods have long been subjected to rigorous examination to estimate the uncertainty of measurement of the various components of each method. Interpretation of this basic concept for water microbiology is under active discussion (BSI, 2003). Virological methods are yet to be scrutinised.

The quality assurance programme of a laboratory will include the use of internal quality control samples in each assay, of known positive and negative material. Qualitative and enumeration assays should have appropriate samples included. Records should be kept of the results of these materials to determine any variation over time so that action may be taken. Appropriate reference organisms should be included in procedures. The laboratory should take part in relevant external quality assessment schemes or proficiency-testing programmes in which unknown samples are sent by independent laboratories for testing.

Water and associated samples to be used or tested within a virological laboratory must be collected with regard to the difficulties of collecting a representative portion of the water under investigation (EN ISO, 1992). This is particularly challenging as water bodies change over time, season and place and the likely numbers of human viruses may be small. Variation in the virological content of the water is likely to be a bigger factor than the uncertainty of measurement within the method in influencing the result of an assay to detect virus. The volume of water that needs to be tested will vary with matrix. A few mililitres of sewage may contain detectable virus but will only represent a tiny fraction of the matrix at one given time. A 1000 l of finished drinking water may contain no detectable viruses but will represent a reasonable portion of a treatment works' output. In Europe samples of recreational water are often 10 l in volume as this was the volume required for testing by the 1976 EU Bathing Water Directive (Council Directive, 1976). Sampling should take place at designated sites that are chosen for the likelihood of producing positive and representative results. Samples should be collected in clean, plastic containers, kept cool and delivered to the laboratory within 24 h. As with all aspects of laboratory processes, the safety of staff undertaking sampling is a vital consideration.

Monitoring

Monitoring of water matrices can serve many purposes and is often an aspect of formal national regulation. The regular and continuing sampling of a body of water provides data, which is the basis of the action, gives confidence in treatment and can form the basis for predictive models of risk to public health. Drinking water, sewage effluent and recreational waters are the common water matrices covered by national regulations as they impact public health. The microbiological parameters recognised worldwide as providing the most effective tools are *E. coli* and intestinal enterococci. These organisms are present in large numbers in faecal material and do not multiply in the environment; the laboratory assays are inexpensive, relatively simple, quick, require only 100 ml of water and so can be repeated often. Repeated sampling is essential to provide an accurate representation of the water body over time and therefore the presence of faecal material. However, the presence of these bacteria is not directly associated to the presence of any human viruses although a higher faecal content of water will increase the likelihood of viruses being also present.

The World Health Organisation (WHO) has reviewed the monitoring and surveillance processes in the full range of water systems for drinking and recreational water (WHO, 1999, 2003, 2006). Human viruses are not often part of formal national monitoring regulations as they are not always present in faeces due to seasonality and are sometimes present in low numbers, the laboratory tests are expensive and need large volumes of water, and until recently were slow to produce results. Much work is currently underway to adapt molecular assays to detect and quantify viruses in water and so that practical, quick and efficient monitoring tools may be produced eventually. Quality control issues are important aspects that need to be addressed (USEPA, 2004). The targets will be adenoviruses and enteroviruses, which are the virus groups most commonly found in environmental waters, or norovirus, a pathogen causing gastroenteritis.

The detection of enteroviruses using a cell culture-based plaque assay is the most well-established virological tool. Although monitoring for human viruses is not often a part of formal regulation, virus assays have been used for many years to assess the performance of treatment technologies in sewage and drinking water treatment plants. For instance, this has been applied to the UV-disinfection process in UK sewage-treatment plants since 1992 as a requirement to be fulfilled for the 'permit to discharge' and has only been withdrawn recently.

Monitoring and surveillance for research purposes has also been done for many projects all over the world on the full range of enteric viruses to provide information on the presence and distribution of those viruses in all of the aquatic environmental matrices. Although less rigorous than formal monitoring, all sampling and testing should follow the same general principles. All the comments in the sampling section of quality control are relevant to monitoring for research or surveillance purposes.

Regulations

WHO (2003, 2006) has maximized the protection of public health through the provision of safe drinking and recreational water as a basic strategy. All aspects of the water environment should be assessed and viewed as a continuous process of which each aspect needs reviewing and action. The quality of drinking water is ensured through the framework of Water Safety Plans (WSP) (DWI, 2005), which are a means to address every aspect of the water supply chain from 'catchment to consumer' using risk assessment and then risk management. In practice this means a first stage of system assessment that identifies hazards at every stage of the water supply, the risks associated with each hazard, control measures to reduce or remove the risk and meeting national standards and regulations. The second stage is operational monitoring which checks each control measure and then plans reactions to problems. Microbiological monitoring is an integral part of both these stages; bacterial indicator organisms, but not viruses are in widespread use as standards.

Most developed countries have used this multiple-barrier approach over many years and the most recent regulations reflect the change to a broad risk-management approach. The EU Drinking Water Directive (European Community Directive, 1998) and the changes to the US Safe Drinking Water Act of 1974 and amendments 1986, 1996 (USEPA, 2006), the Australian Drinking Water Guidelines (2004) and the SANS (2005) for Drinking Water are all examples of this.

Specific reference to viruses is often mentioned as basic information in these Guidelines; that water should contain no viruses but that no numerical standard is in force. The USEPA document (USEPA, 2005) has listed many human viruses on its 'Drinking water contaminant list' although it states that 'the contaminants included are not subject to any proposed or promulgated national primary drinking water regulation but are known or anticipated to occur in public water systems, and may require regulation under the Safe Drinking Water Act'. The South African standard has an action level if cytopathogenic viruses are found during operational monitoring. The National Directive from The Netherlands (Anonymous, 2001) notes a maximum acceptable infection risk of 1 in 10,000 per person per year for enterovirus, which leads to obligatory monitoring of source water.

Worldwide, countries have adopted the WHO (2003) recommendations concerning bathing waters for similar risk assessments to be done. In this case bathing beach profiles and specific regard to sewage disposal are high priorities. *E. coli* and intestinal enterococci are used as indicator organisms for the presence of faecal pollution and therefore potential presence of viruses. One of the few and earliest water regulations involving a virus parameter was the 1976 EU Bathing Water Directive (Council Directive, 1976) which stated that less than one plaque-forming unit (pfu) of enterovirus should be present in 10l of designated bathing water. The assay was to be done if the quality of the water was poor as measured by the bacterial indicators or if circumstances at the site changed. This has now been superseded by the 2006 Directive (European Community Directive, 2006), which does not contain a virus parameter for testing. Detailed profiles of bathing water areas are required instead and virus testing could be used as a part of this survey. A further review of scientific evidence associated with bathing waters hazards and risks will be undertaken in 2008 including that for viruses.

Conclusions

Laboratory organisation for virology laboratories whether in a university research setting or water utility testing facility must now reach the same levels of quality assurance as any other scientific discipline. Training of staff and documentation of a quality system will underpin an effective, well-run laboratory. With this basis, enteric virus assays, when appropriate concentration and detection methods are available, can be included with confidence within national standards and regulations.

Acknowledgement

The author thanks David Sartory for the generous way he shared his expertise and information.

References

Anonymous. Categorisation of pathogens according to hazard and categories of containment. In: Advisory Committee on Dangerous Pathogens. London, UK: The Stationery Office; 1995.

Anonymous. Besluit van 9 januari 2001 tot wijziging van het waterleidingbesluit in verband met de rich tlijn betreffende de kwaliteit van voor menselijke consumptie bestemd water. Staatsblad van het Koninkrijk der Nederlande 2001; 31: 1–53.

Anonymous. The control of substances hazardous to health (5th ed.). Control of substances hazardous to health regulations 2002 (as amended). In: Approved Code of Practice. London, UK: The Stationery Office; 2005.

Australian Drinking Water Guidelines. National Water Quality Management Strategy. Canberra, Australia: Australian Government; 2004.

BS EN 14486. Water Quality: Direct Detection of Human Enterovirus by Monolayer Plaque Assay. London, UK: British Standards Institute; 2005.

BSI DD 260. Water Quality: Enumeration of Micro-organisms in Water Samples. Guidance on the Estimation of Variance of Results with Particular Reference to the Contribution of Uncertainty of Measurement. London, UK: British Standards Institute; 2003.

Council Directive of 8 December 1975 concerning the quality of bathing water (76/440/EEC). Official Journal of the European Communities 5.2.1976; L31: 1–7.

Drinking Water Inspectorate. A brief guide to drinking water safety plans. In: Guidance Notes www.dwi.gov.uk London, UK. 2005.

EN ISO 5667-1. Water quality: sampling part1. In: Guidance on the design of sampling programmes and sampling techniques. Geneva: International Standards Organization; 1992.

European Community Directive 98/83/EC of 3 November 1998 concerning the quality of water intended for human consumption. Official Journal of the European Communities 05/12/1998; L330: p0032–p0054.

European Community Directive 2006/7/EC of 15 February 2006 concerning the management of bathing water quality and repealing Directive 76/160/EEC. Official Journal of the European Communities 2006; L64: 37–51.

ISO 17994:2004 Water quality: Criteria for establishing equivalence between Microbiological Methods. Geneva: International Standards Organization; 2004.

ISO/IEC 17025:2005 General requirements for the competence of testing and calibration laboratories. Geneva: International Standards Organization; 2005.

ISO TR 13843:2000 Water quality: guidance on validation of microbiological methods. Technical Report. Geneva: International Standards Organisation; 2000.

SANS. South African National Standard 241: Drinking water. Pretoria: Standards South Africa; 2005.

Standing Committee of Analysts. Methods for the isolation and identification of human enteric viruses from waters and associated materials.In: Methods for the Examination of Waters and Associated Materials (in this series). Environment Agency, UK; 1995.

Standing Committee of Analysts. The microbiology of drinking water: part 3 practices and procedures. In: Methods for the Examination of Waters and Associated Materials (in this series). Environment Agency, UK; 2002.

US Environmental Protection Agency. Manual for the certification of laboratories analyzing drinking water: criteria and procedures, quality assurance. 4th ed. EPA-815-B-97-001. Cincinnati, OH: US Environmental Protection Agency, Office of Groundwater and Drinking Water; 1997.

US Environmental Protection Agency. Manual of methods for virology. EPA/600/4-84/013. Cincinnati: USEPA; 2001.

US Environmental Protection Agency. Quality assurance/quality control for laboratories performing PCR analyses on environmental samples. Cincinnati, OH: USEPA; 2004.

US Environmental Protection Agency. Drinking water contaminant list 2; final notice. Federal Regulations, 70:36:9071. Cincinnati, OH: USEPA; 2005.

US Environmental Protection Agency. National primary drinking water regulation; long-term 2 enhanced surface water treatment rule. Federal Regulations, 71:2:653. Cincinnati, OH: USEPA; 2006.

WHO. Health-Based Monitoring of Recreational Waters: The Feasibility of a New Approach (The "Annapolis Protocol"). Geneva: World Health Organization; 1999.

WHO. Guidelines for Safe Recreational Water Environments. Volume 1: Coastal and Fresh Waters. Geneva: World Health Organization; 2003.

WHO. Guidelines for Drinking Water Quality, Third Edition, Incorporating First Addendum. Volume 1: Recommendations. Geneva: World Health Organization; 2006.

Human Viruses in Water
Albert Bosch (Editor)
© 2007 Elsevier B.V. All rights reserved
DOI 10.1016/S0168-7069(07)17013-0

Chapter 13

Recent Advances and Future Needs in Environmental Virology

Mark Wong[a], Irene Xagoraraki[b], Joan B. Rose[a]
[a]*Department of Fisheries and Wildlife, Michigan State University, E. Lansing, MI 48824*
[b]*Department of Civil and Environmental Engineering, Michigan State University, E. Lansing, MI 48824*

The global water quality inventory and environmental virology: status and implications

Medically important viruses were first noted as part of the environmental "malaises" early in human history (often described with symptoms such as jaundice) but it was advances in cell culture, electron microscopy, and immunology that spurred the discovery and characterization of human viruses. The first isolations of viruses from water came in the 1950s and 1960s for surface waters and drinking waters, respectively (Gerba and Rose, 1989). Yet it had long been understood that enteric viruses such as poliovirus were shed in feces and thus by association present in sewage and sewage-polluted waters. Our conventional definition of environmental virology has primarily focused on enteric viruses and contaminated drinking water, fecal-oral transmission, and associated person-to-person transmission. Likewise, the management and control of waterborne viruses has focused on disinfection of drinking water and vaccinations. Despite the tremendous improvements in water and sanitation management fueled by a better understanding of the nature of viruses, emerging viruses such as the polyomaviruses and reemerging epidemics of age-old viruses such as poliovirus, as well as concerns associated with intentional use of eradicated viruses such as the smallpox virus, challenge our conventional definition of "environmental virology" and traditional approaches to control.

The global outbreaks of severe acute respiratory syndrome (SARS) and avian influenza (AI) highlight the degree of vulnerability that high-density urban populations face when threatened by novel, unanticipated viral pathogens. This is further underscored by security fears brought on by recent acts of terrorism both in the US and abroad. Less sensational but equally serious outbreaks of many other viruses like norovirus, hantavirus, and West Nile virus have been documented worldwide and are on the rise. Rotavirus-induced diarrhea is still the most prevalent infant killer in many developing nations causing an estimated 140 million cases worldwide and killing almost 600,000 people annually (Parashar et al., 2003).

A United Nations report on the world water crisis situation has highlighted the twin issues of scarcity and impacted quality of the world's drinking water supply (UNESCO, 2003). Scarcity of water in many parts of the world has forced people to turn to increasingly less pristine sources of water for their drinking and other needs. This has given rise to an increased incidence of waterborne disease. The burgeoning world population is also placing a greater strain on the current world water supply. Irrigation currently consumes approximately 70–80% of the world's fresh and groundwater supply while polluting lakes and streams through surface runoff of pesticides, fertilizers, and land-applied biosolids with their associated pathogen loads. The US Food and Agricultural Organization (FAO) anticipates a net expansion of irrigated land of some 45 million hectares in 93 developing countries to a total of 242 million hectares in 2030 and projects that agricultural-water withdrawals will increase by some 14% from 2000 to 2030 to meet future food production needs (FAO, 2002).

The increase in population also means that more wastewater is being generated. Globally, 41% of the population is without adequate sanitation and very little of the global wastewater is treated. (WHO/WSH main site: www.who.int/water_sanitation_health/). According to the US House Transportation Subcommittee on Water Resources (2003), after investing $250 billion in wastewater infrastructure in the US with the passage of the *Clean Water Act* of 1972, our communities' economic well-being, which relies on clean water, and our ability to continue to meet the public health goals are at risk. This is despite there being 16,000 publicly owned wastewater treatment plants, 100,000 major pumping stations, 600,000 miles of sanitary sewers, and 200,000 miles of storm sewers in the US.

Historically, the importance of protecting one's drinking water supply has been well documented and recognized. Poisoning or contaminating an enemies' water supply has been practiced in warfare since at least the fourth century BC. Less deliberate acts of contamination occur more frequently due to industrial accidents, inclement weather, and weak enforcement of regulations or operator neglect. While animal wastes have been implicated in bacterial and parasitic outbreaks, the viruses remain associated with some of the most serious health consequences such as the outbreak of viral hepatitis E in Kanpur, India in 1991 that affected an estimated 79,091 people (Naik et al., 1992) with 30% mortality in pregnant women in the first trimester. The recent widespread poliovirus outbreaks throughout Africa are likely in part due to contaminated water and the inadequate and wastewater treatment (Pavlov et al., 2005).

More recently, attention has also focused on the need to protect recreational water sources (Wade et al., 2003; Standish-Lee and Loboschefsky, 2006). Fresh and salt water sources represent an important recreational resource, especially to economies that rely heavily on tourism. In addition, the increasing scarcity of pristine water sources has meant that the water cycle is being short circuited in order to provide adequate water for drinking, recreation, power generation, agriculture, and industrial processing. The assessment of the impairment of waterways for the various uses based on the "indicator bacteria" and *Escherichia coli* has not provided enough specificity in regard to health risk, sources of the pollution, identification of the responsible party, and control. This has fueled a demand for advanced pathogen detection.

In the twenty-first century, viral diseases have changed the landscape of medicine. Acquired immunodeficiency syndrome (AIDS) now infects millions of people worldwide and up to 30% of the populations in Africa. Waterborne diseases will be particularly devastating to these individuals and the list of potential waterborne viral agents is growing. Certain microbiological advances like the polymerase chain reaction (PCR) and microarray technology may provide the tools necessary for monitoring any new agent of interest. The era of pathogen discovery is not over. We should be prepared to use all the microbiological advances to continue to understand the transmission of diseases through water. This must then be coupled with our knowledge of engineering to determine how to maintain and develop the quantity and quality of our water resources. With the advances made in medicine, microbiology, and engineering we can add to the existing body of knowledge and continue to advance the field of "environmental virology".

Advanced pathogen detection and discovery

The detection of viruses in water and other environmental samples constitutes special challenges. The standard method of detection of viral pathogens in environmental samples uses assays in mammalian cell culture. The infected cell cultures undergo observable morphological changes called cytopathogenic effects (CPEs) that are used for the detection of viruses. Even though many viruses are culturable in several cell lines and are thus detectable by the development of CPEs in cell culture, there are several viruses, like enteric waterborne adenoviruses types 40 and 41, which are difficult to culture and do not produce clear and consistent CPE. Other viruses, like waterborne caliciviruses, have not yet been successfully grown in cell cultures. Conventional cell culture assays for the detection of viruses in environmental samples have limited sensitivity and can be labor-intensive and time-consuming. Two advances, the PCR and microarrays, have spurred the study of viruses and should be further applied to the field of environmental virology.

PCR application for the study of environmental virology

Since only a small percentage of viruses are cultivatable, there is a need to examine in a more systematic fashion the development of standard/consensus methods for the collection and processing of viruses in the water environment via molecular techniques. Table 1 lists some of the critical issues in the application of PCR techniques in environmental virology.

The PCR-detection assay allows for highly sensitive and highly specific detection of virus nucleic acid sequences. PCR was initially used as a research tool for the amplification of nucleic acid products. It is fast gaining acceptance in the clinical diagnostic setting and it has been effectively applied to the detection of viruses from environmental samples. The sensitivity of PCR has been demonstrated to be comparable or superior to cell culture (Raboni et al., 2003; Lee and Jeong, 2004). In order to use PCR one must already know the exact sequence of a given region of the viral nucleic acid. Primers are then designed to amplify that specific region of the genome. Primers for the specific detection of many of the enteric viruses have been published. Nested and multiplex PCR, which are variations of the conventional PCR, have also been applied to virus detection in water samples. The chief drawback of PCR methods is that they are incapable of distinguishing between active and inactive targets.

The latest advancement in molecular methods is the development of quantitative real-time PCR. Quantitative PCR (qPCR) or real-time PCR can be used to quantify the original template concentration in the sample. Following DNA extraction, real-time PCR simultaneously amplifies, detects, and quantifies viral acid in a single tube within a short time. In addition to being quantitative, real-time PCR is also faster than conventional PCR. Real-time PCR requires the use of primers similar to those used in conventional PCR. It also requires oligonucleotide probes labeled with fluorescent dyes, or an alternative fluorescent detection chemistry different from conventional gel electrophoresis, and a thermocycler that can measure fluorescence. For quantification, generation of a standard curve is required from an absolute standard with known quantities of the target nucleic acid or organism. When real-time PCR quantitative results for adenoviruses in environmental samples were compared with conventional cell culture results, it was concluded that the real-time PCR method demonstrated higher quantities of adenoviruses in comparison with conventional techniques. This was likely due to the presence of noninfectious viruses and low sensitivities of tissue culture assays to serotypes present in the environmental samples (Choi et al., 2004; He and Jiang, 2004).

Another modification of the standard PCR method, integrated cell culture (ICC)-PCR makes use of a cell culture step to enhance sensitivity and demonstrate infectivity. The ability to detect only infectious viruses among many noninfectious viruses is important for predictions of public health risk in water and other environmental samples, particularly after disinfection. As described above, molecular methods detect part of the viral DNA or RNA that indicates virus presence but not infectivity. To determine infectivity, molecular methods can be used in conjunction

Table 1

Critical issues for PCR applications to environmental virology

Sample concentration and inhibition control	Environmental samples are less concentrated than clinical samples. Sample pre-concentration, usually by filtration and elution of large volumes of water, is required prior to DNA extraction. PCR in environmental samples is often inhibited by various substances. Sample dilutions and sample pre-treatment methods should be evaluated for inhibition control.
Molecular techniques and quantification	For quantification, the use of real-time PCR is required along with generation of a standard curve from an absolute standard with known quantities of the target nucleic acid or organism. Real-time PCR requires the use of primers similar to those used in conventional PCR. It also requires oligonucleotide probes labeled with fluorescent dyes and a thermocycler that can measure fluorescence.
Molecular techniques and infectivity determination	Instead of CPEs, detection of bacterial mRNA by reverse transcription (RT)-PCR assay can be used to detect infectivity. In the case of DNA viruses, detection of viral mRNA during cultivation is an indication of the presence of infectious viruses. In the case of RNA viruses detection of the double-stranded replicative form in cell cultures inoculated with viruses demonstrates infectivity (Cromeans et al., 2005).
Specificity and primer and probe design	PCR and real-time PCR primers should be designed to amplify only the DNA or RNA of interest. Real-time PCR primers differ from conventional PCR primers. The primer selection in the case of real-time PCR is restricted due to requirements of smaller target amplicon sizes than the PCR target sizes. Also, in the case of real-time PCR there is a need to select probes that meet certain criteria.
QA/QC requirements	The high sensitivity of the molecular techniques requires demanding quality assurance, quality control protocols. PCR techniques are very sensitive to contamination from amplified DNA. Contamination can result in false negatives and false positives. Appropriate lab infrastructure and highly trained personnel are required for effective QA/QC (EPA, 2004). To avoid false-positive results, confirmation of amplification products, such as sequencing is required.

with cell cultures. The ability of viruses to infect cell cultures and to replicate their nucleic acid implies that they are also capable of infecting the human host.

Targeting specific messenger RNA (mRNA), after inoculation of cell cultures with samples containing infectious viruses, cultures are observed for evidence of virus replication. Instead of CPEs, detection of viral mRNA by reverse transcription (RT)-PCR assay can sometimes be used to detect infectivity. DNA viruses, such as waterborne adenoviruses, form mRNA during replication in cells. Only infectious adenoviruses can enter cells and transcribe mRNA during replication; inactivated virus will be unable to do so. Therefore, detection of viral mRNA during cultivation is indicative of the presence of infectious viruses. Positive-strand RNA viruses, such as waterborne enteroviruses and noroviruses, produce a negative, complementary strand of RNA in the early phase of viral replication. This negative strand of RNA is bound to newly formed positive strands of RNA to form a double-stranded replicative form; detection of the replicative form in cell cultures inoculated with viruses can also be used to demonstrate infectivity (Cromeans et al., 2005).

A PCR case study for environmental virology

Human adenoviruses are considered as emerging waterborne viruses. They are currently listed in the Environmental Protection Agency's Contaminate Candidate List.

The importance of adenoviruses and the potential health risks associated with their waterborne transmission has been recognized by the scientific community (Fong and Lipp, 2005). Currently, there are 51 different types of human adenoviruses. Adenoviruses are a common cause of gastroenteritis, upper and lower respiratory system infections, and conjunctivitis (Swenson et al., 2003). Other diseases associated with adenoviruses include acute and chronic appendicitis, cystitis, exanthematous disease, and nervous system diseases (Swenson et al., 2003). Adenoviruses are considered important opportunistic pathogens in immunocompromised patients (Wadell, 1984).

The potential of transmission of adenoviruses through water is suggested by the findings of several researchers. Enriquez et al. (1995) concluded that enteric adenoviruses are more stable in tap water and wastewater than poliovirus. Irving and Smith (1981) reported that adenoviruses are more likely to survive conventional sewage treatment than enteroviruses. In addition, Hurst et al. (1988) estimated that most adenoviruses detected in wastewater may be enteric adenoviruses. Borchardt et al. (2003) associated diarrhea of viral etiology (including adenoviruses) with drinking municipal water. Adenovirus outbreaks have also been associated with recreational exposure and swimming (Foy et al., 1968; Cardwell et al., 1974; D'Angelo et al., 1979; Martone et al., 1980; Turner et al., 1987; Papapetropoulou and Vantarakis, 1998; Harley et al., 2001).

Adenoviruses have been found in wastewater samples in significant numbers in different geographic locations (Irving and Smith, 1981; Krikelis et al., 1985a, b; Hurst et al., 1988, Puig et al., 1994; Greening et al., 2002; He and Jiang 2005). Girones et al. (1995) presented data that showed the prevalence of adenoviruses in sewage samples. The presence of adenovirus has also been reported in river waters (Tani et al., 1995; Cho et al., 2000; Greening et al., 2002; Choi et al., 2004; Lee et al., 2004; Choi and Jiang, 2005; Haramoto et al., 2005; Van Heerden et al., 2005) and a river estuary (Castignolles et al., 1998). Also, Jiang et al. (2001) detected human adenoviruses in urban runoff-impacted coastal waters in southern California, and Fong and Lipp (2004) detected adenoviruses in the coastal reaches of a river in Georgia.

Adenoviruses occurrence has been reported in drinking waters. Chapron et al. (2000) detected adenovirus in untreated surface waters collected and evaluated by the information collection rule. Fourteen of the 29 samples (48%) were positive for adenoviruses and 38% of these samples were determined to be infectious. Van Heerden et al. (2003) reported incidence of adenoviruses in raw and treated drinking water in South Africa. The results indicated human adenoviruses present in (13%) of the raw and (4%) of the treated water samples tested. At a later study, Van Heerden et al. (2005) reported adenovirus in 10 of 188 drinking water samples. Cho et al. (2000) and Lee and Kim (2002) detected adenoviruses in tap water.

To control the problem of adenovirus infections there is a need to further evaluate environmental exposure pathways and assess human exposure risk. Developing reliable and sensitive detection methods is crucial to this effort. It is also critical that the methods be able to quantify and differentiate between serotypes. Advances in molecular techniques have been applied in the detection of adenoviruses. Table 2 presents a summary of molecular techniques that have been applied to this complex group of viruses. Many of these assays have been used for detection in water. There is no approved, universally applied method for the detection of adenoviruses. Further methods development, comparison of existing techniques and evaluation of sensitivity and specificity of techniques is required to advance the state of knowledge relative to human adenoviruses and waterborne infections.

Microarrays and other hybridization-based detection

Microarrays were first described in 1995 by Schena et al. (1995). They are arrays of spots on specially prepared glass or silicon surfaces. Each spot on an array serves as a single test at which a hybridization of DNA, immunological attachment, or chemical reaction can occur. These arrays can be useful for screening multiple samples against multiple targets but they are neither as sensitive nor as specific as PCR or cell culture. Microarray technology is enabled by the ability to deliver submicroliter volumes of material to an attachment surface or matrix. DNA microarrays are arrays in which DNA–DNA/DNA–cDNA or DNA–RNA hybridization reactions are an indication of a positive/negative reaction. Spotted DNA microarrays consist of either PCR generated or synthesized oligonucleotides that have been printed or mechanically spotted on to specially coated glass slides. *In situ* synthesized arrays have their probes chemically synthesized directly on to the support matrix.

Table 2

Summary of molecular methods applied to the detection of human adenoviruses

Species*	Serotype	Target gene	Primer name	Sequence 5'-3'	Length	Amplicon size	Wastewater & water samples	Type of assay	Source
A-F	All 51	Hexon	JTVXF JTVXR	GGA-CGC-CTC-GGA-GTA-CCT-GAG ACI-GTG-GGG-TTT-CTG-AAC-TTG-TT	21 23	96	No	Real-time PCR	Jothikumar et al, 2005
A-F	1-8,11	Hexon	NA NA	GAC-ATG-ACT-TTC-GAG-GTC-GAT-CCC-ATG-GA CCG-GCT-GAG-AAG-GGT-GTG-CGC-AGG-TA	29 26	140	No	Real-time PCR	Watanabe et al, 2005
A-F	All 51	Hexon	NA NA	C(AT)T-ACA-TGC-ACA-TC(GT)-C(CG)G-G C(AG)C-GGG-C(GA)A-A(CT)T-GCA-CCA-G	19 19	68	No	Real-time PCR	Formiga-Cruz et al, 2002
A-F	All 51	Hexon	AD1 AD2	CTG ATG TAC TAC AAC AGC ACT GGC AAC ATG GG GCG TTG CGG TGG TGG TTA AAT GGG TTT ACG TTG TCC AT	32 38	NA	No	PCR	Sarantis et al, 2004
A-F	1-5, 9, 16, 17, 19, 21, 28, 37, 40, 41, 25	Hexon	AD2 AD3	CCC-TGG-TAK-CCR-ATR-TTG-TA GAC-TCY-TCW-GTS-AGY-GGC-C	20 19	NA	Yes	Real-time PCR	He and Jiang, 2005; Choi and Jiang, 2005
A-F	All 51	Hexon	Hex1 Hex2	TTC-CCC-ATG-GCI-CA(CT)-AAC-AC CCC-TGG-TA(GT)-CC(AG)-AT(AG)-TTG-TA	20 20	482	Yes	Nested PCR Multiplex PCR	Ko et al, 2003; Xu et al, 2000
A-F	All 51	Hexon	AQ1 AQ2	GCC-ACG-GTG-GGG-TTT-CTA-AAC-TT GCC-CCA-GTG-GTC-TTA-CAT-GCA-CAT-C	23 25	129	Yes	Real-time PCR	Heim et al, 2003; Haramoto et al, 2005; van Heerden et al, 2005
A-E	2,3,4,7,11,21,11	E1A	AdE1A-F AdE1A-R	GCC-TGC-ACG-ATC-TGT-ATG-AT TCT-CAT-ATA-GCA-AAG-CGC-ACA	20 21	409-446	No	Multiplex PCR	Lin et al, 2004
A-E	2,3,4,5,7,16,21	Fiber	AdFib-F3 AdFib-R3	ACT-GTA-KCW-GYT-TTG-GYT-GT TTA-TTS-YTG-GGC-WAT-GTA-KGA	20 21	430-437	No	Multiplex PCR	Lin et al, 2004

Group	Serotypes	Gene	Primer	Primer sequence	Length	Amplicon	Method	Reference
A–E	3,4,6,7,16,21	Hexon	AdHex-F7 / AdHex-R5	CAC-GAY-GTG-ACC-ACM-GAC-CG / TTK-GGT-CTG-TTW-GGC-ATK-GCY-TG	20 / 23	770–815	No	Multiplex PCR	Lin et al., 2004
C,D,F	(1,2,5,6)$_x$(8,19)$_x$(40,41)	Hexon	JHKXF / JHKXR	GGA-CGC-CTC-GGA-GTA-CCT-GA / CGC-TGI-GAC-CIG-TCT-GTG-G	20 / 19	135	No	Real-time PCR	Ko et al., 2005
A	All	Fiber	NA	TGC ATT TAG TGT TTG ATG AA / ATA GGT TTA GAT GTA TCT CCC TGT AA	20 / 26	NA	No	Multiplex PCR	Pehler-Harrington et al., 2004
A	31	Hexon	31F / 31R	AGA-TAT-GAC-ATT-TGA-AGT-TGA-CCC-C / CGC-AGA-TAG-ACC-GCT-TCA-ATG	25 / 21	NA	No	Multiplex PCR	Gu et al., 2003
A	All	Fiber	AdA1 / AdA2	GCT-GAA-GAA-MCW-GAA-GAA-AAT-GA / CRT-TTG-GTC-TAG-GGT-AAG-CAC	23 / 21	1444–1537	No	Multiplex PCR	Xu et al., 2000
B	All	Fiber	NA	TCT TCC CAA CTC TGG TAC / CCT GGG TTT ATA AAG GGG TG	18 / 20	NA	No	Multiplex PCR	Pehler-Harrington et al., 2004
B	All	Fiber	AdB1 / AdB2	TST-ACC-CYT-ATG-AAG-ATG-AAA-GC / GGA-TAA-GCT-GTA-GTR-CTK-GGC-AT	23 / 23	670–772	No	Multiplex PCR	Xu et al., 2000; Lin et al., 2004
C	All	Fiber	NA	TCA TAT CAT GGG TAA CAG ACA T / CCC ATG TAG GCG TGG ACT TC	22 / 20	NA	No	Multiplex PCR	Pehler-Harrington et al., 2004
C	2,5	E1A	AdC-E1AF / AdC-E1AR	CCA-CCT-ACC-CTT-CAC-GAA-CT / CTC-GTG-GCA-GGT-AAG-ATC-G	20 / 19	260	Yes	PCR	Ko et al., 2003
C	All	Fiber	AdC1 / AdC2	TAT-TCA-GCA-TCA-CCT-CCT-TTC-C / AAG-CTA-TGT-GGT-GGT-GGG-GC	22 / 20	1988–2000	No	Multiplex PCR	Xu et al., 2000
D	All	Fiber	NA	CTC CGG GTG GAA GAT GAC T / ATT GGG TCA GCC AGT TTG AGT	19 / 21	NA	No	Multiplex PCR	Pehler-Harrington et al., 2004

Table 2 (*continued*)

Species*	Serotype	Target gene	Primer name	Sequence 5'-3'	Length	Amplicon size	Wastewater & water samples	Type of assay	Source
D	All	Fiber	AdD1	GAT-GTC-AAA-TTC-CTG-GTC-CAC	21	1205-1221	No	Multiplex PCR	Xu et al., 2000
			AdD2	TAC-CCG-TGC-TGG-TGT-AAA-AAT-C	22				
E	All	Fiber	NA	TTG GCT CAG GTT TAG GAC TCA GT	23	NA	No	Multiplex PCR	Pehler-Harrington et al., 2004
				CTG TTT AGG GTA ATC TTT ATA TTC CCT	27				
E	All	Fiber	AdE1	TCC-CTA-CGA-TGC-AGA-CAA-CG	20	967	No	Multiplex PCR	Xu et al., 2000
			AdE2	AGT-GCC-ATC-TAT-GCT-ATC-TCC	21				
F	NA	Hexon	Adenovirus.est.fwd	ATG-TAT-TCC-TTT-TTC-CGA-AAC-TTC-CA	23	244	No	Real-time PCR	Logan et al., 2006
			Adenovirus.est.rev	GCC-ACA-TGG-TGC-GAT-CGC-A	19				
F	40,41	Fiber	JTVFF	AAC-TTT-CTC-TCT-TAA-TAG-ACG-CC	23	118	No	Real-time PCR	Jothikumar et al., 2005; Ko et al., 2005
			JTVFR	AGG-GGG-CTA-GAA-AAC-AAA-A	19				
F	40	Hexon	f-AD157	ACC-CAC-GAT-GTA-ACC-ACA-GAC-A	22	88	Yes	Real-time PCR	Jiang et al., 2005
			r-AD245	ACT-TTG-TAA-GAG-TAG-GCG-GTT-TCC	24				
F	40,41	Fiber	NA	AAC ATG CTC ATC CAA ATC TCG CCT A	25	NA	No	Multiplex PCR	Pehler-Harrington et al., 2004
				TTC AGT TAT GTA GCA AAA TAC AGC	24				
F	40,41	E1B	4041-1	CTG-ATG-GAG-TTT-TGG-AGT-G	19	NA	No	Multiplex PCR	Rohayem et al., 2004
			4041-2	CCA-TTA-GCC-TGC-TCC-TTA	18				
F	40,41	E1A	AdF-E1AF	GGG-AAC-TGG-GAT-GAC-AT	17	280	Yes	PCR	Ko et al., 2003
			AdF-E1AR	CCS-TCT-TCA-TAG-CAT-TTC	18				
F	40,41	Fiber	AdF1	ACT-TAA-TGC-TGA-CAC-GGG-CAC	21	541-586	No	Multiplex PCR	Xu et al., 2000
			AdF2	TAA-TGT-TTG-TGT-TAC-TCC-GCT-C	22				

F	40,41	Hexon	hexAA1885	GCC-GCA-GTG-GTC-TTA-CAT-GCA-CAT-C	25	308	Yes	PCR, Nested PCR Multiplex PCR	Allard et al., 1990; Puig et al., 1994; Girones et al., 1995; Pina et al., 1998; Castignolles et al., 1998; Cho et al., 2000; Chapron et al., 2000; Greening et al., 2002; Maluquer de Motes et al., 2004; Lee et al., 2004; Rigotto et al., 2005; van Heerden et al., 2005; Choo and Kim, 2006
			hexAA1913	CAG-CAC-GCC-GCG-GAT-GTC-AAA-GT	23				
F	41	E1B	41AA142	TCT-GAT-GGA-GTT-TTG-GAG-TGG-CTA	24	2187	No	PCR	Allard et al., 1990
			41AA358	AGA-AGC-ATT-AGC-GGG-AGG-GTT-AAG	24				
F	40,41	E1A	40AA45	ATT-GCT-GTT-GGC-GCT-TTT-GAC-ATA-G	25	858	No	PCR	Allard et al., 1990
			41AA129	TCA-AGA-GGA-CTT-GGG-GCG-CTT-TAA	24				

Note: A–F species include all 51 human infectious species.

Microarrays have conventionally been used to develop gene expression profiles of certain targets of interest. Increasingly, research has also focused on adapting microarray technology to screen clinical specimens against multiple target pathogens in a highly efficient manner (Zhou, 2003; Bodrossy and Sessitsch, 2004). Microarrays have been designed for the detection and genotyping of hepatitis B virus, adenoviruses, Epstein-Barr virus, herpes simplex virus, influenza virus, and human papillomavirus (Sengupta et al., 2003; Boriskin et al., 2004; Korimbocus et al., 2005; Min et al., 2006; Song et al., 2006).

Proposals have been put forth for using DNA microarrays as an environmental detection tool and possible biodefense tool (Pannucci et al., 2004; Sergeev et al., 2004). Only a few examples exist for the application of microarray technology on environmental samples. For example, Kelly et al. (2005) have used DNA microarrays to analyze the nitrifying bacterial community in a wastewater treatment plant. Wu et al. (2004) developed a community genome array that was able to reveal species and strain differences in microbial community composition in soil, river, and marine sediments.

The use of microarrays as an environmental research tool can be divided into two broad categories: arrays that serve to detect specific gene sequences regardless of source and arrays that target specific pathogens. Straub and Chandler (2003) have proposed that a unified system for the detection of waterborne pathogens would significantly advance public health and microbiological water analysis and have indicated that advances in sample collection, on-line sample processing and purification, and DNA microarray technologies may form the basis of a universal method to detect known and emerging waterborne pathogens. Table 3 summarizes some of the viral microarrays and their applications.

A number of commercially available microarray chip platforms are currently available. Their main differences are the manufacturing method employed and feature density. Affymetrix GeneChip arrays are able to accommodate up to 1.3 million unique features on a $5'' \times 5''$ quartz wafer and are manufactured using a photolithographic masking technique (Pease et al., 1994). Agilent microarrays have a 44,000 feature set and are synthesized using an inkjet printing method (Hughes et al., 2001). Febit's Geniom microarrays contain only 6000 features but hybridization can be carried out with eight chips in parallel allowing multiple sample processing (Guimil et al., 2003). Nimblegen microarrays contain 3,90,000 probes per array and are manufactured using a micromirror focusing technique. Less expensive glass slide arrays for smaller probe sets are also within the in-house fabrication capability of most research institutions (approximately 16,000 features per slide). Current limitations of the technology include issues regarding validation and low starting microbial biomass (Wu et al., 2006) and need to be addressed before the technology may be applied.

A microarray case study: community-level monitoring for viruses

At present, the monitoring of public health occurs at the individual patient level. The highly disseminated nature of the public health system, however, means that it

Table 3

Viral microarrays and their applications

Virus types	Numbers of gene sequences	Application	Major finding	References
Influenza A virus	12,000 features	Subtyping and sequencing	Integrated microfluidic system. Mismatch discrimination is achieved at the enzymatic ligation step.	Liu et al. (2006)
Varicella-zoster virus (VZV)	5 pairs of oligonucleotide probes 18–21 mer long	Distinguish 3 major circulating genotypes of VZV	Evaluated against 6 reference strains and 130 clinical specimens	Sergeev et al. (2006)
Animal pestiviruses DNA suspension microarray	8 probes	Detection and differentiation of animal pestiviruses	40 strains of CSFV, BVDV1, BVDV2 and BDV tested	Deregt et al. (2006)
Pan-viral DNA microarray, virochip ver. 3	Approximately 22,000 oligonucleotide probes	Detected human parainfluenza virus 4 (HPIV-4)	Conventional clinical laboratory testing using an extensive panel of microbiological tests failed to yield a diagnosis. Microarray worked.	Chiu et al. (2006)
Six species of orthopoxvirus (OPV)	110 oligonucleotide probes	Simultaneous detection and identification of six species of OPV including Variola, Monkeypox, Cowpox, Camelpox, Vaccinia, and Ectromelia viruses.	The method allowed us to discriminate OPV species from VZV, herpes simplex 1 virus (HSV-1), and HSV-2 that cause infections with clinical manifestations similar to OPV infections.	Ryabinin et al. (2006)
Respiratory pathogen	No data	20 common respiratory and 6 category A biothreat	The results demonstrate a novel, timely, and unbiased method for	Wang et al. (2006)

Table 3 (*continued*)

Virus types	Numbers of gene sequences	Application	Major finding	References
microarray ver 1 (RPM v.1) Pan-viral DNA microarray, virochip	1592 probes, 25-mer	pathogens known to cause febrile respiratory illness Virus discovery and identification	the molecular epidemiologic surveillance of influenza viruses. Identification of a Novel Gammaretrovirus in Prostate Tumors, and strongly implicate RNase L activity in the prevention or clearance of infection *in vivo*.	Urisman et al. (2006)
Four major serotypes of dengue virus	216 probes 22-mer	Detection and identification	Host-blind probe design	Putonti et al. (2006)
Flu-chip 55 diagnostic microarray	55 capture and label probes	Detection and subtyping of influenza A, B and Avian influenza H5N1	The combined results for two assays provided the absolutely correct types and subtypes for an average of 72% of the isolates, the correct type and partially correct subtype information for 13% of the isolates, the correct type only for 10% of the isolates, false-negative signals for 4% of the isolates, and false-positive signals for 1% of the isolates.	Townsend et al. (2006)
Epstein-Barr virus genome-chip	71 PCR amplified fragments, 12 control DNA fragments	Detects gene expression patterns of EBV in tumor cells	This study demonstrates that the EBV-chip is useful for screening infection with EBV in tumors, which may lead to insights into tumorigenesis associated with this virus.	Li et al. (2006)

Universal viral chip	No data	Characterization of all currently known viruses in Genbank	Have designed virus probes that are used not only to identify known viruses but also for discerning the genera of emerging or uncharacterized ones.	Chou et al. (2006)
Enterovirus microarray	13 probes	Detecting and differentiating EV71 and CA16	144 clinical specimens examined. Diagnostic accuracy of 92.0% for EV71 and 95.8% for CA16. Diagnostic accuracy for other enteroviruses (non-EV71 or -CA16) was 92.0%.	Chen et al. (2006)
Affymetrix resequencing RPM v.1	No data	Species- and strain-level identification of respiratory viruses	Broad-spectrum respiratory pathogen surveillance	Lin et al. (2006)
VZV expression microarray	71 probes 75-mer	Displays gene expression profile of VZV	Was able to show differences in levels of transcriptions among the various VZV ORFs	Kennedy et al. (2005)
Foot-and-mouth disease (FMD) DNA chip	155 probes, 35–45 mer long	Detection and typing of FMDV serotypes and differentiation from other viruses causing vesicular diseases	23 different FMDV strains representing all seven serotypes were detected and typed by the FMD DNA chip.	Baxi et al. (2005)
CNS viral pathogen chip	40,588 probes 20-mer	Identify herpes simplex virus type 1 (HSV-1), HSV-2, and cytomegalovirus; all serotypes of human enteroviruses and five flaviviruses (West Nile virus, dengue viruses and Langat virus)	Able to detect the 3 major CNS disease-causing viruses from a single sample	Korimbocus et al. (2005)

Table 3 (*continued*)

Virus types	Numbers of gene sequences	Application	Major finding	References
Flavivirus microarray	8 probes, 500 nucleotides long	Detect and distinguish between yellow fever (YF), West Nile, Japanese encephalitis (JE), and the dengue 1–4 viruses	Verified on all 7 flavivirus types. Detects and identifies even diverged strains of West Nile and Dengue virus	Nordstrom et al. (2005)
Ligation-detection microarray	6 detection sites	Detection and genotyping of SARS coronavirus (SARS-CoV)	20 samples assayed with the universal microarray were confirmed by DNA sequencing	Long et al. (2004)
Hepatitis B and D virus chip	14 probe fragments	Hepatitis D and Hepatitis B virus detection		Zhaohui et al. (2004)
Pan viral CNS chip	38 gene targets for 13 viral causes of meningitis	Detects and differentiates between echoviruses, HSV-1 and 2, VZV, human herpesvirus 7, human herpesvirus 6A and 6B, Epstein-Barr virus, polyomavirus JC and BK, cytomegalovirus, mumps and measles viruses		Boriskin et al. (2004)

takes either a long time or a massive influx of cases before a disease outbreak is recognized. The trend towards increasingly urbanized and dense city living and the more frequent travel between communities necessitates that community health monitoring adopts a more proactive preventative role instead of merely recording and reporting disease data. In order to do so, there need to be tools that are able to screen for the large panel of possible viral pathogens which are representative of the pathogen loads present in the larger community. To meet this requirement it becomes logical to monitor the community's sewage using microarrays designed to detect the presence of waterborne pathogens.

Using the OligoArray version 2.1 software written by Rouillard et al. (2003), Wong et al. (2006) have designed a total of 780 probes to detect 25 of the common virus families known to cause gastroenteritis. Two additional probes were placed unto the microarray as quality control sequences for fabrication. The 25 virus families targeted in this micoarray are hepatitis A and E virus, human adenovirus A–F, noroviruses, sapoviruses, human enterovirus groups A–E, polioviruses, rotavirus groups A–C, coronaviruses, astroviruses, human cytomegalovirus, torovirus, polyomaviruses, and picobirnaviruses. These oligonucleotide probes were synthesized unto a 63×116 microarray by the Gulari Research Group at the University of Michigan (Gao et al., 2001).

Figure 1 illustrates the processing steps required to prepare a sample for hybridization on an array. Environmental samples are first concentrated using conventional virus concentration methods like organic flocculation, tangential flow filtration, or ultrafiltration. Concentrates are then biologically amplified by passage through cell culture and virus nucleic acid is extracted. Labeling of viral nucleic acid is

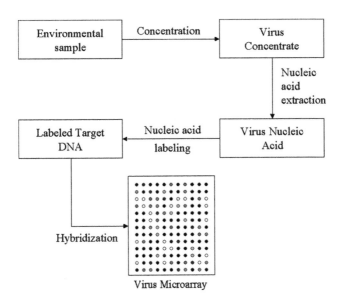

Fig. 1 The processing steps required to prepare a sample for hybridization on an array.

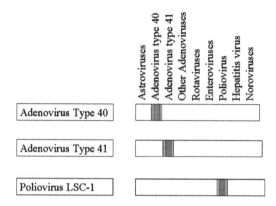

Fig. 2 The hybridization profile obtained from the hybridization of poliovirus, adenovirus type 40 and 41 with the virus microarray.

carried out through the use of DNA polymerase or RT enzymes. The labeled nucleic acid can then be hybridized on the microarray and a distinctive hybridization pattern is produced when target viruses bind to their complementary probes on the array.

Testing of the virus microarray was carried out using poliovirus virus LSC-1 and adenovirus type 40 and 41 as test subjects. Poliovirus was cultured on Buffalo green monkey (BGM) cells and adenoviruses were cultured on MA104 cells and their respective nucleic acid contents were extracted. Poliovirus RNA was labeled using the Superscript Indirect Labelling System (Invitrogen, Inc. Carlsbad, CA). Adenovirus DNA was labeled using the large fragment of DNA polymerase I (Klenow Fragment). Hybridization was carried out using an in-house hybridization wash station at 20°C for approximately 17 h. Subsequently, the microarray was scanned using a Molecular Devices model 4000B Genepix Scanner (Molecular Devices Corporation, Sunnyvale, CA), washed at 25°C, scanned, and subsequently washed and rescanned at 1°C intervals to generate a washing curve profile. Figure 2 illustrates the hybridization profile obtained from the hybridization of poliovirus, adenovirus type 40 and 41 with the virus microarray. Excellent signal to noise was achieved and the viruses were easily identified and distinguished.

Current work demonstrates that the chip can be used with viral concentrates from cell culture systems originating from sewage, with unique virus detection in community wastewater indicative of infections in the population. Thus future applications suggest, as with PCR, it will be a combination of conventional methods used in environmental virology and new techniques that will enhance pathogen discovery.

Prediction of the future: trends and needs in environmental virology

Viruses remain a public health concern and should remain a priority for the water and health community. These bio-nano particles are excreted in high concentrations by infected individuals, have high potency (probability of infection is high with low numbers (Haas et al., 1999)), and are environmentally robust. The ability of both DNA viruses and RNA viruses to rapidly evolve means new and emerging viral pathogens will need to be addressed. Pathogen discovery and characterization, occurrence in the environment, exposure pathways, and health outcomes via environmental exposure need to be addressed. This will likely follow a new microbial risk framework, which will require focused research on some important properties of viral disease transmission. The future will require models that examine community risks and provide explicit links between the models currently under development for environmental exposure and infectious disease.

It is predicted that there will be in each of these areas some key scientific issues that will need to be addressed (Table 4). New advances like qPCR and microarray technology should be applied to wastewater streams for viral pathogen discovery and for characterization of disease in our populations. These data should be used to further an analysis of water quality and health status. Adenoviruses, polyomaviruses, noroviruses, and other respiratory viruses need to be characterized, as does respiratory viral transmission particular via recreational exposure. Greater assessment of

Table 4

Predicted needs for advancing the field of environmental virology

Needs for method development in environmental virology

Needs	Advances
Quantification	Real-time PCR
Infectivity	Cell-culture PCR, mRNA detection
Specificity	Primer sequence databases
Quick response time	Real-time PCR, biosensors
Multiple detection	Microarrays, multiplex PCR

Areas of interest in environmental virology

Area	Focused interest
Virus discovery and characterization	Community assessment
	Cancer viruses
	Obesity viruses
	Heart viruses
	Continued global spread of poliovirus
	Respiratory viruses
	Animal viruses
Human exposure pathways and risk assessment	Viruses in sewage
	Recreational waters
	Storm waters
	Rural waters (ground, septic, agricultural waters)
	Irrigation waters

the occurrence of animal viruses in water, their role in animal health, and application to source tracking are needed. Finally, environmental virology needs to include the detection, occurrence, and transmission of viruses from fomites. The health impacts of these viral pathogens in nontraditional outcomes such as cancer are future areas requiring research.

References

Allard A, Girones R, Juto P, Wadell G. Polymerase chain reaction for detection of adeno-viruses in stool samples. J Clin Microbiol 1990; 28(12): 2659–2667.

Baxi MK, Baxi S, Clavijo A, Burton KM, Deregt D. Microarray-based detection and typing of foot-and-mouth disease virus. Vet J 2005; 172(3): 473–481.

Bodrossy L, Sessitsch A. Oligonucleotide microarrays in microbial diagnostics. Curr Opin Microbiol 2004; 7(3): 245–254.

Borchardt MA, Chyou PH, DeVries EO, Belongia EA. Septic system density and infectious diarrhea in a defined population of children. Environ Health Perspect 2003; 111(5): 742–748.

Boriskin YS, Rice PS, Stabler RA, Hinds J, Al-Ghusein H, Vass K, Butcher PD. DNA microarrays for virus detection in cases of central nervous system infection. J Clin Microbiol 2004; 42(12): 5811–5818.

Cardwell GG, Lindsey NJ, Wulff H, Donnelly DD, Bohl FN. Epidemic with adenovirus type 7 acute conjuctivitis in swimmers. Am J Epidemiol 1974; 99: 230–234.

Castignolles N, Petit F, Mendel I, Simon L, Cattolico L, Buffet-Janvresse C. Detection of adenovirus in the waters of the Seine River estuary by nested-PCR. Mol Cell Probes 1998; 12(3): 175–180.

Chapron CD, Ballester NA, Fontaine JH, Frades CN, Margolin AB. Detection of astro-viruses, enteroviruses, and adenovirus types 40 and 41 in surface waters collected and evaluated by the information collection rule and an integrated cell culture-nested PCR procedure. Appl Environ Microbiol 2000; 66(6): 2520–2525.

Chen TC, Chen GW, Hsiung CA, Yang JY, Shih SR, Lai YK, Juang JL. Combining mul-tiplex reverse transcription-PCR and a diagnostic microarray to detect and differentiate enterovirus 71 and coxsackievirus A16. J Clin Microbiol 2006; 44(6): 2212–2219.

Chiu CY, Rouskin S, Koshy A, Urisman A, Fischer K, Yagi S, Schnurr D, Eckburg PB, Tompkins LS, Blackburn BG, Merker JD, Patterson BK, Ganem D, DeRisi JL. Micro-array detection of human parainfluenzavirus 4 infection associated with respiratory failure in an immunocompetent adult. Clin Infect Dis 2006; 43(8): e71–e76.

Cho HB, Lee SH, Cho JC, Kim SJ. Detection of adenoviruses and enteroviruses in tap water and river water by reverse transcription multiplex PCR. Can J Microbiol 2000; 46(5): 417–424.

Choi S, Jiang SC. Real-time PCR quantification of human adenoviruses in urban rivers indicates genome prevalence but low infectivity. Appl Environ Microbiol 2005; 71(11): 7426–7433.

Choi SB, Chu W, Han J, He J, Jiang S. Application of Real-Time PCR and Tissue Culture Assay for Adenovirus Detection in Two Southern California Urban Rivers. New Orleans, LA: American Society for Microbiology 104th General Meeting; 2004.

Choo YJ, Kim SJ. Detection of human adenoviruses and enteroviruses in Korean oysters using cell culture, intergrated cell culture-PCR, and direct PCR. J Microbiol 2006; 44(2): 162–170.

Chou CC, Lee TT, Chen CH, Hsiao HY, Lin YL, Ho MS, Yang PC, Peck K. Design of microarray probes for virus identification and detection of emerging viruses at the genus level. BMC Bioinformatics 2006; 7: 232.

Cromeans T, Narayanan J, Jung K, Ko G, Wait D, Sobsey M, editors. Development of Molecular Methods to Detect Infectious Viruses in Water. London, UK: IWA Publishing; 2005.

D'Angelo LJ, Hierholzer JC, Keenlyside RA, Anderson LJ, Martone WJ. Pharygoconjunctival fever caused by adenovirus type 4: report of a swimming pool-related outbreak with recovery of virus from pool water. J Infect Dis 1979; 140: 42–47.

Deregt D, Gilbert SA, Dudas S, Pasick J, Baxi S, Burton KM, Baxi MK. A multiplex DNA suspension microarray for simultaneous detection and differentiation of classical swine fever virus and other pestiviruses. J Virol Methods 2006; 136(1–2): 17–23.

Enriquez CE, Hurst CJ, Gerba CP. Survival of the enteric adenoviruses 40 and 41 in tap, sea, and waste water. Water Res 1995; 29(11): 2548–2553.

EPA. Quality assurance/quality control guidance for laboratories performing PCR analyses on environmental samples. US EPA, Office of Water; 2004.

FAO: World agriculture: towards 2015/2030, an FAO study. Rome: Food and Agriculture Organization; 2002.

Fong T, Lipp EK. Molecular Detection of Waterborne Enteric Viruses in the Coastal Reaches of the Altamaha River. New Orleans LA: American Society for Microbiology 104th General Meeting; 2004.

Fong TT, Lipp EK. Enteric viruses of humans and animals in aquatic environments: health risks, detection, and potential water quality assessment tools. Microbiol Mol Biol Rev 2005; 69(2): 357–371.

Formiga-Cruz M, Tofino-Quesada G, Bofill-Mas D, Lees DN, Henshilwood K, Allard AK, Conden-Hansson AC, Hernroth BE, Vantarakis A, Tsibouxi A, Papapetropoulou M, Furones MD, Girones R. Distribution of human virus contamination in shellfish from different growing areas in Greece, Spain, Sweden, and the United Kingdom. Appl Environ Microbiol 2002; 68(12): 5990–5998.

Foy HM, Cooney MK, Hatlen JB. Adenovirus type 3 epidemic associated with intermittent chlorination of a swimming pool. Arch Environ Health 1968; 17: 795–802.

Gao X, LeProust E, Zhang H, Srivannavit O, Gulari E, Yu P, Nishiguchi C, Xiang Q, Zhou X. A flexible light-directed DNA chip synthesis gated by deprotection using solution photogenerated acids. Nucleic Acids Res 2001; 29(22): 4744–4750.

Gerba CP, Rose JB. Viruses in source and drinking water. In: Advances in Drinking Water Microbiology Research (McFeters GA, editor). Madison, WI: Science Tech; 1989.

Girones R, Puig M, Allard A, Lucena F, Wadell G, Jofre J. Detection of adenovirus and enterovirus by PCR amplification in polluted waters. Water Sci Technol 1995; 31(5–6): 351–357.

Greening GE, Hewitt J, Lewis GD. Evaluation of integrated cell culture-PCR (C-PCR) for virological analysis of environmental samples. J Appl Microbiol 2002; 93(5): 745–750.

Gu Z, Belzer SW, Gibson CS, Bankowski MJ, Hayden RT. Multiplexed, real-time PCR for quantitative detection of human adenovirus. J Clin Microbiol 2003; 41(10): 4636–4641.

Guimil R, Beier M, Scheffler M, Rebscher H, Funk J, Wixmerten A, Baum M, Hermann C, Tahedl H, Moschel E, Obermeier F, Sommer I, Buchner D, Viehweger R, Burgmaier J, Stahler CF, Muller M, Stahler PF. Geniom technology–the benchtop array facility. Nucleosides Nucleotides Nucleic Acids 2003; 22(5–8): 1721–1723.

Haas CH, Rose JB, Gerba CP. Quantitative Microbial Risk Assessment. New York, NY: Wiley; 1999.

Haramoto E, Katayama H, Oguma K, Ohgaki S. Application of cation-coated filter method to detection of noroviruses, enteroviruses, adenoviruses, and torque teno viruses in the Tamagawa River in Japan. Appl Environ Microbiol 2005; 71(5): 2403–2411.

Harley D, Harrower B, Lyon M, Dick A. A primary school outbreak of pharyngoconjunctival fever caused by adenovirus type 3. Commun Dis Intell 2001; 25(1): 9–12.

He J, Jiang S. Quantification of human adenoviruses and enterococcus in environmental water samples. New Orleans, LA: American Society for Microbiology 104th General Conference; 2004.

He JW, Jiang S. Quantification of enterococci and human adenoviruses in environmental samples by real-time PCR. Appl Environ Microbiol 2005; 71(5): 2250–2255.

Heim A, Ebnet C, Harste G, Pring-Akerblom P. Rapid and quantitative detection of human adenovirus DNA by real-time PCR. J Med Virol 2003; 70(2): 228–239.

Hughes TR, Mao M, Jones AR, Burchard J, Marton MJ, Shannon KW, Lefkowitz SM, Ziman M, Schelter JM, Meyer MR, Kobayashi S, Davis C, Dai H, He YD, Stephaniants SB, Cavet G, Walker WL, West A, Coffey E, Shoemaker DD, Stoughton R, Blanchard AP, Friend SH, Linsley PS. Expression profiling using microarrays fabricated by an ink-jet oligonucleotide synthesizer. Nat Biotechnol 2001; 19(4): 342–347.

Hurst CJ, McClellan KA, Benton WH. Comparison of cytopathogenicity, immunofluorescence and in situ DNA hybridization as methods for the detection of adenoviruses. Water Res 1988; 22: 1547–1552.

Irving LG, Smith FA. One-year survey of enteroviruses, adenoviruses, and reoviruses isolated from effluent at an activated-sludge purification plant. Appl Environ Microbiol 1981; 41(1): 51–59.

Jiang S, Noble R, Chu W. Human adenoviruses and coliphages in urban runoff-impacted coastal waters of Southern California. Appl Environ Microbiol 2001; 67(1): 179–184.

Jiang S, Dezfulian H, Chu W. Real-time quantitative PCR for enteric adenovirus serotype 40 in environmental waters. Can J Microbiol 2005; 51: 393–398.

Jothikumar N, Cromeans TL, Hill VR, Lu X, Sobsey MD, Erdman DD. Quantitative real-time PCR assays for detection of human adenoviruses and identification of serotypes 40 and 41. Appl Environ Microbiol 2005; 71(6): 3131–3136.

Kelly JJ, Siripong S, McCormack J, Janus LR, Urakawa H, El Fantroussi S, Noble PA, Sappelsa L, Rittmann BE, Stahl DA. DNA microarray detection of nitrifying bacterial 16S rRNA in wastewater treatment plant samples. Water Res 2005; 39(14): 3229–3238.

Kennedy PG, Grinfeld E, Craigon M, Vierlinger K, Roy D, Forster T, Ghazal P. Transcriptomal analysis of varicella-zoster virus infection using long oligonucleotide-based microarrays. J Gen Virol 2005; 86(Pt 10): 2673–2684.

Ko G, Cromeans TL, Sobsey MD. Detection of infectious adenovirus in cell culture by mRNA reverse transcription-PCR. Appl Environ Microbiol 2003; 69(12): 7377–7384.

Ko G, Jothikumar N, Hill VR, Sobsey MD. Rapid detection of infectious adenoviruses by mRNA real-time RT-PCR. J Virol Methods 2005; 127(2): 148–153.

Korimbocus J, Scaramozzino N, Lacroix B, Crance JM, Garin D, Vernet G. DNA probe array for the simultaneous identification of herpesviruses, enteroviruses, and flaviviruses. J Clin Microbiol 2005; 43(8): 3779–3787.

Krikelis V, Markoulatos P, Spyrou N, Serie C. Detection of indigenous enteric viruses and raw sewage effluents of the City of Athens, Greece, during a two-year survey. Water Sci Technol 1985a; 17: 159–164.

Krikelis V, Spyrou N, Markoulatos P, Serie C. Seasonal distribution of enteroviruses and adenoviruses in domestic sewage. Can J Microbiol 1985b; 31: 345–350.

Lee C, Lee SH, Han E, Kim SJ. Use of cell culture-PCR assay based on combination of A549 and BGMK cell lines and molecular identification as a tool to monitor infectious adenoviruses and enteroviruses in river water. Appl Environ Microbiol 2004; 70(11): 6695–6705.

Lee HK, Jeong YS. Comparison of total culturable virus assay and multiplex integrated cell culture-PCR for reliability of waterborne virus detection. Appl Environ Microbiol 2004; 70(6): 3632–3636.

Lee SH, Kim SJ. Detection of infectious enteroviruses and adenoviruses in tap water in urban areas in Korea. Water Res 2002; 36(1): 248–256.

Li C, Chen RS, Hung SK, Lee YT, Yen CY, Lai YW, Teng RH, Huang JY, Tang YC, Tung CP, Wei TT, Shieh B, Liu ST. Detection of Epstein-Barr virus infection and gene expression in human tumors by microarray analysis. J Virol Methods 2006; 133(2): 158–166.

Lin B, Vora GJ, Thach D, Walter E, Metzgar D, Tibbetts C, Stenger DA. Use of oligonucleotide microarrays for rapid detection and serotyping of acute respiratory disease-associated adenoviruses. J Clin Microbiol 2004; 42(7): 3232–3239.

Lin B, Wang Z, Vora GJ, Thornton JA, Schnur JM, Thach DC, Blaney KM, Ligler AG, Malanoski AP, Santiago J, Walter EA, Agan BK, Metzgar D, Seto D, Daum LT, Kruzelock R, Rowley RK, Hanson EH, Tibbetts C, Stenger DA. Broad-spectrum respiratory tract pathogen identification using resequencing DNA microarrays. Genome Res 2006; 16(4): 527–535.

Liu RH, Lodes MJ, Nguyen T, Siuda T, Slota M, Fuji HS, McShea A. Validation of a fully integrated microfluidic array device for influenza a subtype identification and sequencing. Anal Chem 2006; 78(12): 4184–4193.

Logan C, O'Leary JJ, O'Sullivan N. Real-time reverse transcription-PCR for detection of rotavirus and adenovirus as causative agents of acute viral gastroenteritis in children. J Clin Microbiol 2006; 44(9): 3189–3195.

Long WH, Xiao HS, Gu XM, Zhang QH, Yang HJ, Zhao GP, Liu JH. A universal microarray for detection of SARS coronavirus. J Virol Methods 2004; 121(1): 57–63.

Maluquer de Motes C, Clemente-Casares P, Hundesa A, Martin M, Girones R. Detection of bovine and porcine adenoviruses for tracing the source of fecal contamination. Appl Environ Microbiol 2004; 70(3): 1448–1454.

Martone WJ, Hierholzer JC, Keenlyside RA, Fraser DW, D'Angelo LJ, Winkler WG. An outbreak of adenovirus type 3 disease at a private recreation center swimming pool. Am J Epidemiol 1980; 111: 229–237.

Min W, Wen-Li M, Bao Z, Ling L, Zhao-Hui S, Wen-Ling Z. Oligonucleotide microarray with RD-PCR labeling technique for detection and typing of human papillomavirus. Curr Microbiol 2006; 52(3): 204–209.

Naik SR, Aggarwal R, Salunke PN, Mehrotra NN. A large waterborne viral hepatitis E epidemic in Kanpur, India. Bull World Health Organ 1992; 70(5): 597–604.

Nordstrom H, Falk KI, Lindegren G, Mouzavi-Jazi M, Walden A, Elgh F, Nilsson P, Lundkvist A. DNA microarray technique for detection and identification of seven flaviviruses pathogenic for man. J Med Virol 2005; 77(4): 528–540.

Pannucci J, Cai H, Pardington PE, Williams E, Okinaka RT, Kuske CR, Cary RB. Virulence signatures: microarray-based approaches to discovery and analysis. Biosens Bioelectron 2004; 20(4): 706–718.

Papapetropoulou M, Vantarakis AC. Detection of adenovirus outbreak at a municipal swimming pool by nested PCR amplification. J Infect 1998; 36: 101–103.

Parashar UD, Hummelman EG, Bresee JS, Miller MA, Glass RI. Global illness and deaths caused by rotavirus disease in children. Emerg Infect Dis 2003; 9(5): 565–572.

Pavlov DN, Van Zyl WB, Van Heerden J, Grabow WO, Ehlers MM. Prevalence of vaccine-derived polioviruses in sewage and river water in South Africa. Water Res 2005; 39(14): 3309–3319.

Pease AC, Solas D, Sullivan EJ, Cronin MT, Holmes CP, Fodor SP. Light-generated oligonucleotide arrays for rapid DNA sequence analysis. Proc Natl Acad Sci USA 1994; 91(11): 5022–5026.

Pehler-Harrington K, Khanna M, Waters CR, Henrickson KJ. Rapid detection and identification of human adenovirus species by adenoplex, a multiplex PCR-enzyme hybridization assay. J Clin Microbiol 2004; 42(9): 4072–4076.

Pina S, Puig M, Lucena F, Jofre J, Girones R. Viral Pollution in the environment and in shellfish: human adenovirus detection by PCR as an index of human viruses. Appl Environ Microbiol 1998; 64: 3376–3382.

Puig M, Jofre J, Lucena F, Allard A, Wadell G, Girones R. Detection of adenoviruses and enteroviruses in polluted waters by nested PCR amplification. Appl Environ Microbiol 1994; 60(8): 2963–2970.

Putonti C, Chumakov S, Mitra R, Fox GE, Willson RC, Fofanov Y. Human-blind probes and primers for dengue virus identification. FEBS J 2006; 273(2): 398–408.

Raboni SM, Siqueira MM, Portes SR, Pasquini R. Comparison of PCR, enzyme immuno-assay and conventional culture for adenovirus detection in bone marrow transplant patients with hemorrhagic cystitis. J Clin Virol 2003; 27(3): 270–275.

Rigotto C, Sincero TCM, Simoes CMO, Barardi CRM. Detection of adenoviruses in shellfish by means of conventional-PCR, nested-PCR, and integrated cell culture PCR (ICC/PCR). Water Res 2005; 39: 297–304.

Rohayem J, Berger S, Juretzek T, Herchenroder O, Mogel M, Poppe M, Henker J, Rethwilm A. A simple and rapid single-step multiplex RT-PCR to detect norovirus, astrovirus and adenovirus in clinical stool samples. J Virol Methods 2004; 118(1): 49–59.

Rouillard JM, Zuker M, Gulari E. OligoArray 2.0: design of oligonucleotide probes for DNA microarrays using a thermodynamic approach. Nucleic Acids Res 2003; 31(12): 3057–3062.

Ryabinin VA, Shundrin LA, Kostina EB, Laassri M, Chizhikov V, Shchelkunov SN, Chumakov K, Sinyakov AN. Microarray assay for detection and discrimination of orthopoxvirus species. J Med Virol 2006; 78(10): 1325–1340.

Sarantis H, Johnson G, Brown M, Petric M, Tellier R. Comprehensive detection and serotyping of human adenoviruses by PCR and sequencing. J Clin Microbiol 2004; 42(9): 3963–3969.

Schena M, Shalon D, Davis RW, Brown PO. Quantitative monitoring of gene expression patterns with a complementary DNA microarray. Science 1995; 270(5235): 467–470.

Sengupta S, Onodera K, Lai A, Melcher U. Molecular detection and identification of influenza viruses by oligonucleotide microarray hybridization. J Clin Microbiol 2003; 41(10): 4542–4550.

Sergeev N, Distler M, Courtney S, Al-Khaldi SF, Volokhov D, Chizhikov V, Rasooly A. Multipathogen oligonucleotide microarray for environmental and biodefense applications. Biosens Bioelectron 2004; 20(4): 684–698.

Sergeev N, Rubtcova E, Chizikov V, Schmid DS, Loparev VN. New mosaic subgenotype of varicella-zoster virus in the USA: VZV detection and genotyping by oligonucleotide-microarray. J Virol Methods 2006; 136(1–2): 8–16.

Song Y, Dai E, Wang J, Liu H, Zhai J, Chen C, Du Z, Guo Z, Yang R. Genotyping of hepatitis B virus (HBV) by oligonucleotides microarray. Mol Cell Probes 2006; 20(2): 121–127.

Standish-Lee P, Loboschefsky E. Protecting public health from the impact of body-contact recreation. Water Sci Technol 2006; 53(10): 201–207.

Straub TM, Chandler DP. Towards a unified system for detecting waterborne pathogens. J Microbiol Methods 2003; 53(2): 185–197.

Swenson PD, Wadell G, Allard A, Hierholzer JC. Adenoviruses. In: Manual of Clinical Microbiology (Murray, PR, Baron EJ, Jorgenson JH, Pfaller MA, Yolken RH, editors). 8th ed. Washington, DC: ASM Press; 2003.

Tani N, Dohi Y, Kurumatani N, Yonemasu K. Seasonal distribution of adenoviruses, enteroviruses and reoviruses in urban river water. Microbiol Immunol 1995; 39(8): 577–580.

Townsend MB, Dawson ED, Mehlmann M, Smagala JA, Dankbar DM, Moore CL, Smith CB, Cox NJ, Kuchta RD, Rowlen KL. Experimental evaluation of the FluChip diagnostic microarray for influenza virus surveillance. J Clin Microbiol 2006; 44(8): 2863–2871.

Turner M, Istre GR, Beauchamp H, Baum M, Arnold S. Community outbreak of adenovirus type 7a infections associated with a swimming pool. South Med J 1987; 80: 712–715.

UNESCO. Water for people, water for life. World Water Development Report, United Nations Educational, Scientific and Cultural Organizations; 2003.

Urisman A, Molinaro RJ, Fischer N, Plummer SJ, Casey G, Klein EA, Malathi K, Magi-Galluzzi C, Tubbs RR, Ganem D, Silverman RH, Derisi JL. Identification of a novel gammaretrovirus in prostate tumors of patients homozygous for R462Q RNASEL variant. PLoS Pathog 2006; 2(3): e25.

US House Transportation Subcommittee on Water Resources and Environment, "Hearing on Meeting the Nation's Wastewater Infrastructure Needs," 03/19/03.

Van Heerden J, Ehlers MM, Van Zyl WB, Grabow WO. Incidence of adenoviruses in raw and treated water. Water Res 2003; 37(15): 3704–3708.

Van Heerden J, Ehlers MM, Heim A, Grabow WOK. Prevalence, quantification and typing of adenoviruses detected in river and treated drinking water in South Africa. J Appl Microbiol 2005; 99(2): 234–242.

Wade TJ, Pai N, Eisenberg JN, Colford Jr JM. Do U.S. Environmental Protection Agency water quality guidelines for recreational waters prevent gastrointestinal illness? A systematic review and meta-analysis. Environ Health Perspect 2003; 111(8): 1102–1109.

Wadell G. Molecular epidemiology of human adenoviruses. Curr Top Microbiol Immunol 1984; 110: 191–220.

Wang Z, Daum LT, Vora GJ, Metzgar D, Walter EA, Canas LC, Malanoski AP, Lin B, Stenger DA. Identifying influenza viruses with resequencing microarrays. Emerg Infect Dis 2006; 12(4): 638–646.

Watanabe M, Kohdera U, Kino M, Haruta T, Nukuzuma S, Suga T, Akiyoshi K, Ito M, Suga S, Komada Y. Detection of adenovirus DNA in clinical samples by SYBR Green real-time polymerase chain reaction assay. Pediatr Int 2005; 47(3): 286–291.

Wong M, Hashsham SA, Gulari E, Rose JB. Development of a Virulence Factor Biochip and its Validation for Microbial Risk Assessment in Drinking Water. EPA Star Grant Report; 2006.

Wu L, Thompson DK, Liu X, Fields MW, Bagwell CE, Tiedje JM, Zhou J. Development and evaluation of microarray-based whole-genome hybridization for detection of micro-organisms within the context of environmental applications. Environ Sci Technol 2004; 38(24): 6775–6782.

Wu L, Liu X, Schadt CW, Zhou J. Microarray-based analysis of subnanogram quantities of microbial community DNAs by using whole-community genome amplification. Appl Environ Microbiol 2006; 72(7): 4931–4941.

Xu W, McDonough MC, Erdman DD. Species-specific identification of human adenoviruses by a multiplex PCR assay. J Clin Microbiol 2000; 38(11): 4114–4120.

Zhaohui S, Wenling Z, Bao Z, Rong S, Wenli M. Microarrays for the detection of HBV and HDV. J Biochem Mol Biol 2004; 37(5): 546–551.

Zhou J. Microarrays for bacterial detection and microbial community analysis. Curr Opin Microbiol 2003; 6(3): 288–294.

List of Contributors

Robert L. Atmar
Departments of Medicine and Molecular Virology & Microbiology
Baylor College of Medicine
1 Baylor Plaza, MS BCM280
Houston, TX 77030, USA

Jamie Bartram
World Health Organization
Geneva, Switzerland

Soile Blomqvist
Enterovirus Laboratory, National Public Health Institute (KTL)
WHO Collaborating Centre for Poliovirus Surveillance
    and Enterovirus Research
Mannerheimintie 166
00300 Helsinki, Finland

Charles P. Gerba
Department of Soil, Water and Environmental Science
University of Arizona
Tucson, AZ, USA

Wilhelm O. K. Grabow
University of Pretoria
Pretoria
South Africa

Tapani Hovi
Enterovirus Laboratory, National Public Health Institute (KTL)
WHO Collaborating Centre for Poliovirus Surveillance
    and Enterovirus Research
Mannerheimintie 166
00300 Helsinki, Finland

Juan Jofre
Department of Microbiology, School of Biology
University of Barcelona
Diagonal 645
08028-Barcelona, Spain

Françoise S. Le Guyader
Laboratoire de Microbiologie
IFREMER, BP 21105
44311 Nantes cedex 03, France

Kristina D. Mena
University of Texas
Health Science Center at Houston School of Public Health
1100 N. Stanton Street, Suite 110
El Paso, TX 79902 USA

Rosa M. Pintó
Enteric Virus Laboratory, Department of Microbiology
School of Biology, University of Barcelona
Diagonal 645
08028 Barcelona, Spain

Ana Maria de Roda Husman
National Institute of Public Health (RIVM)
Centre of Infectious Disease Control (CIb)
WHO Collaborating Centre for Risk Assessment of
  Pathogens in Food and Water
Antonie van Leeuwenhoeklaan 9
3720 BA Bilthoven, The Netherlands

Merja Roivainen
Enterovirus Laboratory, National Public Health Institute (KTL)
WHO Collaborating Centre for Poliovirus Surveillance
  and Enterovirus Research
Mannerheimintie 166
00300 Helsinki, Finland

Joan B. Rose
Department of Fisheries and Wildlife
Michigan State University
13 Natural Resources
E. Lansing MI, 48824 USA

Juan-Carlos Saiz
Laboratory of Zoonotic and Environmental Virology, Department
of Biotechnology
Instituto Nacional de Investigación Agraria y Alimentaria (INIA)
Ctra.Coruña km. 7.5
28040 Madrid, Spain

Syed A. Sattar
Centre for Research on Environmental Microbiology (CREM)
University of Ottawa
451 Smyth Road
Ottawa, ON, Canada K1 H 8M5

Kellogg Schwab
Johns Hopkins Bloomberg School of Public Health
Department of Environmental Health Sciences
615 N. Wolfe St. Room E6620
Baltimore, MD 21205-2103, USA

Jane Sellwood
Health Protection Agency
Environmental Virology Unit
Microbiology Laboratory Royal Berkshire Hospital
Reading RG1 5AN, UK

Susan Springthorpe
Centre for Research on Environmental Microbiology (CREM)
University of Ottawa
451 Smyth Road
Ottawa, ON, Canada K1 H 8M5

Mark Wong
    Department of Fisheries and Wildlife
    Michigan State University
    13 Natural Resources
    E. Lansing MI, 48824 USA

Peter Wyn-Jones
    Institute of Geography and Earth Sciences
    University of Wales
    Aberystwyth, UK

Irene Xagoraraki
    Department of Civil and Environmental Engineering
    Michigan State University
    E. Lansing MI, 48824 USA

Index

Page numbers suffixed by t and f refer to Tables and Figures respectively.

Printed and bound by CPI Group (UK) Ltd, Croydon, CR0 4YY

03/10/2024

01040430-0013